T0313709

PROCESS CONTROL SYSTEM FAULT DIAGNOSIS

PROCESS CONTROL SYSTEM FAULT DIAGNOSIS

A BAYESIAN APPROACH

Ruben Gonzalez

Fei Qi

Biao Huang

WILEY

Library of Congress Cataloging-in-Publication Data

Names: Gonzalez, Ruben, 1985- author. | Qi, Fei, 1983- author | Huang, Biao,
     1962- author.
Title: Process control system fault diagnosis : a Bayesian approach / Ruben
     Gonzalez, Fei Qi, Biao Huang.
Description: First edition. | Chichester, West Sussex, United Kingdom : John
     Wiley & Sons, 2016. | Includes bibliographical references and index.
Identifiers: LCCN 2016010340| ISBN 9781118770610 (cloth) | ISBN 9781118770597
     (epub)
Subjects: LCSH: Chemical process control–Statistical methods. | Bayesian
     statistical decision theory. | Fault location (Engineering)
Classification: LCC TP155.75 .G67 2016 | DDC 660/.2815–dc23 LC record available at
https://lccn.loc.gov/2016010340

A catalogue record for this book is available from the British Library.

Set in 10/12pt, TimesLTStd by SPi Global, Chennai, India.

Printed and bound in Malaysia by Vivar Printing Sdn Bhd

1   2016

Contents

Preface

Background

Control performance monitoring (CPM) has been and continues to be one of the most active research areas in the process control community. A number of CPM technologies have been developed since the late 1980s. It is estimated that several hundred papers have been published in this or related areas. CPM techniques have also been widely applied in industry. A number of commercial control performance assessment software packages are available off the shelf.

CPM techniques include controller monitoring, sensor monitoring, actuator monitoring, oscillation detection, model validation, nonlinearity detection and so on. All of these techniques have been designed to target a specific problem source in a control system. The common practice is that one monitoring technique (or monitor) is developed for a specific problem source. However, a specific problem source can show its signatures in more than one monitor, thereby inducing alarm flooding. There is a need to consider all monitors simultaneously in a systematic manner.

There are a number of challenging issues:

1. There are interactions between monitors. A monitor cannot be designed to just monitor one problem source in isolation from other problem sources. While each monitor may work well when only the targeted problem occurs, relying on a single monitor can be misleading when other problems also occur.
2. The causal relations between a problem source and a monitor are not obvious for industrial-scale problems. First-principles knowledge, including the process flowchart, cannot always provide an accurate causal relation.
3. Disturbances and uncertainties exist everywhere in industrial settings.
4. Most monitors are either model-based or data-driven; it is uncommon for monitor results to be combined with prior process knowledge.

Clearly, there is a need to develop a systematic framework, including theory and practical guidelines, to tackle the these monitoring problems.

Control Performance Diagnosis and Control System Fault Diagnosis

Control systems play a critical role in modern process industries. Malfunctioning components in control systems, including sensors, actuators and other components, are not uncommon in industrial environments. Their effects introduce excess variation throughout the process, thereby reducing machine operability, increasing costs and emissions, and disrupting final product quality control. It has been reported in the literature that as many as 60% of industrial controllers may have some kind of problem.

The motivation behind this book arises from the important task of isolating and diagnosing control performance abnormalities in complex industrial processes. A typical modern process operation consists of hundreds or even thousands of control loops, which is too many for plant personnel to monitor. Even if poor performance is detected in some control loops, because a problem in a single component can invoke a wide range of control problems, locating the underlying problem source is not a trivial task. Without an advanced information synthesis and decision-support system, it is difficult to handle the flood of process alarms to determine the source of the underlying problem. Human beings' inability to synthesize high-dimensional process data is the main reason behind these problems. The purpose of control performance diagnosis is to provide an automated procedure that aids plant personnel to determine whether specified performance targets are being met, evaluate the performance of control loops, and suggest possible problem sources and a troubleshooting sequence.

To understand the development of control performance diagnosis, it is necessary to review the historical evolution of CPM. From the 1990s and 2000s, there was a significant development in CPM and, from the 2000s to the 2010s, control performance diagnosis. CPM focuses on determining how well the controller is performing with respect to a given benchmark, while CPD focuses on diagnosing the causes of poor performance. CPM and CPD are of significant interest for process industries that have growing safety, environmental and efficiency requirements. The classical method of CPM was first proposed in 1989 by Harris, who used the minimum variance control (MVC) benchmark as a general indicator of control loop performance. The MVC benchmark can be obtained using the filtering and correlation (FCOR) algorithm, as proposed by Huang et al. in 1997; this technique can be easily generalized to obtain benchmarks for multivariate systems. Minimum variance control is generally aggressive, with potential for poor robustness, and is not a suitable benchmark for CPM of model predictive control, as it does not take input action into account. Thus the linear quadratic Gaussian (LQG) benchmark was proposed in the PhD dissertation of Huang in 1997. In order to extend beyond simple benchmark comparisons, a new family of methods was developed to monitor specific instruments within control loops for diagnosing poor performance (by Horch, Huang, Jelali, Kano, Qin, Scali, Shah, Thornhill, etc). As a result, various CPD approaches have appeared since 2000.

To address the CPD problem systematically, Bayesian diagnosis methods were introduced by Huang in 2008. Due to their ability to incorporate both prior knowledge and data, Bayesian methods are a powerful tool for CPD. They have been proven to be useful for a variety of monitoring and predictive maintenance purposes. Successful applications of the Bayesian approach have also been reported in medical science, image processing, target recognition, pattern matching, information retrieval, reliability analysis and engineering diagnosis. It provides a flexible structure for modelling and evaluating uncertainties. In the presence of noise and disturbances, Bayesian inference provides a good way to solve the monitoring and diagnosis problem, providing a quantifiable measure of uncertainty for decision making. It is one of the most widely applied techniques in statistical inference, as well being used to diagnose engineering problems.

The Bayesian approach was applied to fault detection and diagnosis (FDI) in the mechanical components of transport vehicles by Pernestal in 2007, and Huang applied it to CPD in

2008. CPD techniques bear some resemblance to FDI. Faults usually refer to failure events, while control performance abnormality does not necessarily imply a failure. Thus, CPD is performance-related, often focusing on detecting control related problems that affect control system performance, including economic and environmental performance, while FDI focuses on the failure of components. Under the Bayesian framework, both can be considered as an abnormal event or fault diagnosis for control systems. Thus control system fault diagnosis is a more appropriate term that covers both.

Book Objective, Organization and Readership

The main objectives of this book are to establish a Bayesian framework for control system fault diagnosis, to synthesize observations of different monitors with prior knowledge, and to pinpoint possible abnormal sources on the basis of Bayesian theory. To achieve these objectives, this book provides comprehensive coverage of various Bayesian methods for control system fault diagnosis. The book starts with a tutorial introduction of Bayesian theory and its applications for general diagnosis problems, and an introduction to the existing control loop performance-monitoring techniques. Based upon these fundamentals, the book turns to a general data-driven Bayesian framework for control system fault diagnosis. This is followed by presentation of various practical problems and solutions. To extend beyond traditional CPM with discrete outputs, this book also explores how control loop performance monitors with continuous outputs can be directly incorporated into the Bayesian diagnosis framework, thus improving diagnosis performance. Furthermore, to deal with historical data taken from ambiguous operating conditions, two approaches are explored:

- Dempster–Shafer theory, which is often used in other applications when ambiguity is present
- a parametrized Bayesian approach.

Finally, to demonstrate the practical relevance of the methodology, the proposed solutions are demonstrated through a number of practical engineering examples.

This book attempts to consolidate results developed or published by the authors over the last few years and to compile them together with their fundamentals in a systematic way. In this respect, the book is likely to be of use for graduate students and researchers as a monograph, and as a place to look for basic as well as state-of-the-art techniques in control system performance monitoring and fault diagnosis. Since several self-contained practical examples are included in the book, it also provides a place for practising engineers to look for solutions to their daily monitoring and diagnosis problems. In addition, the book has comprehensive coverage of Bayesian theory and its application in fault diagnosis, and thus it will be of interest to mathematically oriented readers who are interested in applying theory to practice. On the other hand, due to the combination of theory and applications, it will also be beneficial to applied researchers and practitioners who are interested in giving themselves a sound theoretical foundation. The readers of this book will include graduate students and researchers in chemical engineering, mechanical engineering and electrical engineering, specializing in process control, control systems and process systems engineering. It is expected that readers will be acquainted with some fundamental knowledge of undergraduate probability and statistics.

Acknowledgements

The material in this book is the outcome of several years of research efforts by the authors and many other graduate students and post-doctoral fellows at the University of Alberta. In particular, we would like to acknowledge those who have contributed directly to the general area of Bayesian statistics that has now become one of the most active research subjects in our group: Xingguang Shao, Shima Khatibisepehr, Marziyeh Keshavarz, Kangkang Zhang, Swanand Khare, Aditya Tulsyan, Nima Sammaknejad and Ming Ma. We would also like to thank our colleagues and collaborators in the computer process control group at the University of Alberta, who have provided a stimulating environment for process control research. The broad range of talent within the Department of Chemical and Materials Engineering at the University of Alberta has allowed cross-fertilization and nurturing of many different ideas that have made this book possible. We are indebted to industrial practitioners Aris Espejo, Ramesh Kadali, Eric Lau and Dan Brown, who have inspired us with practical relevance in broad areas of process control research. We would also like to thank our laboratory support from Artin Afacan, computing support from Jack Gibeau, and other supporting staff in the Department of Chemical and Materials Engineering at the University of Alberta. The support of the Natural Sciences and Engineering Research Council of Canada and Alberta Innovates Technology Futures for this and related research work is gratefully acknowledged. Last, but not least, we would like to acknowledge Kangkang Zhang, Yuri Shardt and Sun Zhou for their detailed review of and comments on the book.

Some of the figures presented in this book are taken from our previous work that has been published in journals. We would like to acknowledge the journal publishers who have allowed us to re-use these figures:

Figures 3.1 and 14.1 are adapted with permission from *AIChE Journal*, Vol. 56, Qi F, Huang B and Tamayo EC, 'A Bayesian approach for control loop diagnosis with missing data', pp. 179–195. ©2010 John Wiley and Sons.

Figures 4.4 and 13.2 are adapted with permission from *Automatica*, Vol. 47, Qi F and Huang B, 'Bayesian methods for control loop diagnosis in the presence of temporal dependent evidences', pp. 1349–1356. ©2011 Elsevier.

Figures 4.1, 4.3, 4.5–4.7, 13.1 and 13.3 are adapted with permission from *Industrial & Engineering Chemistry Research*, Vol. 49, Qi F and Huang B, 'Dynamic Bayesian approach for control loop diagnosis with underlying mode dependency', pp. 8613–8623. © 2010 American Chemical Society.

Figures 8.1–8.4 are adapted with permission from *Journal of Process Control*, Vol. 24, Gonzalez R and Huang B, 'Control loop diagnosis using continuous evidence through kernel density estimation', pp. 640–651. ©2014 Elsevier.

List of Figures

List of Tables

Nomenclature

Symbol	Description
α	Frequency parameter for the Dirichlet distribution
$\alpha\{\frac{\bullet}{\boldsymbol{m}_k}\}$	Frequency parameters pertaining to the ambiguous mode $\boldsymbol{m}_k$
μ	Population mean
Σ	Population covariance
σ	Population standard deviation
Θ	Complete set of probability/proportion parameters
$\Theta\{\frac{\bullet}{\boldsymbol{m}_k}\}$	The set of elements in Θ pertaining to the ambiguous mode $\boldsymbol{m}_k$
$\hat{\Theta}$	Informed estimate of Θ
$\boldsymbol{\Theta}$	Complete set of probability/proportion parameters (matrix form)
$\boldsymbol{\Theta}^*$	Inclusive estimate of Θ (matrix form)
$\boldsymbol{\Theta}_*$	Exclusive estimate of Θ (matrix form)
θ	A probability/proportion parameter
$\theta\{\frac{m}{\boldsymbol{m}}\}$	Proportion of data in ambiguous mode $\boldsymbol{m}$ belonging to mode m
$Bel(M)$	Lower-bound probability of mode M
C	State of the component of interest (random variable)
c	State of the component of interest (observation)
$\mathcal{C}(M)$	The event where mode M was diagnosed
$\mathcal{C}(M)\|M$	The event where mode M was diagnosed and M was true
$\mathcal{C}(\bar{M})\|M$	The event where a mode other than M was diagnosed and M was true
$\mathcal{D}$	Historical record of evidence
D_i	ith element of historical evidence data record $\mathcal{D}$
E	Evidence (random variable)
e	Evidence (observation)
F_N	False negative diagnosis rate
$\boldsymbol{G}$	Generalized BBA
$\boldsymbol{G}[:, m]$	mth column of $\boldsymbol{G}$ (MATLAB notation)
$\boldsymbol{G}[k, :]$	kth row of $\boldsymbol{G}$ (MATLAB notation)
H	Bandwidth matrix (Kernel density estimation)
$\boldsymbol{H}$	Hessian matrix
i.i.d.	Independent and identically distributed
$\boldsymbol{J}$	Jacobian matrix

Symbol	Description
K	Support for conflict (Dempster–Shafer theory)
K	Kernel function (kernel density estimation)
M	Operational mode (random variable)
$\boldsymbol{M}$	Potentially ambiguous operational mode (random variable)
m	Operational mode (observation)
$\boldsymbol{m}$	Potentially ambiguous operational mode (observation)
MIC	Mutual information criterion
CMIC	Conditional mutual information criterion
$n(E)$	Number of times evidence E has been observed
$n(E, M)$	Number of times evidence E and mode M have been jointly observed
$n(M)$	Number of times mode M was observed
ODE $[f(x)]$	Ordinary differential equation solver applied to $f(x)$
$p(E)$	Normalization over evidence (probability of evidence)
$p(E|M)$	Likelihood (probability of evidence given the mode)
$p(M)$	Prior (prior probability of the mode)
$p(M|E)$	Posterior (probability of mode given the evidence)
$Pl(M)$	Upper bound probability of mode M
P	Posterior state covariance (Kalman filter)
Q	Model error covariance (Kalman filter)
R	Observation error covariance (Kalman filter)
S	Sample covariance matrix
$S(E|\boldsymbol{M})$	Support for evidence E given potentially ambiguous mode $\boldsymbol{M}$
$S(\boldsymbol{M})$	Support for potentially ambiguous mode $\boldsymbol{M}$
$S(E|\boldsymbol{M})$	Support for potentially ambiguous mode $\boldsymbol{M}$ given evidence E
UCEM	Underlying complete evidence matrix
UKF $[f(x)]$	Unscented Kalman filter with a model $f(x)$

Part One

Fundamentals

1

Introduction

1.1 Motivational Illustrations

Consider the following scenarios:

Scenario A

You are a plant operator, and a gas analyser reading triggers an alarm for a low level of a vital reaction component, but from experience you know that this gas analyser is prone to error. The difficulty is, however, that if the vital reaction component is truly scarce, its scarcity could cause plugging and corrosion downstream that could cost over $120 million in plant downtime and repairs, but if the reagent is not low, shutting down the plant would result in $30 million in downtime. Now, imagine that you have a diagnosis system that has recorded several events like this in the past, using information from both upstream and downstream, is able to generate a list of possible causes of this alarm reading, and displays the probability of each scenario. The diagnosis system indicates that the most possible cause is a scenario that happened three years ago, when the vital reagent concentration truly dropped, and by quickly taking action to bypass the downstream section of the plant a $120-million incident was successfully avoided. Finally, imagine that you are the manager of this plant and discover that after implementing this diagnosis system, the incidents of unscheduled downtime are reduced by 60% and that incidents of false alarms are reduced by 80%.

Scenario B

You are the head of a maintenance team of another section of the plant with over 40 controllers and 30 actuators. Oscillation has been detected in this plant, where any of these controllers or actuators could be the cause. Because these oscillations can push the system into risky operating regions, caution must be exercised to keep the plant in a safer region, but at the cost of poorer product quality. Now, imagine you have a diagnosis tool that has data recorded from previous incidents, their troubleshooting solutions, and the probabilities of each incident.

Process Control System Fault Diagnosis: A Bayesian Approach, First Edition. Ruben Gonzalez, Fei Qi and Biao Huang.
© 2016 John Wiley & Sons, Ltd. Published 2016 by John Wiley & Sons, Ltd.

With this tool, we see that the most probable cause (at 45%) was fixed by replacing the stem packing on Valve 23, and that the second most probable cause (at 22%) was a tank level controller that in the past was sometimes overtuned by poor application of tuning software. By looking at records, you find out that a young engineer recently used tuning software to re-tune the level controller. Because of this information, and because changing the valve packing costs more, you re-tune the controller during scheduled maintenance, and at startup find that the oscillations are gone and you can now safely move the system to a point that produces better product quality. Now that the problem has been solved, you update the diagnosis tool with the historical data to improve the tool's future diagnostic performance. Now imagine, that as the head engineer of this plant, you find out that 30% of the most experienced people on your maintenance team are retiring this year, but because the diagnostic system has documented a large amount of their experience, new operators are better equipped to figure out where the problems in the system truly are.

Overview

These stories paint a picture of why there has been so much research interest in fault and control loop diagnosis systems in the process control community. The strong demand for better safety practices, decreased downtime, and fewer costly incidents (coupled with the increasing availability of computational power) all fuel this active area of research. Traditionally, a major area of interest has been in detection algorithms (or *monitors* as they will be called in this book) that focus on the behaviour of the system component. The end goal of implementing a monitor is to create an alarm that would sound if the target behaviour is observed. As more and more alarms are developed, it becomes increasingly probable that a single problem source will set off a large number of alarms, resulting in an alarm flood. Such scenarios in industry have caused many managers to develop alarm management protocols within their organizations. Scenarios such as those presented in scenarios A and B can be realized and in some instances have already been realized by research emphasizing the best use of information obtained from monitors and historical troubleshooting results.

1.2 Previous Work

1.2.1 Diagnosis Techniques

The principal objective in this book is to diagnose the operational mode of the process, where the mode consists of the operational state of all components within the process. For example, if a system comprises a controller, a sensor and a valve, the mode would contain information about the controller (e.g. well tuned or poorly tuned), the sensor (e.g. biased or unbiased) and the valve (e.g. normal or sticky). As such, the main problem presented in this book falls within the scope of *fault detection and diagnosis*.

Fault detection and diagnosis has a vast (and often times overwhelming) amount of literature devoted to it for two important reasons:

1. The problem of fault detection and diagnosis is a legitimately difficult problem due to the sheer size and complexity of most practical systems.

2. There is great demand for fault detection and diagnosis as it is estimated that poor fault management has cost the United States alone more than \$20 billion annually as of 2003 (Nimmo 2003).

In a three-part publication, Venkatasubramanian et al. (2003b) review the major contributions to this area and classify them under the following broad families: quantitative model-driven approaches (Venkatasubramanian et al. 2003b), qualitative model-driven approaches (Venkatasubramanian et al. 2003a), and process data-driven approaches (Venkatasubramanian et al. 2003c). Each type of approach has been shown to have certain challenges. Quantitative model-driven approaches require very accurate models that cover a wide array of operating conditions; such models can be very difficult to obtain. Qualitative model-driven approaches require attention to detail when developing heuristics, or else one runs the risk of a spurious result. Process data-driven approaches have been shown to be quite powerful in terms of detection, but most techniques tend to yield results that make fault isolation difficult to perform. In this book, particular interest is taken in the quantitative model-driven and the process data-driven approaches.

Quantitative Model-driven Approaches

Quantitative model-driven approaches focus on constructing the models of a process and using these models to diagnose different problems within a process (Lerner 2002) (Romessis and Mathioudakis 2006). These techniques bear some resemblance to some of the monitoring techniques described in Section 1.2.2 applied to specific elements in a control loop. Many different types of model-driven techniques exist, and have been broken down according to Frank (1990) as follows:

1. *The parity space approach* looks at analytical redundancy in equations that govern the system (Desai and Ray 1981).
2. *The dedicated observer and innovations approach* filters residual errors from the Parity Space Approach using an observer (Jones 1973).
3. *The Fault Detection Filter Approach* augments the State Space models with fault-related variables (Clark et al. 1975; Willsky 1976)
4. *The Parameter Identification Approach* is traditionally performed offline (Frank 1990). Here, modeling techniques are used to estimate the model parameters, and the parameters themselves are used to indicate faults.

A popular subclass of these techniques is deterministic fault diagnosis methods. One popular method in this subclass is the parity space approach (Desai and Ray 1981), which set up parity equations having analytical redundancy to look at error directions that could correspond to faults. Another popular method is the observer-based approach (Garcia and Frank 1997), which uses an observer to compare differences in the predicted and observed states.

Stochastic techniques, in contrast to deterministic techniques, use fault-related parameters as augmented states; these methods enjoy the advantage of being less sensitive to process noise (Hagenblad et al. 2004), being able to determine the size and precise cause of the fault, but are very difficult to implement in large-scale systems and often require some excitement

(Frank 1996). Including physical fault parameters in the state often requires a nonlinear form of the Kalman filter (such as the extended Kalman filer (EKF), unscented Kalman filter (UKF) or particle filter) because these fault-related parameters often have nonlinear relationships with respect to the states. Such techniques were pioneered by Isermann (Isermann and Freyermuth 1991), (Isermann 1993) with other important contributions coming from Rault et al. (1984). The motivation for including fault parameters in the state is the stochastic Kalman filter's ability to estimate state distributions. By including fault parameters in the state, fault parameter distributions are automatically estimated in parallel with the state. Examples of this technique include that of Gonzalez et al. (2012), which made use of continuous augmented bias states, while Lerner et al. (2000) made use of discrete augmented fault states.

Process Data-driven Approaches

A popular class of techniques for process monitoring are data-driven modeling methods, where one of the more popular techniques is principal component analysis (PCA) (Ge and Song 2010). These techniques create black-box models assuming that the data can be explained using a linear combination of independent Gaussian latent variables (Tipping and Bishop 1998); a transformation method is used to calculate values of these independent Gaussian variables, and abnormal operation is detected by performing a significance test. The relationship between abnormal latent variables and the real system variables is then used to help the user determine what the possible causes of abnormality could be. There have also been modifications of the PCA model to include multiple Gaussian models (Ge and Song 2010; Tipping and Bishop 1999) where the best local model is used to calculate the underlying latent variables used for testing.

All PCA models assume that the underlying variables are Gaussian, but more recent methods (Lee et al. 2006) do away with this assumption by first using independent components analysis (ICA) to calculate values of independent latent variables (which are not assumed to be Gaussian under ICA) and then using a kernel density estimation to evaluate the probability density of that value. Low probability densities indicate that the process is behaving abnormally. Even more recent work (Gonzalez et al. 2015) uses Bayesian networks instead of PCA/ICA to break down the system into manageable pieces; this allows the user to define variables of interest for monitoring and determine the causal structures used to help narrow down causes. Abnormality is detected if key process variables take on improbable values or if groups of key process variables take on improbable patterns. Results from this approach are generally easier to interpret than PCA/ICA-based methods.

Bayesian Data-driven Approaches

This book focuses on using the Bayeisan data-driven approach, which is distinct from other fault detection and diagnosis methods, mainly for the reason that the Bayesian approach is a *higher-level diagnosis method* (Pernestal 2007; Qi 2011). This type of approach is not meant to compete with previously mentioned fault detection and diagnosis methods; instead, the Bayesian approach provides a unifying framework to simultaneously use many of these methods at once. As such, it can take input from many different fault detection and diagnosis techniques in order to make a final decision. In this book, other diagnosis methods and even instruments themselves are treated as input sources and are referred to as *monitors*; this term is chosen mainly because previous work (Qi 2011) focused heavily on using input from control loop monitoring techniques (described in Section 1.2.2).

For Bayesian diagnosis, data from monitors must be collected for every scenario that one would wish to diagnose. In this book, such scenarios are referred to as operational *modes*. When new monitor information arrives, the new information is compared to historical data in order to determine which historical mode best fits the new information. The Bayesian diagnosis technique ranks each of the modes based on posterior probability, which is calculated using Bayes' theorem (Bayes 1764/1958):

$$p(M|E) = \frac{p(E|M)p(M)}{p(E)}$$

$$P(E) = \sum_i p(E|m_i)p(m_i)$$

where

- $p(M|E)$ is the posterior probability, or probability of the mode M given evidence E
- $p(E|M)$ is the likelihood of the evidence E given the historical mode M
- $p(M)$ is the prior probability of the historical mode M
- $p(E)$ is the probability of the evidence E (which is a normalizing constant).

In the Bayesian diagnosis technique, the historical data and mode classifications are used to construct the likelihood $p(E|M)$, and prior probabilities of modes are assigned to $p(M)$ using expert knowledge. While collecting data for historical modes may be a challenge, the Bayesian method at least allows us to collect data in a way that is not necessarily representative of the true mode occurrence rate. For example, if mode 1 occurs 90% of the time, then representative sampling would require that 90% of the data come from mode 1. Bayesian methods (which use prior probabilities to cover mode representation) allow us to collect an arbitrary amount of data for each mode, giving us a lot more flexibility in data collection than other methods.

1.2.2 Monitoring Techniques

Much of this work focuses on monitoring and diagnosing control-loops (a schematic for a typical control loop is given in Figure 1.1); for this area of research, there exists abundant

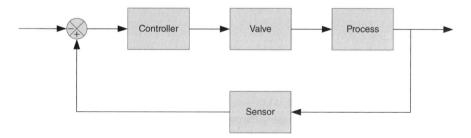

Figure 1.1 Typical control loop

Table 1.1 List of monitors for each system

Simulated	Bench-scale	Industrial-scale
Control performance	Sensor bias	Raw sensor readings
Valve stiction	Process operation	
Process model		

literature on assessing the performance of the entire loop as well as diagnosing problems within the loop's core components. These methods (defined as monitors in this book) can be directly used to create alarms or notification statuses which alert operators and engineers about risky or inefficient operation.

Monitors tend to focus on one or more of the main components in a control system: for example, the controller, the actuator (often a valve), the process and the sensor. The following monitors will be considered in this book as examples but the diagnosis approach as proposed in this book can be applied to other monitors as well.

- **Control performance monitors** are intended to monitor the performance of the controller, but are often affected by other parts of the control loop.
- **Sensor bias monitors** focus on sensor performance.
- **Valve stiction monitors** focus on valve performance, but can sometimes be affected by other sources of oscillation.
- **Process model monitors** evaluate the correctness of the process model, which has utility in diagnosing controller performance and process performance. Deviation from the model can indicate a change in the system operation, and perhaps even a fault. In addition, because control tuning is performed with a model in mind, changes in the model may indicate that the current controller configuration is not suitable for current operation.
- **Process operation monitors** tend to fall under the category of fault detection, and aim to diagnose abnormalities and faults within a process.

The methods in this book are tested on three particular testbed systems: a simulated system, a bench scale system and an industrial scale system. Each type of monitor has been used in at least one of the testbed systems; a summary of monitors for each system is presented in Table 1.1. The simulated system makes use of three monitors (control performance monitors, valve stiction monitors and process model monitors) while the bench scale system makes use of the two remaining monitor types (a process operation monitor and two sensor bias monitors). The industrial-scale system uses no monitors, but instead directly uses data from the various sensors within the facility.

Control Performance Monitoring

Control performance assessment is concerned with the analysis of available control loop performance against certain benchmarks, while control performance monitoring is concerned with monitoring control performance change with respect to certain references. Due to their similarity, the two terminologies have often been used interchangeably and it is commonly accepted that they can represent each other. Research in this areas was pioneered by Harris et al. 1999 for proposing the minimum variance control (MVC) benchmark. Huang et al. (1995) developed a filtering and correlation (FCOR) algorithm to estimate the MVC benchmark that can be easily extended to multivariate systems. A state space framework for the MVC benchmark was proposed by McNabb and Qin (2005). The MVC index was extended to multivariate systems by Harris et al. (1996) and Huang et al. (1997); the latter tackled MIMO MVC benchmark by introducing the unitary interactor matrix. The MVC benchmark provides a readily computable and physically significant bound on control performance.

The theoretical variance lower bound of MVC may not be achievable for most practical controllers. More realistic performance indices are needed. Ko and Edgar (1998) discussed a PID benchmark. An approach was presented by Qin (1998), stating that MVC can be achievable for a PID controller when the process time delay is either small or large, but not medium. Huang and Shah (1999) proposed the linear quadratic Gaussian (LQG) regulator benchmark as an alternative to the MVC benchmark, based on the process model. Model-based approaches also exist for benchmarking model predictive control (MPC) systems (see Shah et al. (2001) and Gao et al. (2003)).

The benchmarks discussed above mainly focus on stochastic performance. However, these benchmarks can also be related to deterministic performances, such as overshoot, decay ratio, settling time, etc. Ko and Edgar (2000) modified the MVC index to include setpoint variations in the inner loop of cascade control. The influence of setpoint changes on the MVC index was discussed by Seppala et al. (2002), who proposed a method to decompose the control error into two components: one that resulted from setpoint changes, and another from a setpoint detrended signal. Thornhill et al. (2003) examined the reasons why performance during setpoint change differs from the performance during operation at a constant setpoint. The extension of the MVC index to the varying setpoint case has also been discussed by McNabb and Qin (2005).

Some other methods have also been proposed for control performance assessment. Kendra and Cinar (1997) applied frequency analysis to evaluate control performance. An r statistic was introduced by Venkataramanan et al. (1997) that detects deviations from setpoint, regardless of the output noise. Li et al. (2003) proposed a relative performance monitor, which compares the performance of a control loop to that of a reference model.

A number of commercial control performance assessment software packages are available on the market, such as the Intune software tools by Control Soft, LoopScout by Honeywell Hi-Spec Solutions, Performance Surveyor by DuPont, etc. (Jelali 2006). Various successful industrial applications have also been reported (Hoo et al. 2003; Jelali 2006).

In this book, the FCOR algorithm (Huang et al. 1995) is used to calculate the MVC benchmark and serves as the control performance monitor for the simulated system.

Valve Stiction Monitors

The undesirable behaviour of control valves is the biggest single contributor to poor control loop performance (Jelali and Huang 2009). According to Jelali and Huang (2009), 20–30% of control loop oscillations are induced by valve nonlinearities, including stiction, deadband, hysteresis, etc. Among these problems, stiction is the most common one in the process industry (Kano et al. 2004). Oscillation in control loops increases the variability of process variables, which in turn affects product quality, increases energy consumption, and accelerates equipment wear. Detecting valve stiction in a timely manner will bring significant economic benefits, and thus there is a strong incentive for valve stiction detection research. A comprehensive review and comparison of valve stiction detection methods can be found in Jelali and Huang (2009).

Singhal and Salsbury (2005) proposed a stiction detection methodology by calculating the ratio of the areas before and after the oscillation peaks of the PV signal. A method for diagnosing valve stiction was developed based on observations of control loop signal patterns by Yamashita (2006). The method determines typical patterns from valve input and valve output/process variables in the control loop, and thus does not allow detection of stiction which shows up in different patterns. Scali and Ghelardoni (2008) improved the work of Yamashita (2006) to allow different possible stiction patterns to be considered. Choudhury et al. (2007) proposed a controller gain change method, which is based on the change in the oscillation frequency due to changes in the controller gain to detect valve stiction. Yu and Qin (2008) showed that this method can fail to detect the presence of the sticky valve in interacting multi-input multi-output systems. A strategy based on the magnitude of relative change in oscillation frequency due to changes in controller gain is proposed to overcome the limitations of the existing method.

Despite the various work regarding stiction detection, valve stiction quantification remains a challenging problem. Choudhury et al. (2008) proposed a method to quantify stiction using the ellipse fitting method. The PV vs. OP plot is fitted to an ellipse and the amount of stiction is estimated as the maximum length of the ellipse in the OP direction. Chitralekha et al. (2010) treated the problem of estimating the valve position as an unknown input estimation problem. The unknown input is estimated by means of an input estimator based on the Kalman filter. Jelali (2008) presented a global optimization based method to quantify valve stiction. A similar method was also proposed by Srinivasan et al. (2005). The approach is based on identification of a Hammerstein model consisting of a sticky valve and a linear process. The stiction parameters and the model parameters are estimated simultaneously with a global grid search optimization method. Jelali and Huang (2009) presented a closed-loop stiction quantification approach using routine operating data. A suitable model structure of valve stiction is chosen prior to conducting valve stiction detection and quantification. Given the stiction model structure, a feasible search domain of stiction model parameters is defined, and a constrained optimization problem is solved in order to determine the stiction model parameters.

The aforementioned stiction qualification methods all assume that the process is linear. Nallasivama et al. (2010) proposed a method to qualify the stiction for closed-loop nonlinear systems. The key idea used in the approach is based on the identification of extra information available in process output, PV, compared to the controller output, OP. Stiction phenomenon leads to many harmonic components compared to the Fourier transform of the Volterra system, which allows stiction detection in nonlinear loops.

In this book, a simple stiction monitoring algorithm is used for the simulated system based on fitting the valve's input–output relationship to an ellipse (Choudhury et al. 2008). If stiction is absent, the data should be easily fitted by an ellipse. However, if the fit is poor, it is likely that stiction is present.

Model–plant Mismatch Monitors

A large volume of work has been published for open-loop model validation. However, the literature has been relatively sparse on studies concerned with on-line model validation using closed-loop data.

Huang (2001) developed a method for the analysis of detection algorithms in the frequency domain under closed-loop conditions. The divergence algorithm is extended to the model validation for the general Box–Jenkins model under closed-loop conditions through the frequency domain approach. Based on the two-model divergence method, Jiang et al. (2009) developed two closed-loop model validation algorithms, which are only sensitive to plant model changes. Of the two algorithms, one is sensitive to changes in both plant and disturbance dynamics, while the other is only sensitive to changes in plant dynamics, regardless of changes in disturbance dynamics and additive process faults, such as sensor bias.

Badwe et al. (2009) proposed a model mismatch detection method based on the analysis of partial correlations between the model residuals and the manipulated variables. The more significant this correlation is, the higher is the possibility that there exists model mismatch. Badwe et al. (2010) further extended their earlier work by analysing the impact of model mismatch on the control performance.

Selvanathan and Tangirala (2010) introduced a plant-model ratio (PMR) as a measure to quantify the model–plant mismatch in the frequency domain. The PMR provides a mapping between its signatures and changes in process models, and thus the changes in model gain, time constant and time delay can be identified. Although it is claimed that the PMR can be estimated from closed-loop operating data, a significant underlying assumption is that the setpoint contains at least a pulse change. This assumption, however, can be restrictive in practice.

In this book the output error (OE) model method is used for model error monitoring. This algorithm focuses on multi-input, single-output (MISO) systems, even though the simulated process is a MIMO (multiple input, multiple-output) system; however, a MIMO system can be easily constructed using several MISO systems.

Bias Monitors

Sensor bias can also be a problem in control loops, as sensors are the main reference for control action. A common method for detecting sensor bias in process industries is the use of data reconciliation and gross error detection (Mah and Tamhane 1982). Most data-reconciliation and gross error detection methods have been proposed for offline implementation (Ozyurt and Pike 2004); recently, Qin and Li. (2001) and Gonzalez et al. (2012) developed on-line versions.

In this book, bias monitors for the bench-scale process focus on the flow meter output versus pump speed. This type of monitor is effective for positive-displacement pumps such as those found in the bench-scale process.

Process Operation Monitors

Process operation monitoring is a broad area of research, mainly because of the large variety of processes that can be monitored and the large number of operation phenomena that can be targeted (such as faults/breakdowns, abnormal/suboptimal operation and violation of operating limits). Literature in this area falls under fault detection and diagnosis literature, which is reviewed in Section 1.2.1.

In this book, for the bench-scale system, a quantitative model-driven technique is used based on the Kalman filter; here, the state is augmented in order to include two fault-related parameters (representing leaks). Under ideal conditions, the parameters have values of zero (no leak), but as leaks are introduced, the parameter values change to values significantly greater than 0.

1.3   Book Outline

This book is broken down into two major parts, *Fundamentals* and *Application*, and each of the major contributions is generally represented in both parts. The fundamentals section focuses on theoretical development and justification of the proposed techniques, while the application section focuses on succinctly conveying all information required to apply these techniques. Since both parts are meant to be stand-alone, there may be some slight overlap between them, namely the parts in the fundamentals section that are directly relevant to applications.

A number of techniques exist in this book that many readers may not be familiar with, namely Bayesian diagnosis, Dempster–Shafer theory, kernel density estimation and bootstrapping. A tutorial is provided which covers fundamental aspects of all four of these techniques.

1.3.1   Problem Overview and Illustrative Example

The main objective of this work is to diagnose the process operating mode (which contains information about the state of each process component of interest, such as sticky valves, biased sensors, inaccurate process models etc.). Before diagnosing modes, we collect historical data from monitors for each mode; this historical data is used to diagnose the mode when new evidence becomes available online. Because it is assumed that corresponding modes are available with the historical data, this book takes a *supervised learning approach* when applying historical data.

In order to easily illustrate the challenges associated with Bayesian diagnosis, we start from a toy problem where the modes consist of two different coins, one with a bias toward heads (probability of heads = 0.6) and one that is fair (probability of heads = 0.5). The probability estimates are obtained through historical data of coin flips. For the diagnosis problem, a coin is randomly selected and we wish to use evidence of coin flipping to determine which coin was selected. The evidence is provided by two people flipping the same coin once.

1.3.2   Overview of Proposed Work

This book aims to address various challenging issues with respect to Bayesian diagnosis. A visual map of these solutions is given in Figure 1.2, where shaded boxes indicate problems, and white boxes indicate solutions proposed by this book; dotted lines indicate a combination of multiple solutions.

Autodependent Modes

For industrial processes, mode changes tend to be quite rare, which means that the mode at time t is highly dependent on the mode at time $t - 1$. Taking this type of dependency into account can significantly increase the precision of the diagnosis results, as consecutive pieces

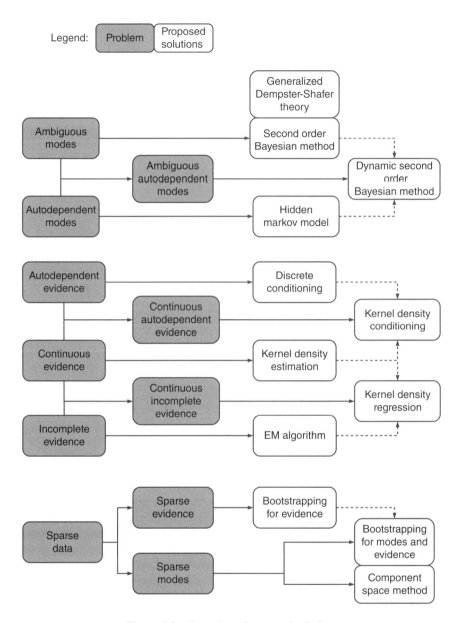

Figure 1.2 Overview of proposed solutions

of evidence contain more information than individual pieces. This type of dependency has been addressed in Qi and Huang (2010b).

Returning to our coin-flipping example, consider the case where after each pair of flips there is a probability of the coin being switched. If that probability is low, the 'mode' has a strong time-wise autodependence. This means that consecutive pieces of evidence contain even more information about the mode than single pieces of evidence themselves.

Autodependent Evidence

Monitor readings often use data from previous time steps in order to calculate a result. If monitor readings are not sampled slowly enough, the evidence will be autodependent. Taking the autodependence of evidence into account was addressed in Qi and Huang (2011a).

Autodependent evidence can also be applied to our coin-flipping example. If the second coin flipper obtained heads at $t - 1$, and the first coin flipper at time t placed the coin on his thumb heads-side-up (with tails being similarly treated) then results would exhibit time-wise dependence.

Incomplete Discrete Evidence

It is not uncommon in process industries that historical records are unavailable during certain time intervals. Since sensors are also used for monitoring, the corresponding monitor will also be unavailable, rendering a data point incomplete (as some elements are missing). Simply discarding incomplete data points will result in a loss of information so Qi and Huang (2010a) proposed using Bayesian methods to recover the useful information from these incomplete data points.

Using our coin-flipping example, consider the case where the evidence from the two people flipping coins is dependent. For example, after the first person flips a coin, whatever side faces up will be placed face up on the second person's thumb. Now the historical data contains the results of both people flipping coins. In some circumstances, the result from one of the two flips will be missing. Because the coin flips are dependent, Qi and Huang (2010a) adopt Bayesian methods to use the information present to make up for the missing information.

Ambiguous Modes from a Bayesian Perspective

Qi and Huang (2010a) addressed the issue where some elements of historical evidence records are missing, causing the evidence to be incomplete. However, just as evidence requires input from multiple monitors, the mode requires information from multiple components. If any information about the components is missing, a number of different modes will be possible, causing the mode to be *ambiguous*.

For example, if some of the historical data from coin-flipping exercises contained no information on which coin was flipped (biased or fair) then either coin could have produced the results (heads or tails) and the corresponding mode (or coin in this case) is *ambiguous*. Since the conditioning variable is unknown, the probability cannot be calculated in a straightforward manner.

Ambiguous Modes from a Dempster–Shafer Perspective

Demspter–Shafer theory (Dempster 1968; Shafer 1976) has been deemed by many to be a generalization of Bayesian diagnosis that is able to handle ambiguity. However, it is shown in this book that Dempster–Shafer theory does not adequately formulate the problem of ambiguous modes in the historical data when likelihoods $p(E|M)$ are used. Some modifications are required in order to properly fit the data-driven diagnosis problem into a Dempster–Shafer framework.

Using Continuous Evidence

Previously, it was assumed that information used by the diagnosis method was discrete (our coin-flipping exercise yields discrete evidence). In reality, however, most monitors yield an output that is continuous (e.g. a monitor that monitors changes in compressor pressure). In order to reduce the amount of information lost through discretizing continuous values, this book proposes the use of *kernel density estimation* (as proposed in Gonzalez and Huang (2014)) to make use of the continuous data directly.

Continuous evidence can also suffer from missing data, but because kernel density estimation is non-parametric, the expectation-maximization (EM) algorithm is not directly applicable. The most common method used to deal with missing evidence in a kernel density estimation framework is kernel density regression (a method used to calculate the expected value of the missing data). The completed data is then included in the data set used for kernel density estimation. Due to the simplicity of this solution, the problem of missing continuous evidence is included in the continuous evidence chapter.

Sparse Evidence given a Mode

If a process has a large number of components, the number of possible modes will be very large. In such cases, it is quite possible for data from a particular mode to be quite sparse. Qi and Huang (2011b) recommended the use of bootstrapping as a method to generate additional data and get a better representation of the monitor distribution.

For the coin-flipping example, consider the case where one of the coins (such as the biased one) does not have a large amount of historical data. Bootstrapping is a method that was suggested in Qi and Huang (2011b) to resolve this issue by simulating more coin flips by randomly drawing from previous results recorded in the historical samples.

Sparse or Missing Modes in the Data

As the number of components in a system increases, the possible modes will increase exponentially. For systems with a large number of components, it is likely that data for a significant number of modes will be missing entirely.

For example, in our coin-flipping exercise, even though we only have two coins (e.g. modes), we might not have any historical data from one of the coins. This book will present techniques on what one can do if certain modes of interest are absent from the historical data.

Dynamic Application of Continuous Evidence and Ambiguous Modes

When accounting for ambiguous modes, the solution for addressing mode autodependence will be affected. Similarly, when accounting for continuous evidence, the solution for autodependent evidence will be affected. In Part One, which deals with fundamentals, the solution to autodependent modes is addressed in Chapter 6, which discusses ambiguous modes. Likewise, the solution to autodependent evidence is addressed in Chapter 8. However, in Part Two, which deals with application, the solutions to both autodependent modes and evidence are dealt with in one chapter (Chapter 18).

References

Badwe A, Gudi R, Patwardhan R, Shah S and Patwardhan S 2009 Detection of model-plant mismatch in mpc applications. *Journal of Process Control* **19**, 1305–1313.

Badwe A, Patwardhan R, Shah S, Patwardhan S and Gudi R 2010 Quantifying the impact of model-plant mismatch on controller performance. *Journal of Process Control* **20**, 408–425.

Bayes T 1764/1958 An essay towards solving a problem in the doctrine of chances. *Biometrika* **45**, 296–315.

Chitralekha SB, Shah SL and Prakash J 2010 Detection and quantification of valve stiction by the method of unknown input estimation. *Journal of Process Control* **20**, 206–216.

Choudhury MAAS, Kariwala V, Thornhill NF, Douke H, Shah SL, Takada H and Forbes JF 2007 Detection and diagnostics of plant-wide oscillations. *Canadian Journal of Chemical Engineering* **85**, 208–219.

Choudhury M, Jain M and Shah S 2008 Stiction – definition, modelling, detection and quantification. *Journal of Process Control* **18**(3-4), 232–243.

Clark RN, Fosth DC and Walton VM 1975 Detection instrument malfunctions in control systems. *IEEE Trans. Aerospace Electron. Syst* **AES-II**, 465–473.

Dempster A 1968 A generalization of Bayesian inference. *Journal of the Royal Statistical Society. Series B* **30**(2), 205–247.

Desai M and Ray A 1981 A fault detection and isolation methodology. *Proc. 20th Conf. on Decision and Control*, pp. 1363–1369.

Frank P 1990 Fault diagnosis in dynamic systems using analytical and knowledge-based redundancy: a survey and some new results. *Automatica* **26**(3), 459–474.

Frank P 1996 Analytical and qualitative model-based fault diagnosis – a survey and some new results. *European Journal of Control* **2**, 6–28.

Gao J, Patwardhan R, Akamatsu K, Hashimoto Y, Emoto G and Shah SL 2003 Performance evaluation of two industrial MPC controllers. *Control Engineering Practice* **11**, 1371–1387.

Garcia E and Frank P 1997 Deterministic nonlinear observer-based approaches to fault diagnosis: a survey. *Control Engineering Practice* **5**(5), 663–670.

Ge Z and Song Z 2010 Mixture Bayesian regularization method of PPCA for multimode process monitoring. *Process Systems Engineering* **56**(11), 2838–2849.

Gonzalez R and Huang B 2014 Control-loop diagnosis using continuous evidence through kernel density estimation. *Journal of Process Control* **24**(5), 640–651.

Gonzalez R, Huang B, Xu F and Espejo A 2012 Dynamic Bayesian approach to gross error detection and compensation with application toward an oil sands process. *Chemical Engineering Science* **67**(1), 44–56.

Gonzalez R, Huang B and Lau E 2015 Process monitoring using kernel density estimation and Bayesian networking with an industrial case study. *ISA Transactions* **58**, 330–347.

Hagenblad A, Gustafsson F and Klein I 2004 A comparison of two methods for stochstic fault detection: the parity space approach and principle component analysis. *13th IFAC Symposium on System Identification (SYSID 2003)*, Rotterdam, The Netherlands, 27–29 August 2003.

Harris T, Boudreau F and MacGregor JF 1996 Performance assessment using multivariable feedback controllers. *Automatica* **32**, 1505–1518.

Harris T, Seppala C and Desborough L 1999 A review of performance monitoring and assessment techniques for univariate and multivariate control systems. *Journal of Process Control* **9**(1), 1–17.

Hoo K, Piovoso M, Sheneller P and Rowan D 2003 Process and controller performance monitoring: overview with industrial applications. *International Journal of Adaptive Control and Signal Processing* **17**, 635–662.

Huang B 2001 On-line closed-loop model validation and detection of abrupt parameter changes. *Journal of Process Control* **11**, 699–715.

Huang B and Shah S 1999 *Performance Assessment of Control Loops*. Springer-Verlag.

Huang B, Shah S and Kwok K 1995 On-line control performance monitoring of MIMO processes. *Proceedings of American Control Conference*, 21–23 June 1995, Seattle, USA.

Huang B, Shah S and Kwok E 1997 Good, bad or optimal? Performance assessment of multivariable process. *Automatica* **33**, 1175–1183.

Isermann R 1993 Fault diagnosis of machines via parameter estimation and knowledge processing. *Automatica* **29**, 815–836.

Isermann R and Freyermuth B 1991 Process fault diagnosis based on process model knowledge, parts i (principles) and ii (case study experiments). *ASME Journal of Dynamic Systems, Measurement Control* **113**, 620–633.

Jelali M 2006 An overview of control performance assessment technology and industrial applications. *Control Engineering Practice* **14**(5), 441–466.

Jelali M 2008 Estimation of valve stiction in control loops using separable least-squares and global search algorithms. *Journal of Process Control* **18**, 632–642.

Jelali M and Huang B (eds) 2009 *Detection and Diagnosis of Stiction in Control Loops: State of the Art and Advanced Methods*. Springer.

Jiang H, Huang B and Shah S 2009 Closed-loop model validation based on the two-model divergence method. *Journal of Process Control* **19**, 644–655.

Jones H 1973 Failure detection in linear systems. PhD thesis. MIT, Cambridge, MA.

Kano M, Maruta H, Kugemoto H and Shimizu K 2004 Practical model and detection algorithm for valve stiction. *Proceedings of the IFAC Symposium on Dynamics and Control of Process Systems*, Boston, USA.

Kendra S and Cinar A 1997 Controller performance assessment by frequency domain techniques. *Journal of Process Control* **7**, 181–194.

Ko B and Edgar T 1998 Assessment of achievable PI control performance for linear processes with dead time. *Proceedings of American Control Conference*, 24–26 June 1998, Philadelphia, Pennsylvania, USA.

Ko B and Edgar TF 2000 Performance assessment of cascade control loops. *AIChE Journal* **46**, 281–291.

Lee JM, Lee IB and Qin S 2006 Fault detection and diagnosis based on modified independent component analysis. *AIChE Journal* **52**, 3501–3514.

Lerner UN 2002 *Hybrid Bayesian networks for reasoning about complex systems*. PhD thesis. Stanford University.

Lerner U, Parr R, Koller D and Biswas G 2000 Bayesian fault detection and diagnosis in dynamic systems. *17th National Conference on Artificial Intelligence (AAAI-00)* , July 30–August 3, 2000, Austin, Texas.

Li Q, Whiteley JR and Rhinehart RR 2003 A relative performance monitor for process controllers. *International Journal of Adaptive Control and Signal Processing* **17**, 685–708.

Mah RSH and Tamhane AC 1982 Detection of gross errors in process data. *AIChE Journal* **28**(5), 828–830.

McNabb CA and Qin SJ 2005 Projection based MIMO control performance monitoring: II – measured disturbances and setpoint changes. *Journal of Process Control* **15**, 89–102.

Nallasivama U, Babjib S and Rengaswamy R 2010 Stiction identification in nonlinear process control loops. *Computers and Chemical Engineering*. **34**(11), 1890–1898.

Nimmo I 2003 Adequately address abnormal situation operations. *Chemical Engineering Progress* **91**(9), 36–45.

Ozyurt D and Pike R 2004 Theory and practice of simultaneous data reconciliation and gross error detection for chemical processes. *Computers and Chemical Engineering* **28**, 381–402.

Pernestal A 2007 *A Bayesian Approach to Fault Isolation with Application to Diesel Engines*. PhD thesis. KTH School of Electrical Engineering, Sweden.

Qi F 2011 *Bayesian Approach for Control Loop Diagnosis*. PhD thesis. University of Alberta.

Qi F and Huang B 2010a A Bayesian approach for control loop diagnosis with missing data. *AIChE Journal* **56**(1), 179–195.

Qi F and Huang B 2010b Dynamic Bayesian approach for control loop diagnosis with underlying mode dependency. *Industrial & Engineering Chemistry Research* **49**, 8613–8623.

Qi F and Huang B 2011a Bayesian methods for control loop diagnosis in the presence of temporal dependent evidences. *Automatica* **47**, 1349–1356.

Qi F and Huang B 2011b Estimation of distribution function for control valve stiction estimation. *Journal of Process Control* **28**(8), 1208–1216.

Qin S 1998 Control performance monitoring – a review and assessment. *Computers and Chemical Engineering* **23**(2), 173–186.

Qin S and Li. W 2001 Detection and identification of faulty sensors in dynamic process. *AIChE Journal* **47**(7), 1581–1593.

Rault A, Jaume D and Verge M 1984 Industrial fault detection and localization *IFAC 9th World Congress, Budapest*, vol. 4, pp. 1789–1792.

Romessis C and Mathioudakis K 2006 Bayesian network approach for gas path fault diagnosis. *Journal of Engineering for Gas Turbines and Power* **128**, 64–72.

Scali C and Ghelardoni C 2008 An improved qualitative shape analysis technique for automatic detection of valve stiction in flow control loops. *Control Engineering Practice* **16**, 1501–1508.

Selvanathan S and Tangirala AK 2010 Diagnosis of poor control loop performance due to model-plant mismatch. *Industrial Engineering Chemistry Research* **49**, 4210–4229.

Seppala CT, Harris TJ and Bacon DW 2002 Time series methods for dynamic analysis of multiple controlled variables. *Journal of Process Control* **12**, 257–276.

Shafer G 1976 *A Mathematical Theory of Evidence*. Princeton University Press.

Shah SL, Patwardhan R and Huang B 2001 Multivariate controller performance analysis: Methods, applications and challenges *Sixth International Conference on Chemical Process Control*, 7–12 January 2001, Tucson, Arizona.

Singhal A and Salsbury T 2005 A simple method for detecting valve stiction in oscillating control loops. *Journal of Process Control* **15**, 371–382.

Srinivasan R, Rengaswamy R, Narasimhan S and Miller R 2005 A curve fitting method for detecting valve stiction in oscillation control loops. *Industrial Engineering Chemistry Research* **44**, 6719–6728.

Thornhill NF, Huang B and Shah SL 2003 Controller performance assessment in set point tracking and regulatory control. *International Journal of Adaptive Control and Signal Processing* **17**, 709–727.

Tipping M and Bishop C 1998 Probabilistic principal component analysis. *Journal of the Royal Statistical Society* **61**(3), 611–622.

Tipping M and Bishop C 1999 Mixtures of probabilistic principal component analysers. *Neural Computation* **11**(2), 443–482.

Venkataramanan G, Shukla V, Saini R and Rhinehart RR 1997 An automated on-line monitor of control system performance. *Proceedings of the American Control Conference*, 4–6 June 1997, Albuquerque, New Mexico, USA.

Venkatasubramanian V, Rengaswamy R and Kavuri SN 2003a A review of process fault detection and diagnosis part ii: Qualitative models and search strategies. *Computers and Chemical Engineering* **27**, 313–326.

Venkatasubramanian V, Rengaswamy R, Yin K and Kavuri SN 2003b A review of process fault detection and diagnosis part i: Quantitative model-based methods. *Computers and Chemical Engineering* **27**, 293–311.

Venkatasubramanian V, Rengaswamy R, Yin K and Kavuri SN 2003c A review of process fault detection and diagnosis part iii: Process history based methods. *Computers and Chemical Engineering* **27**, 327–346.

Willsky A 1976 Detection instrument malfunctions in control systems. *Automatica* **12**, 601–611.

Yamashita Y 2006 An automatic method for detection of valve stiction in process control loops. *Control Engineering Practice* **14**, 503–510.

Yu J and Qin SJ 2008 Statistical MIMO controller performance monitoring. part ii: Performance diagnosis. *Journal of Process Control* **18**, 297–319.

2

Prerequisite Fundamentals

2.1 Introduction

The primary focus of this book is diagnosing the performance of process systems by means of historical data using Bayesian inference. However, there are many practical problems to be considered before such methods can be properly implemented, namely, missing historical evidence, ambiguous historical modes and sparse data (modes and or evidence). In addition, the Bayesian diagnosis method can be enhanced by taking into account dynamic properties of modes and evidence as well as estimating continuous distributions using kernel density estimation. Each of the solutions or enhancements makes use of one of the following five tools:

1. Bayesian inference
2. the EM algorithm
3. techniques for ambiguous modes (including the Dempster–Shafer theory)
4. kernel density estimation
5. bootstrapping.

Working knowledge of each of these tools is useful in understanding the material presented in this book, thus we include a short tutorial for each of them in this chapter. Detailed descriptions and applications of these tools can be found in subsequent chapters.

2.2 Bayesian Inference and Parameter Estimation

Bayesian inference is at the core of every solution mentioned in this book. Philosophically, Bayesian statistics interprets probability in a different manner than the more traditional frequentist approach. The frequentist interpretation discusses problems in dealing with long-term frequencies of data generated from repeated independent random experiments (Venn 1866). However, it does not accommodate the intuitive notion that short-term probabilities exist and have meaning (Korb and Nicholson 2004). By contrast, the Bayesian view of probability asserts that probability represents a subjective degree of belief, a view that was held prior to Venn by de Laplace (1820) and even earlier by Bayes (1764/1958).

In practice, there are two main differences between the Bayesian and frequentist approaches, which mainly addresses parameter estimation and inference.

Process Control System Fault Diagnosis: A Bayesian Approach, First Edition. Ruben Gonzalez, Fei Qi and Biao Huang.
© 2016 John Wiley & Sons, Ltd. Published 2016 by John Wiley & Sons, Ltd.

Parameter Estimation

When estimating parameters, the frequentist approach to parameter estimation assumes that the underlying parameters are not random and hence not subject to chance. By contrast, Bayesian methods assume that the underlying parameters are random and Bayesians must assign prior distributions to these parameters.

Consider an example where we flip a coin 200 times, with 115 results being heads, and 85 results being tails. From these results, we want to know the probability of obtaining heads from a coin flip. The frequentist approach would be to simply estimate the probability parameter from the result ($\theta = 115/200 = 0.575$). There is no distribution associated with this parameter estimation. The Bayesian approach would be to assume a Dirichlet distribution (explained later), which describes the distribution of the probability parameter θ. As we can see in Figure 2.1, the peak probability of this parameter is 0.575, which agrees with the frequentist probability of obtaining heads after flipping a coin. Furthermore, we can see that the distribution is fairly sharp due to the fact that we have performed this experiment about 200 times. Increasing the number of coin flips will increase our confidence in the parameter estimate and make the distribution even sharper.

Inference about One Hypothesis

When performing inference, the frequentist approach yields a probability of being false when assuming each hypothesis. The aim is to select the hypothesis that has the lowest risk of being false. The Bayesian approach will yield probabilities of each hypothesis with the aim of selecting the hypothesis with the largest probability as being true.

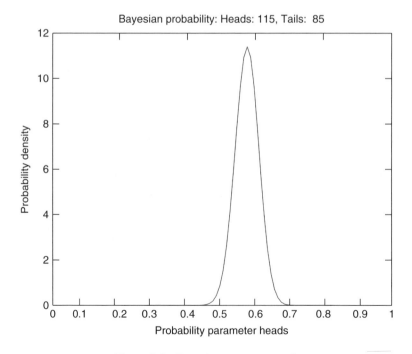

Figure 2.1 Bayesian parameter result

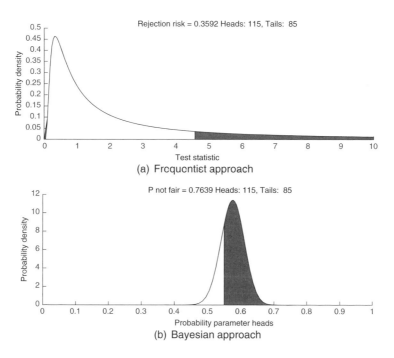

Figure 2.2   Comparison of inference methods

Let us consider the coin-flipping example, where again the heads outcome was observed 115 times, while the tails outcome was observed 85 times and we wish to determine if the coin is fair. When performing inference, the frequentist approach first estimates the distribution of the data. Due to the central limit theorem, we can assume that the mean approaches a normal distribution, enabling us to perform a χ^2 test. Using the χ^2 distribution, we assess the risk of being wrong if we reject the null hypothesis (that the coin is fair, or that $\mu = 0.5$). Figure 2.2(a) shows the corresponding χ^2 distribution, with the shaded area being the risk of rejection (the χ^2 statistic was 0.0229). From integrating the rejection region, we find that we have about a 36% risk of being wrong if we say that the coin is not fair, $\mu \neq 0.5$.

The Bayesian approach to the coin problem is markedly different. First, in order to implement the Bayesian approach, we need to define what a fair coin really is. In this case, let us say that if the probability of heads θ is between 0.45 and 0.55, then the coin is fair. We use the parameter distribution to find out how much of the distribution's area lies outside of this range, which for this case (given in Figure 2.2(b)) is roughly 76%. Thus we can say that based on the observed data, there is a 24% probability that the coin's probability for heads θ lies inside our fairness interval of $[0.45, 0.55]$.

Inference about Many Hypotheses

In the coin-flipping example, the hypothesis was concerned about a continuous hypothesis, that is, the probability parameter θ. Let us consider a new example, where we have two fair coins $\theta = 0.5$ and one coin with a bias toward heads $\theta = 0.6$. Now, instead of θ being continuous (taking on an infinite number of values), it is discrete (taking on the value of either 0.5 or 0.6). For this example, a random coin was selected and 200 trials were performed on this coin, again with heads being observed 115 times and tails being observed 85 times. We would like to determine if the selected coin is one of the fair coins ($\theta = 0.5$) or if it is the biased coin ($\theta = 0.6$).

For the frequentist approach, we calculate the T^2 statistics based on $\mu = 0.5$ and $\mu = 0.6$

$$\chi^2_{\theta=0.5} = \frac{(\hat{\mu} - 0.5)^2}{\hat{\sigma}^2} = \frac{(0.575 - 0.5)^2}{0.2456} = 0.0229$$

$$T^2_{\theta=0.6} = \frac{(\hat{\mu} - 0.6)^2}{\hat{\sigma}^2} = \frac{(0.575 - 0.6)^2}{0.2456} = 0.0025$$

These statistics result in rejection risks of 36% and 84%, respectively. Since $\theta = 0.6$ has the highest rejection risk, we can say that we are more likely to be correct if we say that the biased coin was selected.

Now, in order to use the Bayesian approach, we must familiarize ourselves with Bayes' theorem, which is the fundamental basis for all Bayesian methods

$$p(H|E) = \frac{p(E|H)p(H)}{p(E)}$$

where H represents a hypothesis, E represents evidence and the probability terms are interpreted as follows:

- $p(H)$ is the prior probability of the hypothesis
- $p(E|H)$ is the probability of the evidence given the hypothesis
- $p(E)$ is the probability of the evidence, which can be expressed as

$$p(E) = \int p(E|H)p(H) \, dH$$

- $p(H|E)$ is the posterior probability.

For our applications, H is a random variable representing the possible values of θ (in this example there is a finite number of values, so it is discrete; in the previous examples, θ was continuous).

In order to apply Bayes' theorem, we calculate the likelihoods for $\theta = 0.5$ and $\theta = 0.6$ over the 200 data points.

$$p(E|\theta = 0.5) = 0.5^{115}0.5^{85} = 6.2230 \times 10^{-61}$$

$$p(E|\theta = 0.6) = 0.6^{115}0.4^{85} = 4.5972 \times 10^{-60}$$

We also have the prior probabilities based on the number of coins we have for each hypothesis

$$p(\theta = 0.5) = 2/3$$

$$p(\theta = 0.6) = 1/3$$

The prior probability is based on prior knowledge about the type of coins we have. The resulting Bayesian probabilities are therefore

$$p(\theta = 0.5|E) = \frac{(2/3)6.2230 \times 10^{-61}}{p(E)} = \frac{0.415 \times 10^{-60}}{p(E)} = 0.2131$$

$$p(\theta = 0.6|E) = \frac{(1/3)4.5972 \times 10^{-60}}{p(E)} = \frac{1.532 \times 10^{-60}}{p(E)} = 0.7869$$

From these results, we can say that it is most probable that the biased coin was selected. We can also see that the Bayesian approach has the advantage of being able to use prior probabilities. In addition, the Bayesian approach directly results in probabilities for each hypothesis, which is a more intuitive result than the frequentist result of rejection risks.

Dynamic Inference

One final comment about the difference between frequentist and Bayesian inference is that Bayesian inference can be easily implemented dynamically. Recall that in our Bayesian inference we multiplied the evidence together for the 200 samples in one likelihood calculation step. However, one can obtain the same result by updating the prior probability using a single piece of evidence at a time. For example, let us say that the first observation is heads, then

$$p(\theta = 0.5|E_1) = \frac{(2/3)0.5}{p(E_1)} = \frac{1/3}{p(E_1)} = 5/8$$

$$p(\theta = 0.6|E_1) = \frac{(1/3)0.6}{p(E_1)} = \frac{1/5}{p(E_1)} = 3/8$$

Now let us say the second result was tails, then our previous posterior can be used as a prior for the next inference.

$$p(\theta = 0.5|E_1, E_2) = \frac{p(E_2|\theta = 0.5)p(\theta = 0.5|E_1)}{p(E_2)} = \frac{(5/8)0.5}{p(E_2)} = \frac{5/16}{p(E_2)} = 25/37$$

$$p(\theta = 0.6|E_1, E_2) = \frac{p(E_2|\theta = 0.6)p(\theta = 0.6|E_1)}{p(E_2)} = \frac{(3/8)0.4}{p(E_2)} = \frac{3/20}{p(E_2)} = 12/37$$

If this is continued for the entire 200 observations, we will obtain the same result as the case where all 200 observations were considered at once. For the Bayesian approach, our diagnosis result can be easily updated every time new evidence is made available, and the computational burden is the same for each new data point. Conversely, for the frequentist approach, a new test over the entire dataset has to be calculated every time a new data point is added. In this way, the computational burden increases with each new data point. This property makes the Bayesian approach more practical for on-line applications than the frequentist approach, especially if computational power is limited.

Practical Considerations

In the scientific community, the frequentist approach is generally more popular, understand-ably because parameter estimates and proposed hypotheses are not random, but take fixed values from nature. Furthermore, the aim of scientists is to perform carefully controlled exper-iments where samples can be considered independent and identically distributed, which is a required condition for applying the frequentist approach. Conversely, Bayesian inference is more popular in the artificial intelligence community, especially in areas where real-time decisions have to be made on hypotheses that can change at random, for example diagnosisng problems in diesel engines (Pernestal 2007). Bayesian inference has the advantage that it can be easily implemented in machine learning and on-line diagnosis as it does not require time-wise independence. Data from different scenarios can be collected off-line to estimate their respec-tive distributions, then when implemented online, new evidence can be used to make decisions.

2.2.1 Tutorial on Bayesian Inference

In this tutorial a system with two components is considered (as shown in Figure 2.3):

1. A valve which can be subject to stiction.
2. A sensor which can be subject to bias.

For the sake of illustration, we assume that the two problems do not happen simultaneously, resulting in three operating modes for the process: *normal operation*, *valve stiction*, and *sensor bias*.

Bayesian inference techniques are always applied on top of monitoring techniques. In this way, Bayesian inference was designed to piece together evidence from various sources in order to make a decision. With this in mind, we assume that monitoring algorithms are already in place to detect stiction and bias. The Bayesian technique is simply a layer applied above the monitors in order to make sense of monitoring input. At this point, we will assume that the monitors yield a discrete output. In the case of our example, the output for the stiction monitor is either 0 (stiction not detected) or 1 (stiction detected). Likewise, the output for the bias monitor is either 0 (bias not detected) or 1 (bias detected). The evidence space, therefore, consists of four possible discrete values (as shown in Figure 2.4):

1. $e_1 = [0, 0]$
2. $e_2 = [0, 1]$
3. $e_3 = [1, 0]$
4. $e_4 = [1, 1]$

The goal of using historical evidence is to estimate the *likelihood* $p(E|M)$ for each mode. This can be combined with user-defined prior mode probabilities $p(M)$ in order to obtain a posterior

$$p(M|E) = \frac{p(E|M)p(M)}{\sum_M p(E|M)p(M)} \qquad (2.1)$$

For discrete data, the likelihood $p(E|M)$ can be calculated as

$$p(E|M) = \frac{n(E, M)}{n(M)} \qquad (2.2)$$

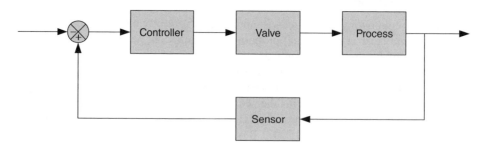

Figure 2.3 Illustrative process

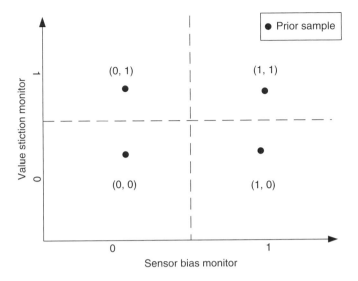

Figure 2.4   Evidence space with only prior samples

where $n(E, M)$ is the number of samples where the evidence E and mode M occur simultaneously, and $n(M)$ is the total number of samples were the mode M occurs.

The motivation for applying the Bayesian technique is alarm management. The monitors themselves are capable of generating alarms, but a problem such as stiction could also affect the sensor bias monitoring alarm. One can see that for systems with a large number components, information on the underlying problem can be obtained from the alarm pattern. Furthermore, the alarms often do not contain information about how certain the alarm is; by contrast, Bayesian techniques assign probabilities to each possible mode, allowing us to ascertain the level of uncertainty about our decision.

Bayesian methods have the added benefit of user-defined priors (denoted as $p(m)$ in Eqn (2.1)). Prior probabilities can be used to assign more weight to modes that are known to occur more frequently. If one does not have any information about prior probabilities, a noninformative flat prior can also be used. This can be applied to inference, but it can also be applied in estimating distributions. For example, we use noninformative prior samples for estimating likelihood distributions by assigning one data point for each discrete evidence, $a(E|m_1) = 1$, $A(m_1) = 4$, as shown in Figure 2.4. Here, all evidence possibilities are shown in a 2×2 grid (2×2 as it contains two monitors with two discrete values each), and a single sample is added to each grid sector representing a possible evidence value. By assigning a point to each grid, we state that for this mode, each possible evidence value was observed once a priori.

After applying the prior samples, historical observations are used to obtain the terms in Eqn (2.2).

$$n(E, M) = \text{Prior}(E, M) + \text{History}(E, M)$$

$$n(M) = \text{Prior}(M) + \text{History}(M)$$

where $\mathrm{Prior}(E, M)$ represents the prior samples of where E and M jointly occur, and $\mathrm{History}(E, M)$ represents the historical data samples where E and M occur. Similarly $\mathrm{Prior}(M)$ and $\mathrm{History}(M)$ represent the prior samples of M and the historical samples of M, respectively. Let us consider the results collected over a number of historical modes, as shown in Table 2.1, which will be used for this example.

When historical data and prior samples have been combined (note: in this example, prior samples placed one sample point for each possible evidence realization under each mode), the result is given in Table 2.2. As a visual example, the evidence space for the *sensor bias* mode is shown in Figure 2.5 with both prior and historical samples included. Likelihoods can be obtained from these samples by normalizing over the frequency of each mode. Results are shown in Table 2.3.

Table 2.1 Counts of historical evidence

E	Normal	Sticky valve	Biased sensor
$e_1 = [0, 0]$	10	0	0
$e_2 = [0, 1]$	0	1	7
$e_3 = [1, 0]$	0	8	1
$e_4 = [1, 1]$	0	1	2

Table 2.2 Counts of combined historical and prior evidence

E	Normal	Sticky valve	Biased sensor
$e_1 = [0, 0]$	11	1	1
$e_2 = [0, 1]$	1	2	8
$e_3 = [1, 0]$	1	9	2
$e_4 = [1, 1]$	1	2	3

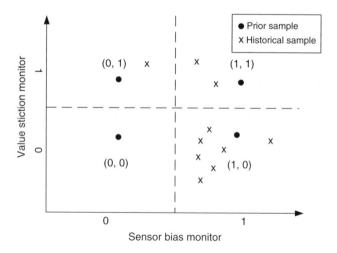

Figure 2.5 Evidence space with prior and historical data

Table 2.3 Likelihoods of evidence

E	Normal	Sticky valve	Biased sensor
$e_1 = [0,0]$	11/14	1/14	1/14
$e_2 = [0,1]$	1/14	1/7	4/7
$e_3 = [1,0]$	1/14	9/14	1/7
$e_4 = [1,1]$	1/14	1/7	3/14

Before performing on-line diagnosis, priors for each mode must be assigned. For instance, if the valve has not been maintained for a considerable amount of time, then a higher prior probability can be assigned to the *sticky valve* mode to reflect our knowledge that the valve has a high chance of being sticky. In such a case, the prior probabilities are assigned as $p(\text{normal}) = 1/4$, $p(\text{sticky valve}) = 1/2$, and $p(\text{biased sensor}) = 1/4$.

With the estimated likelihood probabilities for current evidence E under different modes M, the likelihoods $p(E|M)$ and the user-defined prior probabilities $p(M)$, the posterior probabilities of each mode $m_i \in M$ can be calculated. Among these modes, the one with the largest posterior probability is selected.

As an example, given evidence $[1,0]$ (where the stiction monitor detects a problem and the bias monitor does not) the posterior probabilities can be calculated as

$$p(\text{normal}|[1,0]) \propto p(\text{normal}) \cdot p([1,0]|\text{normal})$$

$$= 1/4 \cdot 1/14 = 1/56 \tag{2.3}$$

$$p(\text{sticky valve}|[1,0]) \propto p(\text{sticky valve}) \cdot p([1,0]|\text{sticky valve})$$

$$= 1/2 \cdot 9/14 = 9/28 \tag{2.4}$$

$$p(\text{biased sensor}|[1,0]) \propto p(\text{biased sensor}) \cdot p([1,0]|\text{biased sensor})$$

$$= 1/4 \cdot 1/7 = 1/28 \tag{2.5}$$

The mode with largest posterior probability, *sticky valve*, is then diagnosed as the underlying process mode. Note that these probabilities do not add up to 1 because they are not normalized by $p(E)$. If proper probabilities are desired, then normalization is required:

$$p(E) = 1/56 + 9/28 + 1/28 = 3/8$$

2.2.2 Tutorial on Bayesian Inference with Time Dependency

Mode Time Dependency

During on-line application, where evidence is obtained at every time instance, unless the time intervals are exceedingly long (which would be undesirable because shorter intervals yield more information), there will be some time dependency with the modes. In general, a mode has a probability of switching $p(M^t|M^{t-1})$ from the mode at time $t - 1$ to the mode at time t and this probability tends to be fairly small, for example it usually takes a while for a valve to become sticky, or for an instrument to be biased, and it takes a while for these problems to be

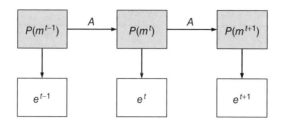

Figure 2.6   Mode dependence (hidden Markov model)

noticed and corrected. Since these switching occurrences can be rare, there tends to be strong autodependence within the modes, thus evidence collected over time yields more information than single pieces of evidence themselves. A visual representation of dependency is available in Figure 2.6, which resembles the well-known *hidden Markov model*.

Mode-time dependency can be taken into account by using switching probabilities. Let us consider our previous example with the following prior probabilities:

$$p(m_1) = 1/2 \qquad \text{normal operation}$$

$$p(m_2) = 1/4 \qquad \text{sticky valve}$$

$$p(m_3) = 1/4 \qquad \text{biased sensor}$$

Now let us consider the modes having the following switching probabilities

$$A \equiv \begin{matrix} p(m_1^t|m_1^{t-1}) = 0.90 & p(m_1^t|m_2^{t-1}) = 0.10 & p(m_1^t|m_3^{t-1}) = 0.10 \\ p(m_2^t|m_1^{t-1}) = 0.05 & p(m_2^t|m_2^{t-1}) = 0.90 & p(m_2^t|m_3^{t-1}) = 0.00 \\ p(m_3^t|m_1^{t-1}) = 0.05 & p(m_3^t|m_2^{t-1}) = 0.00 & p(m_2^t|m_3^{t-1}) = 0.90 \end{matrix}$$

Let us now consider the evidence $e_1 = [0, 0]$ with the following likelihoods obtained from Table 2.3

$$p(e_1|M) = \begin{bmatrix} p(e_1|m_1) = 11/14 \\ p(e_1|m_2) = 1/14 \\ p(e_1|m_3) = 1/14 \end{bmatrix}$$

By combining the likelihood with the prior probabilities we can obtain

$$p(m_1|e_1^{t=1}) = \frac{p(e_1|m_1)p(m_1^{t=1})}{p(e_1^{t=1})}$$

$$= \frac{(11/14)(1/2)}{(11/14)(1/2) + (1/14)(1/4) + (1/14)(1/4)}$$

$$= \frac{11/28}{3/7} = 11/12$$

Similarly, for the other modes, we can obtain

$$p(m_2|e_1^{t=1}) = \frac{p(e_1|m_2)p(m_2^{t=1})}{p(e_1^{t=1})}$$

$$= \frac{(1/14)(1/4)}{3/7} = 1/24$$

$$p(m_3|e_1^{t=1}) = \frac{p(e_1|m_3)p(m_3^{t=1})}{p(e_1^{t=1})}$$

$$= \frac{(1/14)(1/4)}{3/7} = 1/24$$

Now let us consider the probability at $t = 2$ when e_2 is observed. Firstly, the likelihood is shown to be

$$p(e_2|M) = \begin{bmatrix} p(e_2|m_1) = 1/14 \\ p(e_2|m_2) = 1/7 \\ p(e_2|m_3) = 4/7 \end{bmatrix}$$

Now the prior probability for $t = 2$ is the posterior of $t = 1$ with switching probabilities taken into account

$$p(m_1^{t=2}|e_1^{t=1}) = \sum_M p(m_1^t|M^{t-1})p(M|e_1^{t=1})$$

$$= p(m_1^t|m_1^{t-1})p(m_1|e_1^{t=1}) + p(m_1^t|m_2^{t-1})p(m_2|e_1^{t=1}) + p(m_1^t|m_3^{t-1})p(m_3|e_1^{t=1})$$

$$= 0.9 \times (11/12) + 0.1 \times (1/24) + 0.1 \times (1/24) = 5/6$$

$$p(m_2^{t=2}|e_1^{t=1}) = p(m_2^t|m_1^{t-1})p(m_1|e_1^{t=1}) + p(m_2^t|m_2^{t-1})p(m_2|e_1^{t=1}) + p(m_2^t|m_3^{t-1})p(m_3|e_1^{t=1})$$

$$= 0.05 \times (11/12) + 0.9 \times (1/24) + 0.0 \times (1/24) = 1/12$$

$$p(m_3^{t=2}|e_1^{t=1}) = p(m_3^t|m_1^{t-1})p(m_1|e_1^{t=1}) + p(m_3^t|m_2^{t-1})p(m_2|e_1^{t=1}) + p(m_3^t|m_3^{t-1})p(m_3|e_1^{t=1})$$

$$= 0.05 \times (11/12) + 0.0 \times (1/24) + 0.9 \times (1/24) = 1/12$$

where switching probabilities are denoted using t and $t - 1$ due to time invariance (for this time step, $t = 2$ and $t - 1 = 1$). These values are new priors $p(M^{t=2}|e^{t=1})$ which can be combined with the likelihoods $p(e_2^{t=2}|M)$ to obtain a new posterior $p(M^{t=2}|e_2^{t=2})$.

$$p(M^{t=2}|e_2^{t=2}, e_1^{t=1}) = \frac{p(e_2^{t=2}|M)p(M^{t=2}|e^{t=1})}{p(e_2^{t=2})}$$

or more generally

$$p(M^t|E^t, E^{t-1}) = \frac{p(E^t|M)p(M^t|E^{t-1})}{p(E^t)} \tag{2.6}$$

$$p(E^t) = \sum_M p(E^t|M)p(M|E^{t-1})$$

By applying our example, we get the following numerical values for the posterior

$$p(m_1^{t=2}|e_2^{t=2}, e_1^{t=1}) = \frac{p(e_2^{t=2}|m_1)p(m_1^{t=2}|e_1^{t=1})}{p(e_2^{t=2})}$$

$$= \frac{(1/14)(5/6)}{5/42} = 1/2$$

$$p(m_2^{t=2}|e_2^{t=2}, e_1^{t=1}) = \frac{p(e_2^{t=2}|m_2)p(m_2^{t=2}|e_1^{t=1})}{p(e_2^{t=2})}$$

$$= \frac{(1/7)(1/12)}{5/42} = 1/10$$

$$p(m_1^{t=2}|e_2^{t=2}, e_1^{t=1}) = \frac{p(e_2^{t=2}|m_1)p(m_1^{t=2}|e_1^{t=1})}{p(e_2^{t=2})}$$

$$= \frac{(4/7)(1/12)}{5/42} = 2/5$$

Evidence Time Dependency

For evidence time dependency, we consider the case where E^t depends on E^{t-1}, which can happen, for example, if monitors use a window of data which overlaps with the data that monitors use at different time intervals. For the most basic case, where E^t depends on E^{t-1}, the graphical model resembles Figure 2.7.

In such a case, we wish to evaluate $p(M|E^t, E^{t-1})$ as

$$p(M|E^t, E^{t-1}) = \frac{p(E^t|E^{t-1}, M)p(M)}{p(E^t)}$$

We can obtain the required likelihood expression $p(E^t|E^{t-1}, M)$ by the rule of conditioning

$$p(E^t|E^{t-1}, M) = \frac{p(M, E^t, E^{t-1})}{p(E^{t-1}, M)}$$

which yields the estimator

$$p(E^t|E^{t-1}, M) = \frac{n(M, E^t, E^{t-1})}{n(E^{t-1}, M)} \tag{2.7}$$

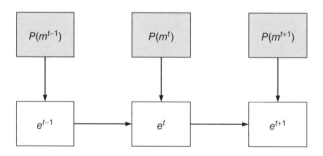

Figure 2.7 Evidence dependence

Table 2.4 Likelihoods of dynamic evidence

E	Normal	Sticky valve	Biased sensor
e_1^t, e_1^{t-1}	0.5	0.01	0.01
e_1^t, e_2^{t-1}	0.07	0.07	.01
e_1^t, e_3^{t-1}	0.07	0.01	0.07
e_1^t, e_4^{t-1}	0.07	0.01	0.01
e_2^t, e_1^{t-1}	0.07	0.07	0.01
e_2^t, e_2^{t-1}	0.01	0.5	0.01
e_2^t, e_3^{t-1}	0.01	0.07	0.07
e_2^t, e_4^{t-1}	0.01	0.06	0.01
e_3^t, e_1^{t-1}	0.07	0.01	0.7
e_3^t, e_2^{t-1}	0.01	0.07	0.7
e_3^t, e_3^{t-1}	0.01	0.01	0.5
e_3^t, e_4^{t-1}	0.01	0.01	0.7
e_4^t, e_1^{t-1}	0.06	0.01	0.01
e_4^t, e_2^{t-1}	0.01	0.07	0.01
e_4^t, e_3^{t-1}	0.01	0.01	0.06
e_4^t, e_4^{t-1}	0.01	0.01	0.01

where $n(M^t, E^t, E^{t-1})$ is the number of times M^t, E^t, E^{t-1} jointly occur, and $n(E^{t-1}, M^t)$ is the number of times E^{t-1}, M^t jointly occur. This will mean that the number of evidence possibilities will be squared. For example, for evidence presented in Table 2.3, the dependent evidence solution will resemble Table 2.4.

This table of evidence can be used in the same manner as the previous table (Table 2.3). For example, if the evidence e_1^t, e_1^{t-1} was observed, the posterior would be

$$p(m_1^t | e_1^t, e_1^{t-1}) = \frac{p(e_1^t | e_1^{t-1}, m_1^t) p(m_1^t)}{p(e_1)}$$

$$= \frac{(0.5)(0.5)}{51/200} = 50/51$$

$$p(m_2^t | e_1^t, e_1^{t-1}) = \frac{p(e_1^t | e_1^{t-1}, m_2^t) p(m_1^t)}{p(e_1)}$$

$$= \frac{(0.01)(0.25)}{51/200} = 1/102$$

$$p(m_3^t | e_1^t, e_1^{t-1}) = \frac{p(e_1^t | e_1^{t-1}, m_3^t) p(m_1^t)}{p(e_1)}$$

$$= \frac{(0.01)(0.25)}{51/200} = 1/102$$

Dynamic Evidence and Modes

The solution for dynamic modes can be easily combined with the solution for dynamic evidence. The dynamic evidence solution only modifies the likelihood, while the dynamic modes

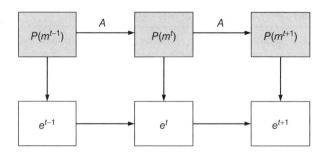

Figure 2.8 Evidence and mode dependence

solution only modifies the prior. Since the dynamic evidence solution only modifies the likeli-hood, we can simply substitute the dynamic evidence likelihood $p(E^t|E^{t-1}, M)$ in Eqn (2.7) for the static likelihood $p(E|M)$ which showed up in the hidden Markov model solution in Eqn (2.6). Thus, when both dynamic evidence and dynamic mode solutions are applied, one can calculate the results using

$$p(M^t|E^t, E^{t-1}) = \frac{p(E^t|E^{t-1}, M^t)p(M^t|E^{t-1}, E^{t-2})}{p(E^t|E^{t-1})} \tag{2.8}$$

$$p(E^t|E^{t-1}) = \sum_{M^t} p(E^t|E^{t-1}, M^t)p(M^t|E^{t-1}, E^{t-2})$$

As one might observe, this solution is applied in the same manner as the dynamic modes solution, but now we replace the evidence (such as Table 2.3) with dynamic evidence (such as Table 2.4). This solution solves the problem depicted in Figure 2.8.

2.2.3   Bayesian Inference vs. Direct Inference

Let us revisit the example where we are monitoring sensor bias and valve stiction for the system in Figure 2.3. Now let us say that we have collected some larger samples of data with the evidence history being summarized in Table 2.5.

By looking at this table, if we observed evidence e_1, an intuitive way to evaluate the probability of the modes would be to count the occurrences of evidence e_1 for each mode

Table 2.5 Counts of combined historical and prior evidence

E	Normal	Sticky valve	Biased sensor	Evidence total
$e_1 = [0,0]$	30	7	4	41
$e_2 = [0,1]$	10	2	10	22
$e_3 = [1,0]$	7	16	2	25
$e_4 = [1,1]$	3	5	4	12
Mode total	50	30	20	100

$(m_1 \rightarrow 30, m_2 \rightarrow 7, m_3 \rightarrow 4)$ and divide it by the total occurrence of evidence $e_1 \rightarrow 41$ so that

$$p(m_1|e_1) = 30/41$$

$$p(m_2|e_1) = 7/41$$

$$p(m_3|e_1) = 4/41$$

This approach is called the *direct approach* as it directly evaluates the evidence based on historical counts. This method, however, assumes that the evidence collected is representative of the mode probability. Thus, it assumes that the prior probability can be obtained from the mode totals at the bottom of Table 2.5:

$$p(m_1) = 50/100 = 0.5$$

$$p(m_2) = 30/100 = 0.3$$

$$p(m_3) = 20/100 = 0.2$$

If these values were used for prior probabilities when applying the Bayesian method, the results obtained would be identical to the results from the direct method:

$$p(m_1|e_1) = \frac{p(e_1|m_1)p(m_1)}{p(e_1)} = \frac{(30/50)(50/100)}{41/100} = 30/41$$

$$p(m_2|e_1) = \frac{p(e_1|m_2)p(m_2)}{p(e_1)} = \frac{(7/30)(30/100)}{41/100} = 7/41$$

$$p(m_3|e_1) = \frac{p(e_1|m_3)p(m_3)}{p(e_1)} = \frac{(4/20)(20/100)}{41/100} = 4/41$$

Thus, as we can see, the Bayesian method is much more flexible as it allows us to select prior probabilities that are not represented in the data. This property of allowing us to use different prior probabilities means that the Bayesian method enjoys certain advantages over the direct probability method:

Bayesian methods allow us to collect arbitrary amounts of data for each mode
When using Bayesian methods, priors take care of the mode probabilities so that the data does not have to.

Bayesian methods allow for easy on-line implementation
When implementing solutions on-line, the prior probabilities change over time when evidence becomes available. This can be easily taken into account when using the Bayesian method, but this is much more difficult to represent when using the direct method.

2.2.4   *Tutorial on Bayesian Parameter Estimation*

In addition to Bayesian inference, some chapters in this book deal with Bayesian parameter estimation. In our previous example, we performed some parameter estimation with respect to the probability parameters

$$p(E|M) = \theta_{E|M} = \frac{n(E, M)}{n(M)}$$

where $n(E, M)$ and $n(M)$ take into account both prior and historical samples of E and M.

However, this estimation does not yield the uncertainty behind the parameter estimate. For example, let us consider a coin-flipping experiment. If we had a prior belief of one sample for heads $C = h$ and one sample for tails $C = t$, and we flipped a coin once and observed heads, our probability of heads and tails would resemble the following

$$p(h) = \theta_h = \frac{1+1}{1+1+1+0} = \frac{2}{3}$$

$$p(t) = \theta_t = \frac{1+0}{1+1+1+0} = \frac{1}{3}$$

From our intuition, however, we know that this result is not very reliable because there are so few data points. What we would like to have is a distribution over $\Theta = \{\theta_h, \theta_t\}$ denoted as $p(\Theta)$ that could be updated using evidence in a Bayesian manner

$$p(\Theta|E) = \frac{p(E|\Theta)p(\Theta)}{p(E)} \tag{2.9}$$

We know that $p(E|\Theta)$ is a categorical distribution, so that in our case

$$p(h|\Theta) = \theta_h = p(h)$$

$$p(t|\Theta) = \theta_t = p(t)$$

The distribution $p(\Theta)$ is known as a *conjugate prior*, which is a distribution that can be updated by $p(E|\Theta)$. Conjugate priors are used in Bayesian parameter estimation problems in order to incorporate prior information about the parameters. These priors can be updated as information becomes available so that they reflect the uncertainty behind the updated parameter. As data becomes more available, the updated parameter distributions become narrower, depicting higher confidence in the estimates.

One of the most important properties of a conjugate prior $p(\Theta)$ to a likelihood distribution $p(E|\Theta)$ is that the posterior $p(\Theta|E)$ must be of the same family of distributions as the prior $p(\Theta)$ after combination with the likelihood according to Eqn (2.9); this allows for easy derivation of the posterior, resulting in a computationally light updating scheme. Finding a conjugate prior is not an easy task and not only depends on the distribution that produces E, but also on the particular parameter Θ of interest. Fortunately, the conjugate priors for many likelihood distributions have been found (Fink 1997) and some results summarized in Table 2.6.

For discrete probabilities, the conjugate prior is identified to be the Dirichlet distribution (Wilks 1962).

$$f(\Theta|\boldsymbol{\alpha}) = \frac{\Gamma\left(\sum_i \alpha_i\right)}{\prod_i \Gamma(\alpha_i)} \prod_i \theta_i^{\alpha_i - 1} \tag{2.10}$$

$$\theta \in [0, 1] \tag{2.11}$$

$$\sum_i \theta_i = 1 \tag{2.12}$$

Table 2.6 List of conjugate priors (Fink 1997)

Random process	Parameter of interest Θ	Conjugate prior $p(\Theta)$
	Discrete univariate processes	
Bernoulli	Binary probability (0 vs. 1)	Beta
Hypergeometric	Binary probability (0 vs. 1)	Beta-binomial
Poisson	Poisson parameter	Gamma
	Discrete multivariate processes	
Multinomial/categorical	Probability parameters	Dirichlet
Multivariate hypergeometric	Probability parameters	Dirichlet–multinomial
	Continuous univariate processes	
Uniform	Uniform probability	Pareto
Pareto	Precision parameter	Pareto
Pareto	Shape parameter	Gamma
Exponential	Mean	Gamma
Gamma	Rate parameter	Gamma
Normal	Mean	Normal
Normal	Inverse variance	Gamma
Lognormal	Normal mean	Normal
Lognormal	Normal inverse variance	Gamma
	Continuous multivariate processes	
Normal	Mean vector	Normal
Normal	Covariance matrix	Wishart
Normal	Mean vector and covariance matrix	Normal–Wishart
Normal regression	Regression coefficients	Normal
Normal regression	Regression coefficients and precision	Normal–gamma

where Γ is called the *gamma function*

$$\Gamma(t) = \int_0^\infty x^{t-1} e^{-x} \; dx \tag{2.13}$$

which is a generalization of the *factorial* function (as it can support continuous values) so that

$$\Gamma(k) = (k-1)! \qquad k \in \text{positive integers} \tag{2.14}$$

$$\Gamma(x+1) = x\Gamma(x) \qquad x \in \text{real numbers} \tag{2.15}$$

The parameter set α can be defined as

$$\alpha_i = n(e_i)$$

so that α is simply a record of the samples. Thus, according to our coin-flipping exercise, the conjugate prior (or the function that can be updated by evidence) is given as

$$f(\Theta|\alpha) = \frac{\Gamma(\alpha_h + \alpha_t)}{\Gamma(\alpha_h)\Gamma(\alpha_t)} \theta_h^{\alpha_h-1} \theta_t^{\alpha_t-1}$$

The motivation behind the Dirichlet distribution lies in its two desirable properties

1. The expected value of θ_i is given as the intuitive fraction of data samples

$$E_{f(\Theta|\alpha)}[\theta_i] = \frac{\alpha_i}{\sum_k \alpha_k} = \frac{n(e_i)}{\sum_k n(e_k)} \qquad (2.16)$$

2. Updating $f(\Theta|\alpha)$ with the observation of e_i simply adds the value of 1 to α_i

$$f(\Theta|\alpha[k+1]) = \frac{p(E_i|\Theta)f(\Theta|\alpha[k])}{p(E_i)} \qquad (2.17)$$

$$\alpha_i[k+1] = \alpha_i[k] + 1 \qquad (2.18)$$

To better understand the Dirichlet distribution, examples are given below.

Example: Expected values of a Dirichlet distribution

Let us say we have observed enough coin flips so that

$$\alpha_h = n(h) = 5$$
$$\alpha_t = n(t) = 7$$

We now wish to calculate the probability of heads or, equivalently, the expected value of θ_h.

$$E_{f(\Theta|\alpha)}[\theta_h] = \int_\Theta \theta_h f(\Theta|\alpha) \, d\Theta$$

$$= \int_\Theta \theta_h \frac{\Gamma(\alpha_h + \alpha_t)}{\Gamma(\alpha_h)\Gamma(\alpha_t)} \theta_h^{\alpha_h - 1} \theta_t^{\alpha_t - 1} \, d\Theta$$

$$= \frac{\Gamma(\alpha_h + \alpha_t)}{\Gamma(\alpha_h)\Gamma(\alpha_t)} \int_\Theta \theta_h \theta_h^{\alpha_h - 1} \theta_t^{\alpha_t - 1} \, d\Theta$$

$$= \frac{\Gamma(\alpha_h + \alpha_t)}{\Gamma(\alpha_h)\Gamma(\alpha_t)} \int_\Theta \theta_h^{(\alpha_h + 1) - 1} \theta_t^{\alpha_t - 1} \, d\Theta$$

Now it should be noted that *Dirichlet's integral* states that

$$\int_\Theta \prod_i \theta_i^{\alpha_i - 1} \, d\Theta = \frac{\prod_i \Gamma(\alpha_i)}{\Gamma\left(\sum_i \alpha_i\right)}$$

which results in

$$E_{f(\Theta|\alpha)}[\theta_h] = \frac{\Gamma(\alpha_h + \alpha_t)}{\Gamma(\alpha_h)\Gamma(\alpha_t)} \times \frac{\Gamma(\alpha_h + 1)\Gamma(\alpha_t)}{\Gamma((\alpha_h + 1) + \alpha_t)}$$

$$= \frac{\Gamma(\alpha_h + \alpha_t)}{\Gamma(\alpha_h)\Gamma(\alpha_t)} \times \frac{[\alpha_h \Gamma(\alpha_h)]\Gamma(\alpha_t)}{\Gamma(\alpha_h + 1 + \alpha_t)}$$

$$= \frac{\Gamma(\alpha_h + \alpha_t)}{\Gamma(\alpha_h)\Gamma(\alpha_t)} \times \frac{[\alpha_h \Gamma(\alpha_h)]\Gamma(\alpha_t)}{[(\alpha_h + \alpha_t)\Gamma(\alpha_h + \alpha_t)]}$$

$$= \frac{1}{\Gamma(\alpha_h)\Gamma(\alpha_t)} \times \frac{\alpha_h \Gamma(\alpha_h)\Gamma(\alpha_t)}{(\alpha_h + \alpha_t)}$$

$$= \frac{\alpha_h}{(\alpha_h + \alpha_t)} = \frac{5}{5+7} = \frac{5}{12}$$

yielding the value that was intuitively expected.

Example: Updating a Dirichlet distribution with new observations

Again, let us say that historically we observed coin flips so that

$$\alpha_h = n(h) = 5$$

$$\alpha_t = n(t) = 7$$

Now let us say that after another coin flip, we observed heads $(E = h)$ and we would like to update our results. According to Bayes' theorem

$$f(\Theta|\alpha, E_i) = f(\Theta|\alpha[k+1]) = \frac{p(E_i|\Theta)f(\Theta|\alpha[k])}{p(E_i)}$$

When applying the updating rule to our example, we have

$$f(\Theta|\alpha[k+1]) = \frac{p(h|\Theta)f(\Theta|\alpha[k])}{p(E_i)}$$

$$= \frac{\theta_h f(\Theta|\alpha[k])}{\int_\Theta \theta_h f(\Theta|\alpha[k])\, d\Theta}$$

Now, from our previous expected value example, we have already shown that

$$\int_\Theta \theta_h f(\Theta|\alpha[k])\, d\Theta = E_{f(\Theta|\alpha[k])}(\theta_h)$$

$$= \frac{\alpha_h}{\alpha_h + \alpha_t}$$

Thus

$$f(\Theta|\alpha[k+1]) = \frac{p(h|\Theta)f(\Theta|\alpha[k])}{p(E_i)}$$

$$= \frac{\theta_h f(\Theta|\alpha[k])}{\frac{\alpha_h}{\alpha_h + \alpha_t}}$$

$$= \frac{\alpha_h + \alpha_t}{\alpha_h} f(\Theta|\alpha[k])\theta_h$$

$$= \frac{\alpha_h + \alpha_t}{\alpha_h} \theta_h \frac{\Gamma(\alpha_h + \alpha_t)}{\Gamma(\alpha_h)\Gamma(\alpha_t)} \theta_h^{\alpha_h - 1} \theta_t^{\alpha_t - 1}$$

$$= \frac{\alpha_h + \alpha_t}{\alpha_h} \frac{\Gamma(\alpha_h + \alpha_t)}{\Gamma(\alpha_h)\Gamma(\alpha_t)} \theta_h^{(\alpha_h + 1) - 1} \theta_t^{\alpha_t - 1}$$

Recalling in Eqn (2.15) that $\Gamma(x + 1) = x\Gamma(x)$,

$$f(\Theta|\boldsymbol{\alpha}[k+1]) = \frac{\Gamma(\alpha_h + \alpha_t + 1)}{\Gamma(\alpha_h + 1)\Gamma(\alpha_t)}\theta_h^{(\alpha_h+1)-1}\theta_t^{\alpha_t - 1}$$

$$= \frac{\Gamma((\alpha_h + 1) + \alpha_t)}{\Gamma(\alpha_h + 1)\Gamma(\alpha_t)}\theta_h^{(\alpha_h+1)-1}\theta_t^{\alpha_t - 1}$$

From this, we can see that the posterior probability $f(\Theta|\boldsymbol{\alpha}[k+1])$ has $(\alpha_h + 1)$ in every place that (α_h) occurred in $f(\Theta|\boldsymbol{\alpha}[k])$. Therefore,

$$\alpha_h[k+1] = \alpha_h + 1 = 5 + 1 = 6$$

when a new outcome of heads $(E = h)$ is observed.

2.3   The EM Algorithm

The expectation–maximization (EM) algorithm is a technique pioneered by Dempster et al. (1977); variations of the EM algorithm technique were used and proposed previously, but this work was the first to present it in a general manner with rigorous proof of convergence. The principal reason for implementing the EM algorithm is to learn relationships and distributions when the data for estimation are incomplete.

Dempster et al. (1977) presented the EM algorithm in three different forms having increased generality:

1. exponential family distributions with closed-form maximum-likelihood solutions
2. exponential family distributions without closed-form maximum-likelihood solutions
3. general distributions.

However, since all solutions can be obtained from the general solution, we set our focus on the most general solution.

Tutorial Problem

For the EM algorithm tutorial, we revisit the simple control loop example as presented in Figure 2.3. Again, we have two monitors, one which monitors instrument bias, while another monitors valve stiction. Consider a mode *biased sensor* where we have some missing values from certain monitors, as shown in Table 2.7. When a monitor's value is missing, we denote it by ×.

We would like to estimate $p(E|m_3)$ when pieces of data are missing from the evidence.

The most general type of solution does not rely on any notion of exponential families. However, because it is the most general, the other two solutions can be derived as a special case of this solution. Thus, the general solution is most often used as a starting point despite its complexity.

Table 2.7 Biased sensor mode

Evidence	Frequency
$[0, 0]$	5
$[0, 1]$	12
$[1, 0]$	4
$[1, 1]$	6
$[0, \times]$	8
$[1, \times]$	2
$[\times, 0]$	3
$[\times, 1]$	10
Total	50

1. **Expectation:** This step involves the construction of the Q function

$$Q(\Phi|\Phi^{[k]}) = E[\log f(Z|\Phi)|y, \Phi^{[k]}] = \int_Z \log\left(f(Z|\Phi)\right)p(Z|y, \Phi^{[k]})\, dZ$$

where Z represents the unobserved part of the dataset (notation is upper case due to its random nature) and y represents the observed part of the dataset (notation is lower case due to its nonrandom nature). Furthermore, Φ is the current (variable) parameter set and $\Phi^{[k]}$ is the previously obtained (constant) parameter set obtained at iteration k.

2. **Maximization:** This step involves numerical maximization of the Q function over the variable parameter set Φ

$$\Phi^{[k+1]} = \arg \max_\Phi E(\log f(Z|\Phi)|y, \Phi^{[k]}))$$

where $\Phi^{[k+1]}$ is the estimate of parameters Φ after iteration k

In order to apply this solution to our tutorial problem, we must first take note of the following notation.

- $\mathcal{D}_c$ is a matrix of complete data entries.
- $\mathcal{D}_{ic}$ is a matrix of incomplete data entries (containing missing $\times$ elements).
- Z is a random vector consisting of the missing (or $\times$) elements within $\mathcal{D}_{ic}$.
- z is a realization of Z.
- y is a vector which represents all of the observed elements within $\mathcal{D}_{ic}$.

When referring to the historical data in our sample problem (shown in Table 2.7), $\mathcal{D}_c$ represents the data points without any missing entries (e.g. data summarized in the first four rows), while $\mathcal{D}_{ic}$ represents the data points that have missing entries (e.g. data summarized in the last four rows). In addition, Z represents all occurrences of missing elements ($\times$) in $\mathcal{D}_{ic}$, while y represents all occurrences of observed elements (not $\times$) in $\mathcal{D}_{ic}$.

Then, when implementing the EM algorithm in our example problem, one goes through the following steps:

1. **Initialization:** As an initialization point, we calculate the $\Theta^{[0]}$ parameters using D_c

$$\theta_{[0,0]} = 5/27$$

$$\theta_{[0,1]} = 12/27$$

$$\theta_{[1,0]} = 4/27$$

$$\theta_{[1,1]} = 6/27$$

2. **Expectation:** This is the step that involves the bulk of the work. We must take the Q function, which is initially expressed as

$$Q(\Theta|\Theta^{[k]}) = \sum_{\mathbf{Z}} p(\mathbf{z}|\mathbf{y}, D_c, \Theta^p) \log p(\mathbf{y}, \mathbf{z}, D_c|\Theta)$$

and derive a usable expression from it. The reason this current expression is hard to implement is that there are $2^{n(\mathbf{z})}$ possible realizations of $\mathbf{Z}$, thus, summation becomes an infeasible exercise. The first simplification can be made by assuming the time-independence of the data. In this way

$$D_c \perp D_{ic} \qquad \therefore D_c \perp \{\mathbf{z}, \mathbf{y}\}$$

This essentially removes D_c from the conditions and allows us to separate the log term

$$Q(\Theta|\Theta^{[k]}) = \sum_{\mathbf{Z}} p(\mathbf{z}|\mathbf{y}, \Theta^{[k]})[\log p(\mathbf{y}, \mathbf{z}|\Theta) + \log p(D_c|\Theta)]$$

Now, we have already established that D_c is independent of $\mathbf{Z}$, which is the summation term. Since

$$\sum_{\mathbf{Z}} p(\mathbf{z}|\mathbf{y}, \Theta^{[k]}) = 1$$

we can separate out the log $p(D_c|\Theta)$ from the summation

$$Q(\Theta|\Theta^{[k]}) = \log p(D_c|\Theta) + \sum_{\mathbf{Z}} p(\mathbf{z}|\mathbf{y}, \Theta^{[k]}) \log p(\mathbf{y}, \mathbf{z}|\Theta)$$

Now, because of time-independence, the summation over $\mathbf{Z}$ can be broken down into time steps $t_1, t_2, \ldots, t_{n(D_{ic})}$. In terms of $\mathbf{Z}$ this means that the realizations are broken down into

$$\mathbf{Z} = \{Z^{t_1}, Z^{t_2}, \ldots Z^{t_{n(D_{ic})}}\} \tag{2.19}$$

where Z^t is a random variable denoting the realizations of all missing ($\times$) values in the tth incomplete data sample. For example,

- If the first element of D_{ic} was $[0, \times] = [0, Z^{t_1}]$ then $Z^{t_1} = \{0, 1\}$.
- If the second element of D_{ic} was $[\times, 1] = [Z^{t_2}, 1]$ then $Z^{t_2} = \{0, 1\}$.
- If the third element of D_{ic} was $[\times, \times] = [Z^{t_3}]$ then $Z^{t_3} = \{[0, 0], [0, 1], [1, 0], [1, 1]\}$.

Note that in being consistent with previous notation, z^t is one of the possible realizations of Z^t so that

$$\sum_i p(Z^t = z_i^t) = 1$$

When time-independence is taken into account, we can assert that

$$p(\boldsymbol{Z}|\boldsymbol{y}, \Theta^{[k]}) = \prod_{t=1}^{n(\mathcal{D}_{ic})} p(Z^t|y^t, \Theta^{[k]})$$

$$\log p(\boldsymbol{Z}, \boldsymbol{y}|\Theta^{[k]}) = \sum_{t=1}^{n(\mathcal{D}_{ic})} \log p(Z^t, y^t|\Theta^{[k]})$$

This results in a summation that is broken down as follows:

$$Q(\Theta|\Theta^{[k]}) = \log p(\mathcal{D}_c|\Theta) +$$

$$\sum_{Z^{t_1}} \cdots \sum_{Z^{n(\mathcal{D}_{ic})}} \left[\prod_{t=1}^{n(\mathcal{D}_{ic})} p(z^t|y^t, \Theta) \right] \left[\sum_{t=1}^{n(\mathcal{D}_{ic})} \log p(z^t, y^t|\Theta^{[k]}) \right]$$

We can simplify this expression by noting that each term of $\log p(z^t, y^t|\Theta^{[k]})$ is front-multiplied by all values of $p(z^t|y^t, \Theta)$, but the first term $\log p(z^{t_1}, y^{t_1}|\Theta^{[k]})$ is independent of all terms in $p(z^t|y^t, \Theta)$ except $p(z^{t_1}|y^{t_1}, \Theta)$. Because the summation of all realizations $p(z^t|y^t, \Theta)$ over Z^t is equal to one, the summations signs all cancel out, except for the one occurring at $t = 1$

$$\sum_{Z^{t_1}} \cdots \sum_{Z^{n(\mathcal{D}_{ic})}} \left[\prod_{t=1}^{n(\mathcal{D}_{ic})} p(z^t|y^t, \Theta) \right] \log p(z^{t_1}, y^t|\Theta^{[k]}) =$$

$$\sum_{Z^{t_1}} p(z^{t_1}|y^{t_1}, \Theta) \log p(z^{t_1}, y^{t_1}|\Theta^{[k]})$$

This property can be generalized to yield

$$\sum_{Z^{t_1}} \cdots \sum_{Z^{n(\mathcal{D}_{ic})}} \left[\prod_{t=1}^{n(\mathcal{D}_{ic})} p(z^t|y^t, \Theta) \right] \log p(z^i, y^i|\Theta^{[k]}) =$$

$$\sum_{Z^i} p(z^i|y^i, \Theta) \log p(z^i, y^i|\Theta^{[k]})$$

When applied to our Q function, the simplification yields the following result:

$$Q(\Theta|\Theta^{[k]}) = \log p(\mathcal{D}_c|\Theta) + \sum_{t=1}^{n(\mathcal{D}_{ic})} \sum_{Z^t} p(z^t|y^t, \Theta) \log p(z^t, y^t|\Theta^{[k]})$$

This is a more workable solution. For example, if each incomplete data entry $\mathcal{D}_{ic}$ has only one element missing, instead of assessing the probability of $2^{n(\mathcal{D}_{ic})}$ realizations of Z and summing them up, we only have to go through two realizations for every Z^t and summing up the results. Thus the computational burden of evaluating the $Q(\Theta|\Theta^{[k]})$ function only increases linearly instead of exponentially.

3. **Maximization:** The maximization step can be performed by numerical optimization

$$\Theta^{[k+1]} = \arg \max_{\Theta} \left[\log p(\mathcal{D}_c|\Theta) + \sum_{t=1}^{n(\mathcal{D}_{ic})} \sum_{Z^t} p(z^t|y^t, \Theta) \log p(z^t, y^t|\Theta^{[k]}) \right]$$

However, the solution to this maximization can be analytically obtained, negating the need for numerical optimization. In order to obtain an analytical result that relates more directly to our original problem, we first note that

$$\log p(\mathcal{D}_c|\Theta) = \sum_{t=1}^{n(\mathcal{D}_c)} \log p(e_c^t|\Theta)$$

$$= n_c[0,0] \log (\theta_{[0,0]}) + n_c[0,1] \log (\theta_{[0,1]}) +$$
$$n_c[1,0] \log (\theta_{[1,0]}) + n_c[1,1] \log (\theta_{[1,1]})$$

where $n_c[0,0]$ is the number of times the evidence $[0,0]$ has been observed in the complete evidence (with other evidence patterns being similarly treated). Now recall that z^t, y^t will correspond to some e_{ic}^t. For example, let us say that $[0, \times]$ is the first element of D_{ic}. Then

$$[0, z^{t_1}] = [0,0] = e_1$$
$$[0, z^{t_1}] = [0,1] = e_2$$

thus

$$p(z^{t_1} = 0, y^{t_1}|\Theta^{[k]}) = p(e_{ic}^{t_1} = [0,0]|\Theta^{[k]})$$
$$p(z^{t_1} = 1, y^{t_1}|\Theta^{[k]}) = p(e_{ic}^{t_1} = [0,1]|\Theta^{[k]})$$

In addition, note the rule of conditioning which can be stated as follows

$$p(z^t|y^t, \Theta) = \frac{p(z^t, y^t|\Theta)}{p(y^t|\Theta)}$$

From this expression, $p(y^t|\Theta)$ is the probability that y_t takes its value, or equivalently it is a normalization constant to ensure that all probabilities of $p(z^t|y^t, \Theta)$ sum to 1 (with respect to all possible values of z^t). Thus according to our example, we can say that

$$p(z^{t_1} = 0|y^{t_1}, \Theta^{[k]}) = \frac{p(E_{ic}^{t_1} = [0,0]|\Theta^{[k]})}{p(E_{ic}^{t_1} = [0,0]|\Theta^{[k]}) + p(E_{ic}^{t_1} = [0,1]|\Theta^{[k]})}$$

$$= \frac{\theta_{[0,0]}}{\theta_{[0,0]} + \theta_{[0,1]}}$$

$$p(z^{t_1} = 1 | y^{t_1}, \Theta^{[k]}) = \frac{p(E_{ic}^{t_1} = [0,1]|\Theta^{[k]})}{p(E_{ic}^{t_1} = [0,0]|\Theta^{[k]}) + p(E_{ic}^{t_1} = [0,1]|\Theta^{[k]})}$$

$$= \frac{\theta_{[0,1]}}{\theta_{[0,0]} + \theta_{[0,1]}}$$

so that in general,

$$p(z^t | y^t, \Theta^{[k]}) = \frac{\theta_{E_{ic}^t}}{\sum_{e \subset E_{ic}^t} \theta_e}$$

where $\sum_{e \subset E_{ic}^t} \theta_e$ sums the θ parameters associated with all possible evidence realizations in E_{ic}^t. For example,

$$\sum_{e \subset [\times, 0]} \theta_e = \theta_{[1,0]} + \theta_{[0,0]}$$

Now $Q(\Theta | \Theta^{[k]})$ can be rewritten as

$$Q(\Theta | \Theta^{[k]}) =$$

$$n_c[0,0] \log(\theta_{[0,0]}) + n_c[0,1] \log(\theta_{[0,1]})$$

$$+ n_c[1,0] \log(\theta_{[1,0]}) + n_c[1,1] \log(\theta_{[1,1]})$$

$$+ n_{ic}[0,\times] \left[\log(\theta_{[0,0]}) \frac{\theta_{[0,0]}}{\theta_{[0,0]} + \theta_{[0,1]}} + \log(\theta_{[0,1]}) \frac{\theta_{[0,1]}}{\theta_{[0,0]} + \theta_{[0,1]}} \right]$$

$$+ n_{ic}[1,\times] \left[\log(\theta_{[1,0]}) \frac{\theta_{[1,0]}}{\theta_{[1,0]} + \theta_{[1,1]}} + \log(\theta_{[1,1]}) \frac{\theta_{[1,1]}}{\theta_{[1,0]} + \theta_{[1,1]}} \right]$$

$$+ n_{ic}[\times,0] \left[\log(\theta_{[1,0]}) \frac{\theta_{[1,0]}}{\theta_{[1,0]} + \theta_{[0,0]}} + \log(\theta_{[0,0]}) \frac{\theta_{[0,0]}}{\theta_{[1,0]} + \theta_{[0,0]}} \right]$$

$$+ n_{ic}[\times,1] \left[\log(\theta_{[1,1]}) \frac{\theta_{[1,1]}}{\theta_{[1,1]} + \theta_{[0,1]}} + \log(\theta_{[0,1]}) \frac{\theta_{[0,1]}}{\theta_{[1,1]} + \theta_{[0,1]}} \right]$$

which can be rearranged as

$$Q(\Theta | \Theta^{[k]}) = \log \theta_{[0,0]} \left[n_c[0,0] + n_{ic}[\times,0] \frac{\theta_{[0,0]}}{\theta_{[0,0]} + \theta_{[1,0]}} + n_{ic}[0,\times] \frac{\theta_{[0,0]}}{\theta_{[0,0]} + \theta_{[0,1]}} \right]$$

$$+ \log \theta_{[0,1]} \left[n_c[0,1] + n_{ic}[\times,1] \frac{\theta_{[0,1]}}{\theta_{[0,1]} + \theta_{[1,1]}} + n_{ic}[0,\times] \frac{\theta_{[0,1]}}{\theta_{[0,0]} + \theta_{[0,1]}} \right]$$

$$+ \log \theta_{[1,0]} \left[n_c[1,0] + n_{ic}[\times,0] \frac{\theta_{[1,0]}}{\theta_{[0,0]} + \theta_{[1,0]}} + n_{ic}[1,\times] \frac{\theta_{[1,0]}}{\theta_{[1,0]} + \theta_{[1,1]}} \right]$$

$$+ \log \theta_{[1,1]} \left[n_c[1,1] + n_{ic}[\times,1] \frac{\theta_{[1,1]}}{\theta_{[0,1]} + \theta_{[1,1]}} + n_{ic}[1,\times] \frac{\theta_{[1,1]}}{\theta_{[1,0]} + \theta_{[1,1]}} \right]$$

By applying the numbers in Table 2.7, $Q(\Theta|\Theta^{[k]})$ is expressed as

$$Q(\Theta|\Theta^{[k]}) = \log \theta_{[0,0]} \left[5 + 3\frac{5/27}{5/27 + 4/27} + 8\frac{5/27}{5/27 + 12/27}\right]$$

$$+ \log \theta_{[0,1]} \left[12 + 10\frac{12/27}{12/27 + 6/27} + 8\frac{12/27}{12/27 + 5/27}\right]$$

$$+ \log \theta_{[1,0]} \left[4 + 3\frac{4/27}{4/27 + 5/27} + 2\frac{4/27}{4/27 + 6/27}\right]$$

$$+ \log \theta_{[1,1]} \left[6 + 10\frac{6/27}{6/27 + 12/27} + 2\frac{6/27}{6/27 + 4/27}\right]$$

resulting in the following expression:

$$Q(\Theta|\Theta^{[k]}) = 9.0196\log \theta_{[0,0]} + 24.314\log \theta_{[0,1]} + 6.133\log \theta_{[1,0]} + 10.533\log \theta_{[1,1]}$$

When maximized over all values in Θ, in light of the constraint

$$\theta_{[0,0]} + \theta_{[0,1]} + \theta_{[1,0]} + \theta_{[1,1]} = 1 \qquad (2.20)$$

the result is

$$\theta_{[0,0]}^{[1]} = 9.0196/50$$

$$\theta_{[0,1]}^{[1]} = 24.314/50$$

$$\theta_{[1,0]}^{[1]} = 6.133/50$$

$$\theta_{[1,1]}^{[1]} = 10.533/50$$

which is an intuitive result. The *expectation* and *maximization* procedures can be repeated until the parameters in Θ converge.

2.4 Techniques for Ambiguous Modes

In Section 2.3 we discussed the problem of missing evidence and how the EM algorithm could be used to infer likelihoods even if some of the historical evidence is incomplete. In this section, we will discuss how to infer likelihoods if some of the modes are incomplete.

We return to our example system presented in Figure 2.3, where we have a sensor that could be biased, and a valve that could become sticky. Let us say that in this case, it is possible for more than one problem to exist at a given time. In this way, there are four modes m_1, m_2, m_3, m_4 as shown in Table 2.8.

As in the case of evidence, it is possible for information about the mode to be missing in the history. In such a case, the mode is ambiguous. For example, let us say that we are sure that there is no bias, but we are unsure as to whether stiction exists or not. The binary label for such a mode would be $[0, \times]$, which would indicate that the modes m_1 and m_2 were possible

Table 2.8 Modes and their corresponding labels

Mode label	Meaning	Binary label
m_1	No bias, no stiction	$[0,0]$
m_2	No bias, stiction	$[0,1]$
m_3	Bias, no stiction	$[1,0]$
m_4	Bias, stiction	$[1,1]$

Table 2.9 Ambiguous modes and their corresponding labels

Mode label	Meaning	Binary label
$\{m_1, m_2\}$	No bias, stiction uncertain	$[0,\times]$
$\{m_3, m_4\}$	Bias, stiction uncertain	$[1,\times]$
$\{m_1, m_3\}$	Bias uncertain, no stiction	$[\times,0]$
$\{m_2, m_4\}$	Bias uncertain, stiction uncertain	$[\times,1]$

Table 2.10 Historical data for all modes

Evidence	Modes								Total
	$[0,0]$	$[0,1]$	$[1,0]$	$[1,1]$	$[0,\times]$	$[1,\times]$	$[\times,0]$	$[\times,1]$	
$e_1, [0,0]$	11	1	1	1	6	2	6	1	29
$e_2, [0,1]$	1	2	8	2	1	2	4	2	22
$e_3, [1,0]$	1	8	2	2	5	4	2	5	29
$e_4, [1,1]$	1	2	3	9	2	6	2	6	31
Total	14	14	14	14	14	14	14	14	112

(resulting in a mode label $\{m_1, m_2\}$). Let us now assume that additional modes were seen in the data according to Table 2.9.

For the tutorials to follow, let us consider some historical data as seen in Table 2.10.

From Table 2.10 we can see that 14 data points were collected from each mode (including the ambiguous ones). If we assume that the mode frequency in the data represents the true mode frequency (so that all modes have an equal chance of happening as presented in the data) we could use the EM algorithm (Dempster et al. 1977) to solve this problem. However, because we are using Bayesian diagnosis methods

$$p(M|E) = \frac{p(E|M)p(M)}{p(E)}$$

the term $p(M)$ indicates that we have a prior probability of modes that is *not obtained from historical data*. The Bayesian evaluation of $p(M|E)$ is convenient because it allows us more freedom in terms of how we select data (e.g. if mode 1 occurs 95% of the time, we do not need to ensure that 95% of the historical data comes from mode 1). Now, the EM algorithm in this case would use the data to find the mode probability $p(M)$, but if we assume that we cannot use the historical data to estimate the mode frequency, as is done in Bayesian diagnosis, the EM algorithm should not be applied as it would attempt to estimate the ambiguous mode statistics (defined as Θ) using historical data.

Instead, we can use unknown parameters Θ to express the likelihood based on different outcomes of the ambiguous modes. These parameters can be used to express a probability range in the diagnosis. This type of approach (using ambiguity to express probability ranges) has been introduced in the area of Dempster–Shafer theory but as we will see in Section 2.4.2 the expression of $p(E|\Theta)$ is somewhat complicated. Furthermore, we will see that in Section 2.4.3 the problem of *Bayesian probability* (with ambiguous modes) cannot be well represented by the formulation given by Shafer (1976), although the problem of *direct probability* can be represented quite well by this same formulation.

2.4.1 Tutorial on Θ Parameters in the Presence of Ambiguous Modes

Previously, we used Θ to express the probability $\theta_i = p(e_i|M)$. In the case of ambiguous modes, Θ is also used to express probability, but now we are concerned about the probability of the mode

$$p(m_i) = \theta\{m_i\}$$

More specifically, however, we are focused on the probability of an unambiguous mode m_i, given a historical mode $\boldsymbol{m}_k$ which could potentially be ambiguous (a boldface $\boldsymbol{m}$ here indicates that the mode can be ambiguous).

$$\theta\{\tfrac{m_i}{\boldsymbol{m}_k}\} \equiv p(m_i|\boldsymbol{m}_k)$$

For example, let us consider the historical mode $\{m_1, m_2\}$ which is ambiguous (refer to Table 2.9). The parameter θ for mode m_1 given this ambiguous mode is

$$\theta\{\tfrac{m_1}{m_1,m_2}\} = p(m_1|\{m_1, m_2\})$$

In other words, $\theta\{\tfrac{m_1}{m_1,m_2}\}$ is the amount of data in $\{m_1, m_2\}$ that actually belongs to m_1. In general, the values of θ are unknown parameters except in the following cases

$$(1) \quad \theta\{\tfrac{m_i}{\boldsymbol{m}_k}\} = p(m_i|\boldsymbol{m}_k) = 0 \qquad\qquad m_i \not\subseteq \boldsymbol{m}_k$$

$$\text{e.g.} \quad \theta\{\tfrac{m_3}{m_1,m_2}\} = p(m_3|\{m_1, m_2\}) = 0$$

$$(2) \quad \theta\{\tfrac{m_i}{\boldsymbol{m}_k}\} = p(m_i|\boldsymbol{m}_k) = 1 \qquad\qquad m_i = \boldsymbol{m}_k$$

$$\text{e.g.} \quad \theta\{\tfrac{m_1}{m_1}\} = p(m_1|m_1) = 1$$

One can see that in these special cases, logic forces the probability to be 1 or 0, hence these cases are *logically forced*.

2.4.2 Tutorial on Probabilities Using Θ Parameters

Now that an interpretation of Θ has been given, we can proceed to express probabilities of unambiguous modes given the data (which includes ambiguous modes).

Direct Probability

Let us first assume that we are not using Bayes' theorem to include prior probabilities $p(M)$ in order to evaluate the posterior $p(M|E)$, but that we are using the data $p(M|E)$ to directly evaluate the probability. In order to do this properly, one must assume that the mode frequencies in the data are representative of the true mode frequency (or that the number of times the mode happens in the data is proportional to its probability).

Let us say that from our system in Figure 2.3 we observe the evidence $E = [0,0]$ and we wish to evaluate the probability of mode $m_1 = [0,0]$. Now in addition to the 11 data points we have observed for this evidence under $m_1 = [0,0]$, we must also consider the data points under $M = [0,\times], [\times,0]$ or equivalently $M = \{m_1, m_2\}, \{m_1, m_3\}$ wherein some of those data points could also belong to $m_1 = [0,0]$. The amount of data in $M = \{m_1, m_2\}, \{m_1, m_3\}$ that does belong to m_1 is given by the parameters $\theta\{\frac{m_1}{m_1,m_2}\}, \theta\{\frac{m_1}{m_1,m_3}\}$ which are unknown. The probability of m_1 is then expressed as

$$p(m_1|e_1, \Theta) = \frac{\left[n(m_1, e_1) + n(\{m_1, m_2\}, e_1)\theta\{\frac{m_1}{m_1,m_2}\} + n(\{m_1, m_3\}, e_1)\theta\{\frac{m_1}{m_1,m_3}\} \right]}{n(e_1)}$$

where $n(m_1, e_1)$, $n(\{m_1, m_2\}, e_1)$ and $n(\{m_1, m_3\}, e_1)$ are the number of observations of the modes $[0,0], [0,\times]$ and $[\times,0]$ when $E = [0,0]$. These values can be obtained from Table 2.10 as 11, 6 and 6, respectively. Similarly $n(e_1)$ is likewise found to be 29 so that

$$p(m_1|e_1, \Theta) = \frac{\left[11 + 6\theta\{\frac{m_1}{m_1,m_2}\} + 6\theta\{\frac{m_1}{m_1,m_3}\} \right]}{29}$$

In general, the probability of $p(M|E, \Theta)$ is given as

$$P(M|E, \Theta) = \frac{1}{n(E)} \sum_{m_k \supseteq M} \theta\{\frac{M}{m_k}\} n(m_k, E)$$

$$= \sum_{m_k \supseteq M} \theta\{\frac{M}{m_k}\} S(m_k|E) \tag{2.21}$$

where the summation condition $m_k \supset M$ indicates that we search through all historical modes and sum over the terms m_k when it contains or is equal to M. The term $S(m_k|E)$ was coined by Shafer (Shafer 1976) as *support* or, in later terminology, the *basic belief assignment* (BBA), which is functionally the same as probability, but makes allowances for ambiguous modes, for example

$$S(m_k|E) = \frac{n(m_k, E)}{n(E)}$$

Bayesian Probability

When using the Bayesian technique for diagnosis, we combine the prior probability $p(M)$ with the likelihood $p(E|M)$, where the likelihood is calculated from historical data

$$p(E|M) = \frac{n(E, M)}{n(M)}$$

When ambiguous modes are in the history, we have to consider not only data points from mode M but also data points from ambiguous modes that could also belong to M. For example, if we wish to consider the likelihood of $E = e_1 = [0, 0]$ given $M = m_1$, the likelihood $p(e_1|m_1, \Theta)$ is expressed as

$$p(e_1|m_1, \Theta) = \frac{n(e_1, m_1) + \theta\{\frac{m_1}{m_1, m_2}\}n(e_1, \{m_1, m_2\}) + \theta\{\frac{m_1}{m_1, m_3}\}n(e_1, \{m_1, m_3\})}{n(m_1) + \theta\{\frac{m_1}{m_1, m_2}\}n(\{m_1, m_2\}) + \theta\{\frac{m_1}{m_1, m_3}\}n(\{m_1, m_3\})}$$

Now $n(m_1, e_1)$, $n(\{m_1, m_2\}, e_1)$ and $n(\{m_1, m_3\}, e_1)$ are again obtained from Table 2.10 as $11, 6$ and 6, respectively. Similarly, $n(m_1)$, $n(\{m_1, m_2\})$ and $n(\{m_1, m_3\})$ are obtained from the bottom row, which indicates the total number of observations from each mode. All three of these values are equal to 14.

$$p(e_1|m_1, \Theta) = \frac{11 + \theta\{\frac{m_1}{m_1, m_2}\}6 + \theta\{\frac{m_1}{m_1, m_3}\}6}{14 + \theta\{\frac{m_1}{m_1, m_2}\}14 + \theta\{\frac{m_1}{m_1, m_3}\}14}$$

In more general terms, the expression for the likelihood can be given as

$$p(E|M, \Theta) = \frac{\sum_{m_k \supseteq M} \theta\{\frac{M}{m_k}\}n(m_k, E)}{\sum_{m_k \supseteq M} \theta\{\frac{M}{m_k}\}n(m_k)}$$

$$= \frac{\sum_{m_k \supseteq M} \theta\{\frac{M}{m_k}\}n(m_k)S(E|m_k)}{\sum_{m_k \supseteq M} \theta\{\frac{M}{m_k}\}n(m_k)} \qquad (2.22)$$

where the support function $S(E|m_k)$ can be obtained for discrete data as

$$S(E|m_k) = \frac{n(m_k, E)}{n(m_k)}$$

2.4.3 Dempster–Shafer Theory

Dempster–Shafer theory was first proposed as a generalization to Bayesian methods in a manner that can account for uncertainty about the hypotheses (or, for our purposes, the modes). Dempster–Shafer theory has been developed as a framework for artificial intelligence with uncertain reasoning and many developments (particularly combination rules and methods of interpretation) have been made in this area. In this tutorial, the basics of Dempster–Shafer theory from Shafer (1976) will be covered.

Dempster–Shafer theory has two main proponents: the rule of conditioning and the rule of combination.

Dempster's Rule of Conditioning

Dempster's rule of conditioning aims to find probability ranges as a method of describing the results. When the probability and likelihood were formulated earlier with Θ parameters, it was admitted that these Θ parameters were unknown. Since these parameters are probabilities and must be contained on the interval of $[0, 1]$, it is possible to find the probability bounds by maximizing and minimizing over Θ. In Shafer (1976), Dempster and Shafer never made use of parameters which would represent allocation of ambiguous mode data. Instead they were more concerned about probability ranges for the final result.

Dempster and Shafer first define the *support function* or *BBA*, which was briefly mentioned before when parametrizing probabilities using Θ. The support function $S(X)$ is defined in the same manner as probability but can be directly applied to ambiguous modes as well as unambiguous ones.

$$S(\boldsymbol{M}) = \frac{n(\boldsymbol{M})}{\sum_k n(\boldsymbol{m}_k)}$$

$$S(\boldsymbol{M}|E) = \frac{n(\boldsymbol{M}, E)}{n(E)}$$

$$S(E|\boldsymbol{M}) = \frac{n(\boldsymbol{M}, E)}{n(\boldsymbol{M})}$$

From the support function, we can calculate the lower bound and upper bound probabilities. In Demspter–Shafer theory, they are known as belief and plausibility.

1. $Bel(X)$ is the *belief* or *lower bound probability* of X.
2. $Pl(X)$ is the *plausibility* or *upper bound probability* of X.

The belief and plausibility can be obtained using *Demspter's rule of conditioning*:

$$Bel(\boldsymbol{M}) = \sum_{\boldsymbol{m}_k \subseteq \boldsymbol{M}} S(\boldsymbol{m}_k) \tag{2.23}$$

$$Pl(\boldsymbol{M}) = \sum_{\boldsymbol{m}_k \cap \boldsymbol{M} \neq \emptyset} S(\boldsymbol{m}_k) \tag{2.24}$$

This rule of conditioning is consistent with our direct probability scenario. If we took the expression in Eqn (2.21) and minimized/maximized it over Θ, the result would be

$$Bel(\boldsymbol{M}|E, \Theta) = \min_{\Theta} \left[\sum_{\boldsymbol{m}_k \supseteq \boldsymbol{M}} \theta\{\tfrac{\boldsymbol{M}}{\boldsymbol{m}_k}\} S(\boldsymbol{m}_k|E) \right] = \sum_{\boldsymbol{m}_k = \boldsymbol{M}} S(\boldsymbol{m}_k) \tag{2.25}$$

$$Pl(\boldsymbol{M}|E, \Theta) = \max_{\Theta} \left[\sum_{\boldsymbol{m}_k \supseteq \boldsymbol{M}} \theta\{\tfrac{\boldsymbol{M}}{\boldsymbol{m}_k}\} S(\boldsymbol{m}_k|E) \right] = \sum_{\boldsymbol{m}_k \supseteq \boldsymbol{M}} S(\boldsymbol{m}_k) \tag{2.26}$$

The summation limits are slightly changed, but only because M is unambiguous. As one can see,

- If M is unambiguous $m_k \subseteq M$ is only true when $m_k = M$.
- If M is unambiguous, $m_k \cap M \neq \emptyset$ is only true when $m_k \supseteq M$.

The reason why Eqn (2.25) minimizes $P(M|E, \Theta)$ (expressed in Eqn (2.21)) is because it is linear with respect to Θ and all of the coefficients on Θ are nonnegative support functions $S(m_k)$. Thus, by minimizing the nonforced values of Θ, $P(M|E, \Theta)$ is also minimized. The only time the condition $\theta \equiv 1$ is forced is when θ is expressed as $\theta \left\{ \frac{M}{M} \right\}$; all other terms can be zero if need be, making $m_k = M$ the only term included in the belief summation condition. By the same reasoning, $P(M|E, \Theta)$ is maximized when Θ is maximized. Thus, we include in our summation every case where Θ is not forced to be zero. As long as $M \subseteq m_k$ then $\theta \{ \frac{M}{m_k} \}$ is not forced to be zero, making $\theta \{ \frac{M}{m_k} \}$ the summation term for plausibility.

As an example, consider our previous case where we found that

$$p(m_1|e_1, \Theta) = S(m_1|e_1)\theta\{\tfrac{m_1}{m_1}\} + S(\{m_1, m_2\}|e_1)\theta\{\tfrac{m_1}{m_1, m_2}\}$$
$$+ S(\{m_1, m_3\}|e_1)\theta\{\tfrac{m_1}{m_1, m_3}\}$$
$$= \left[\frac{11}{29}\theta\{\tfrac{m_1}{m_1}\} + \tfrac{6}{29} \quad \theta\{\tfrac{m_1}{m_1, m_2}\} + \tfrac{6}{29} \quad \theta\{\tfrac{m_1}{m_1, m_3}\} \right]$$

We can see that the value $\theta\{\frac{m_1}{m_1}\} = 1$ is forced by logic, but the other values $\theta\{\frac{m_1}{m_1, m_2}\}$ and $\theta\{\frac{m_1}{m_1, m_3}\}$ can be anywhere between 0 and 1. Since the coefficients on these two values are positive (they both equal $\frac{6}{29}$),

- $p(m_1|e_1, \Theta)$ is minimized when $\theta\{\frac{m_1}{m_1, m_2}\}$ and $\theta\{\frac{m_1}{m_1, m_3}\}$ are set to 0
- $p(m_1|e_1, \Theta)$ is maximized when $\theta\{\frac{m_1}{m_1, m_2}\}$ and $\theta\{\frac{m_1}{m_1, m_3}\}$ are set to 1.

It should be noted, however, that while Dempster's rule of conditioning functions for direct probabilities, it does not function for likelihoods. An adaptation of Dempster–Shafer theory that can be applied to likelihoods is formulated in Chapters 7 and 16, which pertain to generalized Dempster–Shafer theory.

Dempster's Rule of Combination

In addition to the rule of conditioning, Dempster–Shafer theory also has a method to combine information from multiple *independent* sources, much like Bayesian methods. Dempster's rule of combination can be written as

$$S_{12}(m_k) = \frac{1}{1 - K} \sum_{m_k = m_i \cap m_j} S_1(m_i)S_2(m_j) \qquad m_k \neq \emptyset \qquad (2.27)$$

$$K = \sum_{\emptyset = m_i \cap m_j} S(m_i)S(m_j)$$

where $1 - K$ is a normalization constant to ensure that $S(m_k)$ sums to 1. Here, S_1 and S_2 must be support functions that come from independent sources and S_{12} is a support function that combines the two independent sources. If, for example, the bias monitor and stiction monitors could be considered independent, they could each be used to independently construct their own $S(m_k|e)$. The two support functions could then be combined according to Eqn (2.27).

As a combination example, let us consider our system in Figure 2.3 with independent monitors. Let us say that $S(m_k|e)$ is given as

$$
\begin{aligned}
S_1(m_1) &= 4/16 & S_2(m_1) &= 3/16 \\
S_1(m_2) &= 2/16 & S_2(m_2) &- 2/16 \\
S_1(m_3) &= 1/16 & S_2(m_3) &= 2/16 \\
S_1(m_4) &= 1/16 & S_2(m_4) &= 1/16 \\
S_1(m_1, m_2) &= 3/16 & S_2(m_1, m_2) &= 2/16 \\
S_1(m_1, m_3) &= 2/16 & S_2(m_1, m_3) &= 3/16 \\
S_1(m_2, m_4) &= 2/16 & S_2(m_2, m_4) &= 2/16 \\
S_1(m_3, m_4) &= 1/16 & S_2(m_3, m_4) &= 1/16
\end{aligned}
$$

where the subscript of S denotes which set of evidence the support is from.

If we wanted to combine these two BBAs to form S_{12} and assess $S_{12}(m_1)$, we would have

$$
S_{12}(m_1) = \frac{1}{1-K} \sum_{m_1 = m_i \cap m_j} S_1(m_i) S_2(m_j)
$$

$$
= S_1(m_1)S_2(m_1) + S_1(m_1)S_2(m_1, m_2) + S_1(m_1, m_2)S_2(m_1)
$$

$$
+ S_1(m_1)S_2(m_1, m_3) + S_1(m_1, m_3)S_2(m_1)
$$

$$
+ S_1(m_1, m_2)S_2(m_1, m_3) + S_1(m_1, m_3)S_2(m_1, m_2)
$$

$$
= \frac{1}{1-K}\left[\frac{4}{16}\cdot\frac{3}{16} + \frac{4}{16}\cdot\frac{2}{16} + \ldots + \frac{3}{16}\cdot\frac{3}{16} + \frac{2}{16}\cdot\frac{2}{16}\right]
$$

As one can see, each multiplied pair contains sets that intersect to yield m_1.

2.5   Kernel Density Estimation

Up to this point, we have considered evidence E to be discrete. In most cases, the outputs of monitors are actually continuous, but discretization is performed in order to create individual alarms for each monitor. However, we could use smaller and smaller discretization regions to describe the likelihood $p(E|M)$

$$
\lim_{n(E)\to\infty} p(M|E) = \frac{f(E|M)p(M)}{\sum_M f(E|M)p(M)}
$$

where $f(E|M)$ is the probability density function of the likelihood. If the type of distribution is known (such as Gaussian), one could fit the data to this distribution using a parametric approach (such as maximum likelihood estimation). However, in most cases monitor results

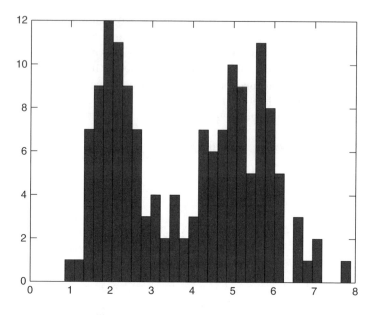

Figure 2.9   Histogram of distribution

do not follow a well-defined distribution, thus nonparametric methods need to be used, namely, discretization or kernel density estimation (Parsen 1962).

2.5.1   From Histograms to Kernel Density Estimates

When trying to estimate a distribution from a dataset without any knowledge of the distribution, most practitioners would turn to the histogram. A histogram partitions the data's domain into bins, and counts how many data points lie within each bin, as shown in Figure 2.9.

The histogram is useful for obtaining a rough idea of what the distribution should look like. A slight adaptation to the histogram is to center the counts around each data point. When evaluating the probability at a new point x, this centred histogram approach simply counts how many data points lie within a bin centred around x. Since the bin positions are more flexible, we can observe in Figure 2.10 that the distribution estimate is much smoother.

The function represented by the centred histogram essentially places a block with area $1/n$ around each data point d_i (where n is the number of data points). When evaluating the probability, we use the criterion

$$\hat{f}(x) = \frac{1}{n} \sum_{\substack{d_i \leq x+h \\ d_i \geq x-h}} \frac{1}{h}$$

where h is a boundary width set by the user. This is a basic form of a kernel density estimate. The general form of a kernel density estimate is

$$\hat{f}(x) = \frac{1}{n} \sum_i \frac{1}{h} K(x, d_i)$$

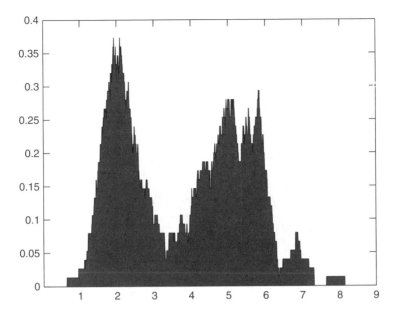

Figure 2.10 Centered histogram of distribution

where in our example the kernel function K is actually a block

$$K(x, d_i, h) = \begin{cases} 0 & \{d_i < x - h\} \\ 1/h & \{x - h \leq d_i \leq x + h\} \\ 0 & \{d_i > x + h\} \end{cases}$$

Finally, after using this form, one realizes that the kernel function does not need to take the shape of a block, but can take on any desired shape, as long as it integrates to unity, for example a standard Gaussian shape

$$K(x, d_i, h) = \frac{1}{h\sqrt{2\pi}} \exp\left[\frac{1}{2}\left(\frac{x - d_i}{h}\right)^2\right]$$

so that

$$\hat{f}(x) = \frac{1}{n} \sum_i \frac{1}{h\sqrt{2\pi}} \exp\left[\frac{1}{2}\left(\frac{x - d_i}{h}\right)^2\right] \tag{2.28}$$

If this kernel were applied to our data, the result would be a function resembling the one shown in Figure 2.11, which, as one might observe, is much smoother.

Kernel density estimation thus far has been presented for univariate applications, but multivariate kernel density estimation is also possible. For example, a multivariate kernel density estimate could use the p-dimensional standard normal distribution to yield

$$\hat{f}(x) = \frac{1}{n} \sum_i \frac{1}{\sqrt{(2\pi)^p |H|}} \exp\left[\frac{1}{2}(x - d_i)^T H^{-1}(x - d_i)\right] \tag{2.29}$$

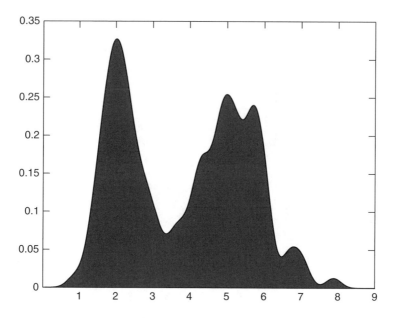

Figure 2.11 Gaussian kernel density estimate

2.5.2 Bandwidth Selection

Kernel density estimation itself is a simple procedure of summing up kernel functions centred around each data point, a procedure that has gone unchanged since its introduction in Parsen (1962). However, the main difficulty lies within the selection of the *bandwidth*, an area where research is ongoing. The bandwidth is synonymous to the bin width in a histogram. Larger bandwidths will result in smoother kernel density estimates, while smaller bandwidths will result in rougher, more jagged estimates. Intuitively, a simple method of bandwidth selection is to start with a small bandwidth that yields a noisy, rough distribution and to use larger bandwidths until the distribution becomes smooth. This is a fairly labour-intensive approach and hard to implement in dimensions larger than two (due to difficulties in visualization).

The goal of selecting a bandwidth is to minimize the error between the kernel density estimate and the true distribution. If the underlying distribution is known, an optimal kernel can be selected. The fundamentals in bandwidth selection have been presented quite nicely in Scott (1992), and the performance of adaptive bandwidth smoothing has been analysed in depth in Terrel and Scott (1992). While the optimal bandwidth can depend on the underlying distribution which we wish to estimate and the kernel type, it depends more strongly on the spread of the data points and the number of data points. Thus, fairly good results can be obtained by simply selecting a distribution (such as the Gaussian distribution) and using its optimal bandwidth estimate for other distributions. Due to this, the optimal Gaussian bandwidth estimator is commonly used

$$H_N = \left(\frac{4}{n(p+2)}\right)^{\frac{2}{p+4}} S \tag{2.30}$$

where S is the sample covariance estimate.

2.5.3 Kernel Density Estimation Tutorial

Let us consider a set of generated data:

$$\mathcal{D} = \begin{bmatrix} 1.90 \\ 0.12 \\ 1.05 \\ -0.23 \\ -0.16 \\ 0.69 \\ 0.56 \\ -1.12 \\ -1.53 \\ -1.09 \end{bmatrix} \qquad \mathrm{cov}(\mathcal{D}) = 1.1594$$

which is shown in in Figure 2.12, where the data points lie along the x axis.

Using the optimal Gaussian bandwidth selector in Eqn (2.30), we obtain a bandwidth of

$$H = \left(\frac{4}{10(1+2)} \right)^{\frac{2}{1+4}} 1.1594$$

$$= \left(\frac{4}{30} \right)^{\frac{2}{5}} 1.1594$$

$$= 0.5179$$

The individual kernels are shown around each data point in Figure 2.13.

The bandwidth parameter can then be used in the kernel density estimate

$$\hat{f}(x) = \frac{1}{n} \sum_i \frac{1}{\sqrt{(2\pi)^p 0.5179}} \exp \left[\frac{1}{2} \frac{(x - d_i)^2}{0.5179} \right]$$

where d_i is each element of $\mathcal{D}$. This final kernel density estimate is visualized in Figure 2.14.

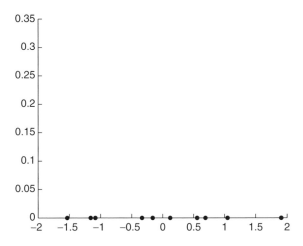

Figure 2.12   Data for kernel density estimation

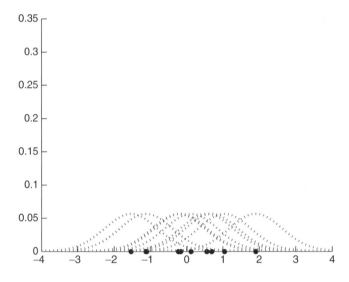

Figure 2.13   Data points with kernels

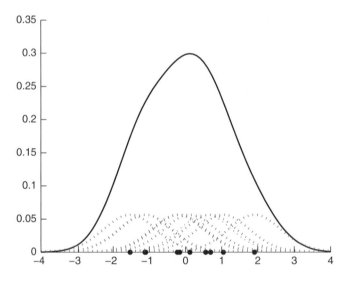

Figure 2.14   Kernel density estimate from data

2.6   Bootstrapping

Bootstrapping is a *resampling* method introduced by Efron (1979) as a modification on jack-knifing (another type of resampling technique). Resampling technques aim to construct new sets of sampled data either by sampling subsets (as in jackknifing) or by sampling with replacement (as in bootstrapping). Bootstrapping is particularly useful if one wishes to obtain a distribution about an estimated parameter. When bootstrapping, it is important that the data are independent and identically distributed (IID).

2.6.1 Bootstrapping Tutorial

Let us consider a case where we have 10 data points and we wish to estimate the mean.

$$\mathcal{D} = \begin{bmatrix} 6.68 \\ 3.83 \\ 4.14 \\ 5.51 \\ 5.33 \\ 4.56 \\ 4.05 \\ 4.77 \\ 4.16 \\ 5.06 \end{bmatrix} \qquad \hat{\mu} = 4.8090$$

However, we would also like to be able to determine a distribution around this estimate $\hat{\mu}$. Bootstrapping enables us to perform this task. Firstly, one can create a new dataset by drawing from the data at random with replacement. Equivalently, one can generate 10 numbers between 1 and 10

$$\mathrm{IND} = [5, 2, 3, 3, 7, 8, 1, 4, 7, 9]$$

and then use these indices to define the new sample. For example, the first element of IND is 5. Thus the first element of the new dataset is $\mathcal{D}(5) = 5.33$. The second element of IND is 2, thus the second element of the new dataset is $\mathcal{D}(2) = 3.83$. This is continued until an entirely new dataset of 10 entries is constructed

$$\mathcal{D}_1 = \begin{bmatrix} 5.33 \\ 3.83 \\ 4.14 \\ 4.14 \\ 4.05 \\ 4.77 \\ 6.68 \\ 5.51 \\ 4.05 \\ 4.16 \end{bmatrix} \qquad \hat{\mu}_1 = 4.6660$$

$\mathcal{D}_1$ is a resampling of $\mathcal{D}$ with replacement. Note that resampling yielded a slightly different value of $\hat{\mu}$. The resampling of $\mathcal{D}$ is then repeated until we have enough values of $\hat{\mu}$ to assess its distribution. For example, after resampling 1000 times, we have a distribution of $\hat{\mu}$ as shown in Figure 2.15.

2.6.2 Smoothed Bootstrapping Tutorial

Resampling straight from the historical data can sometimes have strange and undesirable properties, such as there being a limited number of possible bootstrap sets. For example, from our

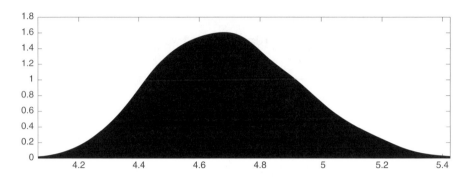

Figure 2.15   Distribution of $\hat{\mu}$ estimate

10 datasets, we could only have 10^{10} different datasets, but since ordering does not matter (because $\hat{\mu}$ involves a sum), there would only be about

$$\frac{(n+r-1)!}{r!(n-1)!} = \frac{(10+10-1)!}{10!(10-1)!} = \frac{21!}{10!9!} = 38\ 798\ 760$$

unique values of $\hat{\mu}$ that could be sampled from. This may seem like a large number, but it would not take very many samples before there is a number of exact repetitions of bootstrapped values of $\hat{\mu}$.

In this example, ordinary bootstrapping would be sampling from a distribution shown in Figure 2.16. As one might see, the resampling can only take on a finite number of values, and the probability is zero for all other legitimate values that the data could have taken.

In order to increase the number of values that $\hat{\mu}$ could take, adding Gaussian noise to each resampled piece of data would be an intuitive solution. This is called the *smoothed bootstrap*. However, adding Gaussian noise to the resampled data begs the question, what variance should this Gaussian noise have? Fortunately, this can be answered using kernel density estimation.

As mentioned in Section 2.5, kernel density estimation (or kernel smoothing) is a method to estimate a smoothed distribution from sampled data. By performing kernel smoothing on the data shown in Figure 2.16 we can obtain an estimate of the distribution, which is obtained by summing Gaussian distributions centred at each data point

$$\hat{f}(x) = \frac{1}{n} \sum_i \frac{1}{\sqrt{(2\pi)^p |H|}} \exp\left[\frac{1}{2}(x - d_i)^T H^{-1}(x - d_i)\right] \tag{2.31}$$

where H is the bandwidth matrix, which can be calculated as

$$H = \left(\frac{4}{n(p+2)}\right)^{\frac{2}{p+4}} S \tag{2.32}$$

Here, p is the dimension of the data, n is the number of data points and S is the sample covariance from the data. This results in the smoothed distribution shown in Figure 2.17. Note that this distribution (from Eqn (2.31)) can be sampled by randomly selecting one of the historical

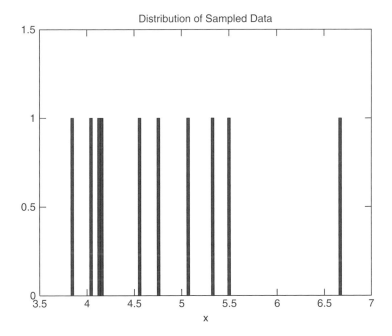

Figure 2.16   Sampling distribution for bootstrapping

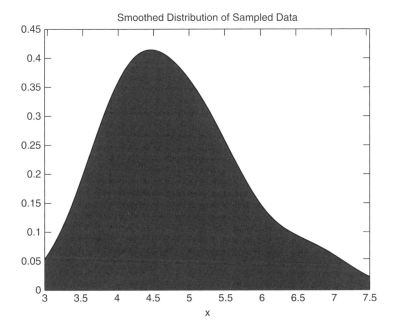

Figure 2.17   Smoothed sampling distribution for bootstrapping

data points, and adding Gaussian noise with covariance H. This is exactly the same procedure as the smoothed bootstrap.

Hence, by using the optimal bandwidth in Eqn (2.32), we can obtain appropriate covariance for the Gaussian noise to be added from our bootstrapped samples.

As an example, let us say we observed the same 10 data points for which we bootstrapped previously, and again we generate the selection indices

$$\text{IND} = [5, 2, 3, 3, 7, 8, 1, 4, 7, 9]$$

The covariance of the data was found to be $\text{var}(\mathcal{D}) = 0.75481$, resulting in a bandwidth of

$$H = \left(\frac{4}{10(1+2)} \right)^{\frac{2}{1+4}} S$$

$$= \left(\frac{4}{30} \right)^{\frac{2}{5}} 0.75481$$

$$= 0.33714$$

We can then add Gaussian noise to the bootstrapped sample

$$\mathcal{D}_1 = \begin{bmatrix} 5.33 \\ 3.83 \\ 4.14 \\ 4.14 \\ 4.05 \\ 4.77 \\ 6.68 \\ 5.51 \\ 4.05 \\ 4.16 \end{bmatrix} + \sqrt{0.33714} \times \texttt{randn}(10, 1)$$

where $\texttt{randn}(10, 1)$ generates a 10×1 vector of Gaussian random variables. The mean of $\mathcal{D}_1$ can be calculated to generate a resampled estimate of $\hat{\mu}_1$, but this time $\hat{\mu}_1$ can essentially take on an infinite number of different values. The resulting distribution of $\hat{\mu}$ is shown in Figure 2.18, from which confidence intervals of $\hat{\mu}$ can be obtained.

2.7   Notes and References

This chapter takes basic concepts from many different sources, including Pernestal (2007), Dempster et al. (1977), Scott (1992) and Efron (1979), and seeks to organize them in a manner that the reader can use to gain a basic understanding of the key concepts discussed later in the book. Similar explanations for many of these topics can be found in Gonzalez (2014).

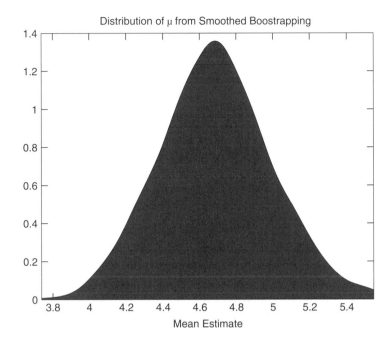

Figure 2.18   Distribution of $\hat{\mu}$ estimate

References

Bayes T 1764/1958 An essay towards solving a problem in the doctrine of chances. *Biometrika* **45**, 296–315.

de Laplace P 1820 *A Philosophical Essay on Probabilities*. Dover.

Dempster A, Laird N and Rubin D 1977 Maximum likelihood from incomplete data via the em algorithm. *Journal of the Royal Statistical Society, Series B (Methodological)* **39**(1), 1–38.

Efron B 1979 Bootstrap methods: Another look at the jackknife. *Annals of Statistics* **7**, 1–26.

Fink D 1997 *A compendium of conjugate priors*. Technical report, Montana State University, Department of Biology, Bozeman Montana, 59717.

Gonzalez R 2014 *Bayesian Solutions to Control Loop Diagnosis*. PhD thesis. University of Alberta.

Korb KB and Nicholson AE 2004 *Bayesian Artificial Intelligence*, 1st edn. Chapman & Hall/CRC.

Parsen E 1962 On estimation of a probability density function and mode. *Annals of Mathematical Statistics* **33**(3), 1065–1076.

Pernestal A 2007 *A Bayesian approach to fault isolation with application to diesel engines*. PhD thesis. KTH School of Electrical Engineering, Sweden.

Scott D 1992 *Multivariate Density Estimation: Theory, Practice, and Visualization*, 1st edn. Wiley.

Shafer G 1976 *A Mathematical Theory of Evidence*. Princeton University Press.

Terrel G and Scott D 1992 Variable kernel density estimation. *Annals of Statistics* **20**(3), 1236–1265.

Venn J 1866 *The Logic of Chance*. MacMillan.

Wilks S 1962 *Mathematical Statistics*. Wiley.

3

Bayesian Diagnosis

3.1 Introduction

The biggest challenge associated with control performance diagnosis is the synthesis of information from multiple sources. Various research has been done on detecting specific problems within a control loop, such as valve stiction, sensor bias, and process model changes. However, there is usually an implicit assumption made that the other unattended components are in good shape. Clearly, this assumption does not always hold. Different problems can produce similar symptoms, thus one alarm can be triggered by different problems. Similarly, one problem source can also affect several monitors simultaneously.

In light of the problems above, the Bayesian approach provides an information synthesis framework by incorporating information from multiple monitors. In this chapter, we will present a description of the control loop diagnosis problem. Following that, a systematic Bayesian approach for data-driven control loop diagnosis is also presented.

3.2 Bayesian Approach for Control Loop Diagnosis

A typical control loop consists of a controller, an actuator, a process, and a sensor. In this book, we assume that the process monitors are designed for the components of interest. For example, if the valve is suspected of having a stiction problem, the output of a valve stiction detection monitor will be available. A control loop diagram and associated monitors are shown in Figure 3.1.

In order to describe and apply the Bayesian control loop diagnosis framework, the associated notation will be introduced in the remainder of this subsection.

3.2.1 Mode M

A system composed of many components will have patterns consisting of the behaviour of each component. Consider a control loop consisting of P components of interest: $C_1, C_2, \ldots, C_P$. A faulty operating mode will indicate that one or more of these components is exhibiting faulty behaviour, while a normal operating mode will have no faulty behaviour in any component.

Process Control System Fault Diagnosis: A Bayesian Approach, First Edition. Ruben Gonzalez, Fei Qi and Biao Huang.
© 2016 John Wiley & Sons, Ltd. Published 2016 by John Wiley & Sons, Ltd.

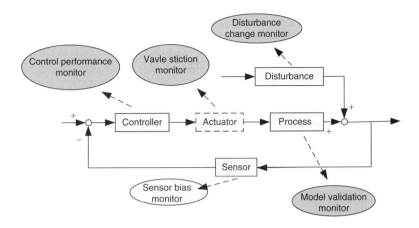

Figure 3.1   Typical control system structure

As an example, a typical process control loop consists of a process, actuator (usually valve), instrumentation or sensor, and a controller. Each component may be subject to different faults: valves may be sticky, controllers may be poorly tuned, and instrumentation could be biased. In addition, the process may have many different components, each with its own set of possible faults. The mode M for such a system would include information on all components within the system. Now suppose that each component C_i has q_i different status values. Then the total number of possible modes is

$$Q = \prod_{i=1}^{P} q_i$$

and the set of all possible modes can be denoted as

$$\mathcal{M} = \{m_1, m_2, \cdots, m_Q\}$$

The objective of control loop diagnosis is to obtain the operating mode or, equivalently, the state of all components of interest in order to isolate the underlying problem.

3.2.2   Evidence E

The diagnosis problem is approached by using evidence E. Evidence can consist of raw process data, but is often more informative if it is processed in such a way as to make it sensitive to the underlying feature or fault we wish to detect. In this book, such processing algorithms are called *monitors*. For example, a simple monitor for instrument bias may be to pass mass balance errors through a low-pass filter. Evidence is the main input to the diagnosis system and can be denoted $E = (\pi_1, \pi_2, \cdots, \pi_L)$, where π_i is the ith evidence source and L is the total number of evidence sources.

Monitor readings are usually continuous, but in order to generate alarms or status descriptions for operators, these readings are often discretized according to predefined thresholds. In this chapter, all monitor readings are assumed to be discretized. For example, control performance monitors can be discretized in order to categorize their values into descriptive ratings such as 'poor', 'acceptable' or 'optimal'.

In order to maintain consistency with other chapters, lower case e represents a specific value of a random variable E. Since E is discrete, it has a finite number of values. For example, let us consider evidence from a control performance monitor and a stiction monitor. One of the possible values could be $e = (\pi_1 = optimal\ control\ performance, \pi_2 = no\ sensor\ bias, \cdots)$.

Now consider a monitor output π_i that has k_i different discrete values. Then the total number of values the evidence can take is

$$K = \prod_{i=1}^{L} k_i$$

and the set of these different possible evidence values is denoted

$$\mathcal{E} = \{e_1, e_2, \cdots, e_K\}$$

where e_i is the ith possible evidence value of E.

3.2.3   Historical Dataset $\mathcal{D}$

Historical data $\mathcal{D}$ includes evidence (in particular, discrete monitor readings) that has been collected from different modes over the desired periods of time. Each data sample d^t, collected at time t consists of one of the possible evidence values e in $\mathcal{E}$ under mode m in $\mathcal{M}$; this can be denoted $d^t = (e^t, m^t)$. The entire data record can be denoted

$$\mathcal{D} = \{d^1, d^2, \cdots, d^{\hat{N}}\}$$

where $\hat{N}$ is the number of historical data samples.

The historical data is divided into sets collected under different operating modes

$$\mathcal{D} = \{\mathcal{D}_{m_1}, \mathcal{D}_{m_2}, \cdots, \mathcal{D}_{m_Q}\}$$

and

$$\mathcal{D}_{m_i} = \{d^1_{m_i}, \cdots, d^{N_{m_i}}_{m_i}\}$$

includes all historical samples within the mode m_i, where N_{m_i} is the number of historical samples corresponding to mode m_i.

Diagnosis is simplest when the evidence is not autodependent, and the methods in this chapter assume auto-independence. In order to avoid autodependence, the data should be sampled sufficiently far apart in time. If each monitor reading is calculated from a separate time segment (or window), then there should be a sufficient gap between these windows in order to ensure that autodependence does not occur. If this assumption is valid, one can obtain the joint probability of the evidence as

$$p(\mathcal{D}) = p(d^1, d^2, \cdots, d^{\hat{N}}) = p(d^1)p(d^2)\cdots p(d^{\hat{N}}) \tag{3.1}$$

Note that it is possible to deal with autodependent data, but this is discussed in a later chapter.

The intent of Bayesian diagnosis is to assign probabilities to operational modes in order to guide the user through the process of investigating the underlying issues. This methodology falls under the broader class of supervised, data-driven classification methods. One of the strengths of this particular method is the ability to combine evidence from multiple sources.

When the current evidence E is observed, having already collected the historical dataset $\mathcal{D}$, Bayes' rule can be applied as follows:

$$p(M|E,\mathcal{D}) = \frac{p(E|M,\mathcal{D})p(M|\mathcal{D})}{p(E|\mathcal{D})} \tag{3.2}$$

Here, $p(M|E,\mathcal{D})$ is the *posterior probability* or the conditional probability of mode M given the evidence E and the historical dataset $\mathcal{D}$. Additionally, $p(E|M,\mathcal{D})$ is the *likelihood* or the probability of observing the evidence E, given the mode M and the historical data $\mathcal{D}$; $p(M|\mathcal{D})$ is the *prior probability* of mode M. Finally, $p(E|\mathcal{D})$ is a scaling factor, and can be calculated as

$$p(E|\mathcal{D}) = \sum_M p(E|M,\mathcal{D})p(M|\mathcal{D})$$

Note that the historical data is collected on the basis of mode, and the number of samples for each mode is not necessarily representative of that mode's frequency. For example, if mode m_1 occurs twice as frequently as mode m_2, it does not require that twice as much data be collected under mode m_1 as under mode m_2. Because of this, it is assumed that the data itself does not offer any information on the underlying mode frequency. This is in fact a strength of the Bayesian method, as it does not force us to collect profuse amounts of data for commonly occurring modes (Pernestal 2007).

As a result, Eqn (3.2) is often written as

$$p(M|E,\mathcal{D}) \propto p(E|M,\mathcal{D})p(M) \tag{3.3}$$

3.3   Likelihood Estimation

Since prior probability is determined by a priori information, the main task of building a Bayesian diagnostic system is the estimation of the likelihood probability $p(E|M,\mathcal{D})$, based on the historical data $\mathcal{D}$. Pernestal (2007) presented a data-driven Bayesian algorithm to estimate the likelihood in diesel engine diagnosis. This method is adopted here for control loop diagnosis.

Suppose that the likelihood of evidence $E = e_i$ under mode $M = m_j$ is to be calculated, where

$$e_i \in \mathcal{E} = \{e_1, \cdots, e_K\}$$

and

$$m_j \in \mathcal{M} = \{m_1, \cdots, m_Q\}$$

The likelihood $p(e_i|m_j,\mathcal{D})$ can only be estimated from the historical data subset $\mathcal{D}_{m_j}$ where the mode $M = m_j$

$$p(e_i|m_j,\mathcal{D}) = p(e_i|m_j,\mathcal{D}_{m_j},\mathcal{D}_{\neg m_j}) = p(e_i|m_j,\mathcal{D}_{m_j}) \tag{3.4}$$

where $\mathcal{D}_{\neg m_j}$ is the dataset whose underlying mode is not m_j.

The likelihood probability can be computed by marginalization over all possible likelihood parameters,

$$p(e_i|m_j, \mathcal{D}_{m_j}) = \int_\Omega p(e_i|\Theta_{m_j}, m_j, \mathcal{D}_{m_j}) f(\Theta_{m_j}|m_j, \mathcal{D}_{m_j}) d\Theta_{m_j} \tag{3.5}$$

where $\Theta_{m_j} = \{\theta_{1|m_j}, \theta_{2|m_j}, \cdots, \theta_{K|m_j}\}$ are the likelihood parameters for all possible evidence values of mode m_j, and K is the total number of possible evidence values. For example, $\theta_{i|m_j} = p(e_i|m_j)$ is the likelihood of evidence e_i when the underlying mode is m_j. Ω is the space of all likelihood parameters Θ_{m_j}. In Eqn (3.5), $f(\Theta_{m_j}|m_j, \mathcal{D}_{m_j})$ can be calculated according to Bayes' rule:

$$f(\Theta_{m_j}|m_j, \mathcal{D}_{m_j}) = \frac{p(\mathcal{D}_{m_j}|\Theta_{m_j}, m_j) f(\Theta_{m_j}|m_j)}{p(\mathcal{D}_{m_j}|m_j)} \tag{3.6}$$

In Eqn (3.6), the Dirichlet distribution is commonly used to model priors of the likelihood parameters (Pernestal 2007), with Dirichlet parameters $a_{1|m_j}, \cdots, a_{K|m_j}$,

$$f(\Theta_{m_j}|m_j) = \frac{\Gamma(\sum_{i=1}^K a_{i|m_j})}{\prod_{i=1}^K \Gamma(a_{i|m_j})} \prod_{i=1}^K \theta_{i|m_j}^{a_{i|m_j}-1} \tag{3.7}$$

where $a_{i|m_j}$ can be interpreted as the number of prior samples for evidence e_i under mode m_j, which will be elaborated shortly; $\Gamma(\cdot)$ is the gamma function,

$$\Gamma(x) = \int_0^\infty t^{x-1} e^{-t} dt \tag{3.8}$$

Here all the independent variables x of the gamma functions are prior numbers of evidence values, which take positive integers, so

$$\Gamma(x) = (x-1)! \tag{3.9}$$

The likelihood of historical data subset $\mathcal{D}_{m_j}$ can be written as

$$p(\mathcal{D}_{m_j}|\Theta_{m_j}, m_j) = \prod_{t=1}^{N_{m_j}} p(d_{m_j}^t|\Theta_{m_j}, m_j) \tag{3.10}$$

The data sample at time t in the historical data subset $\mathcal{D}_{m_j}$ includes the underlying mode m_j and the evidence e^t,

$$d_{m_j}^t = (e^t, m_j)$$

Thus when $e^t = e_i$,

$$p(d_{m_j}^t|\Theta_{m_j}, m_j) = \theta_{i|m_j} \tag{3.11}$$

Combining Eqn (3.10) and Eqn (3.11), we have

$$p(\mathcal{D}_{m_j}|\Theta_{m_j}, m_j) = \prod_{i=1}^K \theta_{i|m_j}^{n_{i|m_j}} \tag{3.12}$$

where $n_{i|m_j}$ is the number of historical samples, and where the evidence $E = e_i$, and the underlying mode $M = m_j$.

Substituting Eqn (3.12) and Eqn (3.6) in Eqn (3.5), the following result can be obtained for the likelihood (Pernestal 2007):

$$p(E = e_i | M = m_j, \mathcal{D}) = \frac{n_{i|m_j} + a_{i|m_j}}{N_{m_j} + A_{m_j}} \tag{3.13}$$

where $n_{i|m_j}$ is the number of historical samples with the evidence $E = e_i$ occurring under mode m_j, $a_{i|m_j}$ is the number of prior samples that is assigned to evidence e_i under mode m_j, $N_{m_j} = \sum_i n_{i|m_i}$, and $A_{m_j} = \sum_i a_{i|m_j}$. To simplify the notation, the subscript m_j will be omitted when it is clear from the context.

With the estimated likelihood probabilities for current evidence E under different modes m_i, $P(E|m_i, \mathcal{D})$, and the user-defined prior probabilities $p(m_i)$, posterior probabilities of each mode $m_i \in \mathcal{M}$ can be calculated according to Eqn (3.3). Among these modes, the one with largest posterior probability is typically picked up as the underlying mode based on the maximum *a posteriori* (MAP) principle, and the abnormality associated with this mode is then diagnosed as the problem source. The users can then use the diagnostic result as a reference to further verify the status of the problematic components as identified in the mode.

3.4 Notes and References

The data-driven Bayesian diagnosis framework presented in this chapter stems from Pernestal (2007) and Qi and Huang (2008). The derivation process of the data-driven approach can be found in Qi and Huang (2008).

References

Pernestal A 2007 *A Bayesian approach to fault isolation with application to diesel engines*. PhD thesis. KTH School of Electrical Engineering, Sweden.

Qi F and Huang B 2008 Data-driven Bayesian approach for control loop diagnosis. *Proceedings of American Control Conference*, Seattle, USA.

4

Accounting for Autodependent Modes and Evidence

4.1 Introduction

The Bayesian methods discussed previously assume that the evidence and underlying modes are temporally independent. The corresponding graphic model is shown in Figure 4.1. This condition, however, may not hold for many engineering problems. With the evidence and mode-transition information being considered, the temporal information can be synthesized within the Bayesian framework to improve diagnostic performance.

In this chapter, the problem of temporal dependence in the evidence is solved by a data-driven Bayesian approach by considering evidence-transition probabilities. A hidden Markov model is built to address temporal dependence in the modes and a data-driven algorithm is developed to estimate the mode transition probability. The mode-dependence solution is then combined with the evidence-dependence solution in order to develop a recursive hidden Markov model that can be used for online control-loop diagnosis.

4.2 Temporally Dependent Evidence

4.2.1 Evidence Dependence

Depending on the sampling process and the external disturbance, the evidence can be temporally dependent or independent. If there is sufficient space between evidence samples, and the evidence is not affected by slow-moving unmeasured disturbances, the evidence can be considered independent. A common example of a slow-moving disturbance is ambient temperature. Consider a system where the evidence source is a monitor reading based on hourly averages with no overlap. Due to cyclical temperature changes that are roughly repeated every 24 samples, the monitor readings may follow a predictable pattern if they depend on temperature. In such a case, if the monitor is affected by the temperature, stronger diagnosis results could arguably be achieved if the model considered temporal dependence to explain the predictable monitor variation.

Process Control System Fault Diagnosis: A Bayesian Approach, First Edition. Ruben Gonzalez, Fei Qi and Biao Huang.
© 2016 John Wiley & Sons, Ltd. Published 2016 by John Wiley & Sons, Ltd.

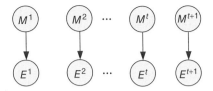

Figure 4.1 Bayesian model with independent evidence data samples

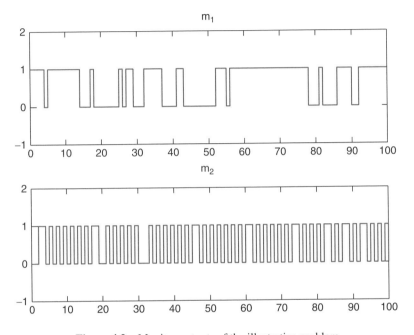

Figure 4.2 Monitor outputs of the illustrative problem

Aside from the practical issues, another limitation with the conventional Bayesian approach is its inability to capture all time-domain information. As an illustration, suppose that the system under diagnosis has two modes m_1 and m_2, with one available discrete monitor π, having two possible values, 0 and 1. A set of 100 monitor output samples is shown in Figure 4.2. The title in each plot indicates the underlying operating mode under which the evidence data is collected. In addition, the likelihood probability of evidence being 0 or 1 is summarized in Table 4.1.

As seen in Table 4.1, the probabilities of the evidence taking values of $e = 0$ and $e = 1$ are almost the same for the two modes; the difference between these modes is that the evidence is much less likely to switch under mode m_1. The Bayesian diagnosis approach discussed in the previous chapter is not able to distinguish the two modes, since it solely relies on the 'point' evidence of the likelihood. However, if information regarding temporal dependence is included, the diagnosis accuracy would be improved significantly.

With evidence autodependence being considered, the current evidence will depend on both the current underlying mode and the previous evidence. By assuming that the evidence

Table 4.1   Likelihood estimation of the illustrative problem

e	0	1
m_1	0.46	0.54
m_2	0.48	0.52

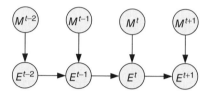

Figure 4.3   Bayesian model considering dependent evidence

dependence follows a Markov process, the corresponding graphic model is as shown in Figure 4.3, and the mode posterior probability is calculated as

$$p(M^t|E^{t-1}, E^t, \mathcal{D}) \propto p(E^t|M, E^{t-1}, \mathcal{D})p(M^t) \tag{4.1}$$

4.2.2   Estimation of Evidence-transition Probability

The evidence-transition probability $p(E^t|M^t, E^{t-1})$ is a crucial parameter when considering the dependence of evidence in a Bayesian diagnosis problem and can be obtained using historical data $\mathcal{D}$. The likelihood term to be obtained from history is $p(E^t|M^t, E^{t-1}, \mathcal{D})$, which represents the likelihood of evidence E^t given current underlying mode M^t and previous evidence E^{t-1}. Consequently, an evidence-transition sample should include the three items: E^t, M^t, and E^{t-1}.

$$d_E^{t-1} = \{M^t, E^{t-1}, E^t\} \tag{4.2}$$

We define the new evidence transition dataset $\mathcal{D}_E$ as

$$\mathcal{D}_E = \{d_E^1, \cdots, d_E^{\hat{N}-1}\}$$
$$= \{(M^2, E^1, E^2), \cdots, (M^{\hat{N}}, E^{\hat{N}-1}, E^{\hat{N}})\} \tag{4.3}$$

Figure 4.4 shows how evidence-transition samples are formed from the historical dataset. Figure 4.4 shows the overlapping nature of these samples. The diagonally hatched regions

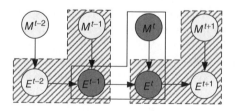

Figure 4.4   Illustration of evidence transition samples

show two time samples, while the outlined region with shaded nodes represent another time sample. All three samples are valid candidates as separate historical samples.

Let us assume that the evidence transition probability from $E^{t-1} = e_s$ to $E^t = e_r$ under mode $M^t = m_c$ is to be estimated from the evidence transition dataset

$$p(E^t | E^{t-1}, M^t, \mathcal{D}_E) = p(e_r | e_s, m_c, \mathcal{D}_E) \tag{4.4}$$

The evidence-transition probability $p(e_r | e_s, m_c, \mathcal{D}_E)$ can only be estimated from the subset $\mathcal{D}_{E|m_c}$ where the underlying mode $M^t = m_c$,

$$p(e_r | e_s, m_c, \mathcal{D}_E) - p(e_r | e_s, m_c, \mathcal{D}_{E|m_c}, \mathcal{D}_{E|\neg m_c})$$
$$= p(e_r | e_s, m_c, \mathcal{D}_{E|m_c}) \tag{4.5}$$

where $\mathcal{D}_{E|\neg m_c}$ is the historical transition dataset when the underlying mode M^t is a mode other than m_c. Without causing confusion, the subscript m_c will be omitted in the following derivation.

By marginalizing over all possible evidence-transition probability parameters, the evidence-transition probability can be calculated as

$$p(e_r | e_s, m_c, \mathcal{D}_E) = \int_{\Psi_1, \cdots, \Psi_K} p(e_r | \Phi_1, \cdots, \Phi_K, e_s, m_c, \mathcal{D}_E)$$
$$f(\Phi_1, \cdots, \Phi_K | e_s, m_c, \mathcal{D}_E) d\Phi_1 \cdots \Phi_K. \tag{4.6}$$

where K is the number of all possible discrete evidence values and $\Phi_i = \{\phi_{i,1}, \phi_{i,2}, \cdots, \phi_{i,K}\}$ represents the probability parameter set for every possible evidence transition of e_i under mode m_c. For instance, $\phi_{s,r} = p(e_r | e_s, m_c)$ is the probability of evidence transition from $E^{t-1} = e_s$ to $E^t = e_r$ with underlying mode $M^t = m_c$. According to the definition of $\phi_{s,r}$, the transition probabilities of evidence e_i under mode m_c must sum to 1.

$$\sum_{i=1}^{K} \phi_{s,i} = \sum_{i=1}^{K} p(e_i | e_s, m_c) = 1 \tag{4.7}$$

where Ψ_i is the space of all possible evidence transition parameter sets Φ_i, subject to Eqn (4.7). The value of $p(e_r | \Phi_1, \cdots, \Phi_K, e_s, m_c, \mathcal{D}_E)$ is determined by the parameter set Φ_s only.

$$p(e_r | \Phi_1, \cdots, \Phi_K, e_s, m_c, \mathcal{D}_E) = p(e_r | \Phi_s, e_s, m_c, \mathcal{D}_E) \tag{4.8}$$

Substituting Eqn (4.8) in Eqn (4.6) generates

$$p(e_r | e_s, m_c, \mathcal{D}_E)$$
$$= \int_{\Psi_1, \cdots, \Psi_K} p(e_r | \Phi_s, e_s, m_c, \mathcal{D}_E) \cdot f(\Phi_1, \cdots, \Phi_K | e_s, m_c, \mathcal{D}_E) d\Phi_1 \cdots \Phi_K$$
$$= \int_{\Psi_1, \cdots, \Psi_K} \phi_{s,r} \cdot f(\Phi_1, \cdots, \Phi_K | e_s, m_c, \mathcal{D}_E) d\Phi_1 \cdots \Phi_K \tag{4.9}$$

The second term, $f(\Phi_1, \cdots, \Phi_K | e_s, m_c, \mathcal{D}_E)$, can be calculated from a Bayesian perspective. It can be viewed as the posterior probability of parameter sets $\{\Phi_1, \cdots, \Phi_K\}$,

$$f(\Phi_1, \cdots, \Phi_K | e_s, m_c, \mathcal{D}_E) =$$
$$\frac{p(\mathcal{D}_E | e_s, m_c, \Phi_1, \cdots, \Phi_K) f(\Phi_1, \cdots, \Phi_K | e_s, m_c)}{p(\mathcal{D}_E | e_s, m_c)} \tag{4.10}$$

where

$$p(\mathcal{D}_E | e_s, m_c) = \int_{\Psi_1, \cdots, \Psi_K} p(\mathcal{D}_E | e_s, m_c, \Phi_1, \cdots, \Phi_K)$$
$$f(\Phi_1, \cdots, \Phi_K | e_s, m_c) d\Phi_1 \cdots \Phi_K \tag{4.11}$$

In Eqn (4.10), the first term in the numerator, $p(\mathcal{D}_E | e_s, m_c, \Phi_1, \cdots, \Phi_K)$, is the likelihood of transition data with parameter sets $\{\Phi_1, \cdots, \Phi_K\}$. Once the mode and parameter sets $\{\Phi_1, \cdots, \Phi_K\}$ are fixed, the likelihood of transition dataset $\mathcal{D}_E$ is determined, and the current evidence e_s may be considered irrelevant.

$$p(\mathcal{D}_E | e_s, m_c, \Phi_1, \cdots, \Phi_K) = p(\mathcal{D}_E | m_c, \Phi_1, \cdots, \Phi_K)$$
$$= \prod_{i=1}^{K} \prod_{j=1}^{K} \phi_{i,j}^{\tilde{n}_{i,j}} \tag{4.12}$$

where $\tilde{n}_{i,j}$ is the number of evidence transitions from e_i to e_j in the transition dataset.

Similarly, the prior probability of transition parameter set Φ_i only depends on the underlying mode m_c, and is independent of current evidence e_s.

$$f(\Phi_1, \cdots, \Phi_K | e_s, m_c) = f(\Phi_1, \cdots, \Phi_K | m_c)$$

With the common assumption that the priors for different parameter sets Φ_i and $\Phi_{j \neq i}$ are independent (Pernestal 2007),

$$f(\Phi_1, \cdots, \Phi_K | m_c) = f(\Phi_1 | m_c) \cdots f(\Phi_K | m_c) \tag{4.13}$$

The Dirichlet distribution is selected as it allows us to include priors of the likelihood parameters

$$f(\Phi_i | m_c) = \frac{\Gamma(\sum_{j=1}^{K} b_{i,j})}{\prod_{j=1}^{K} \Gamma(b_{i,j})} \prod_{j=1}^{K} \phi_{i,j}^{b_{i,j}-1} \tag{4.14}$$

where $b_{i,1}, \cdots, b_{i,K}$ are the Dirichlet parameters. Thus

$$f(\Phi_1, \cdots, \Phi_K | e_s, m_c) = \prod_{i=1}^{K} \frac{\Gamma(\sum_{j=1}^{K} b_{i,j})}{\prod_{j=1}^{K} \Gamma(b_{i,j})} \prod_{j=1}^{K} \phi_{i,j}^{b_{i,j}-1} \tag{4.15}$$

where $b_{i,j}$ can be regarded as the number of prior samples of the evidence transition from e_i to e_j. $\Gamma(\cdot)$ is the gamma function,

$$\Gamma(x) = \int_0^\infty t^{x-1} e^{-t} dt \tag{4.16}$$

In this chapter, x represents counts of evidence transitions. Because these counts are positive integers, the gamma function can be replaced with its equivalent factorial expression:

$$\Gamma(x) = (x-1)! \tag{4.17}$$

Substituting Eqn (4.15) and Eqn (4.12) in Eqn (4.10),

$$f\left(\Phi_1, \cdots, \Phi_K | e_s, m_c, \mathcal{D}_E\right)$$

$$= \frac{p(\mathcal{D}_E | e_s, m_c, \Phi_1, \cdots, \Phi_K) f(\Phi_1, \cdots, \Phi_K | e_s, m_c)}{p(\mathcal{D}_E | e_s, m_c)}$$

$$= \frac{1}{p(\mathcal{D}_E | e_s, m_c)} \cdot \prod_{i=1}^{K} \frac{\Gamma(\sum_{j=1}^{K} b_{i,j})}{\prod_{j=1}^{K} \Gamma(b_{i,j})} \prod_{j=1}^{K} \phi_{i,j}^{b_{i,j}-1} \cdot \prod_{i=1}^{K}\prod_{j=1}^{K} \phi_{i,j}^{\tilde{n}_{i,j}} \tag{4.18}$$

Let

$$\mu = \prod_{i=1}^{K} \frac{\Gamma(\sum_{j=1}^{K} b_{i,j})}{\prod_{j=1}^{K} \Gamma(b_{i,j})} \tag{4.19}$$

and then Eqn (4.18) can be re-orgnized as

$$f\left(\Phi_1, \cdots, \Phi_K | e_s, m_c, \mathcal{D}_E\right)$$

$$= \frac{\mu}{p(\mathcal{D}_E | e_s, m_c)} \cdot \prod_{i=1}^{K}\prod_{j=1}^{K} \phi_{i,j}^{b_{i,j}-1} \cdot \prod_{i=1}^{K}\prod_{j=1}^{K} \phi_{i,j}^{\tilde{n}_{i,j}}$$

$$= \frac{\mu}{p(D | e_s, m_c)} \prod_{i=1}^{K}\prod_{i=1}^{K} \phi_{i,j}^{\tilde{n}_{i,j}+b_{i,j}-1} \tag{4.20}$$

The transition probability from evidence e_s to e_r can be calculated as

$$p(e_r | e_s, m_c, \mathcal{D}_E)$$

$$= \int_{\Psi_1, \cdots, \Psi_K} \phi_{s,r} \cdot f(\Phi_1, \cdots, \Phi_K | e_s, m_c, \mathcal{D}_E) d\Phi_1 \cdots \Phi_K$$

$$= \int_{\Psi_1, \cdots, \Psi_K} \phi_{s,r} \cdot \frac{\mu}{p(\mathcal{D}_E | e_s, m_c)} \cdot \prod_{i=1}^{K}\prod_{j=1}^{K} \phi_{i,j}^{\tilde{n}_{i,j}+b_{i,j}-1} d\Phi_1 \cdots \Phi_K$$

$$= \frac{\mu}{p(\mathcal{D}_E | e_s, m_c)} \int_{\Psi_1} \prod_{j=1}^{K} \phi_{1,j}^{\tilde{n}_{1,j}+b_{1,j}-1} d\Phi_1 \cdots \int_{\Psi_s} \phi_{s,r}^{\tilde{n}_{s,r}+b_{s,r}} \prod_{j \neq r} \phi_{s,j}^{\tilde{n}_{s,j}+b_{s,j}-1} d\Phi_s$$

$$\cdots \int_{\Psi_K} \prod_{j=1}^{K} \phi_{K,j}^{\tilde{n}_{K,j}+b_{K,j}-1} d\Phi_K \tag{4.21}$$

In Eqn (4.21), $p(\mathcal{D}_E | e_s, m_c)$ is the scaling factor as defined in Eqn 4.10. According to Eqn (4.11),

$$p(\mathcal{D}_E|e_s, m_c)$$

$$= \int_{\Psi_1, \cdots, \Psi_K} p(\mathcal{D}_E|e_s, m_c, \Phi_1, \cdots, \Phi_K) \cdot f(\Phi_1, \cdots, \Phi_K|e_s, m_c)d\Phi_1 \cdots \Phi_K$$

$$= \int_{\Psi_1, \cdots, \Psi_K} \prod_{i=1}^{K} \frac{\Gamma(\sum_{j=1}^{K} b_{i,j})}{\prod_{j=1}^{K} \Gamma(b_{i,j})} \cdot \prod_{i=1}^{K}\prod_{j=1}^{K} \phi_{i,j}^{\tilde{n}_{i,j}+b_{i,j}-1} d\Phi_1 \cdots \Phi_K$$

$$= \mu \cdot \int_{\Psi_1} \prod_{j=1}^{K} \phi_{1,j}^{\tilde{n}_{1,j}+b_{1,j}-1} d\Phi_1 \cdots \int_{\Psi_K} \prod_{j=1}^{K} \phi_{K,j}^{\tilde{n}_{K,j}+b_{K,j}-1} d\Phi_K$$

$$= \mu \cdot \prod_{i=1}^{K} \frac{\prod_{j=1}^{K} \Gamma(\tilde{n}_{i,j}+b_{i,j})}{\Gamma(\tilde{N}_i+B_i)} \tag{4.22}$$

where $\tilde{N}_i = \sum_j \tilde{n}_{i,j}$ is the total number of evidence transition samples, from previous evidence e_i to all possible current evidence e_j under mode m_c, and $B_i = \sum_j b_{i,j}$ is the corresponding total number of evidence transition samples under mode m_c.

Similarly, we have

$$\int_{\Psi_1} \prod_{j=1}^{K} \phi_{1,j}^{\tilde{n}_{1,j}+b_{1,j}-1} d\Phi_1 \cdots \int_{\Psi_s} \phi_{s,r}^{\tilde{n}_{s,t}+b_{s,r}} \cdot \prod_{j \neq r} \phi_{s,j}^{\tilde{n}_{s,j}+b_{s,j}-1} d\Phi_s \cdots \int_{\Psi_K} \prod_{j=1}^{K} \phi_{K,j}^{\tilde{n}_{K,j}+b_{K,j}-1} d\Phi_K$$

$$= \frac{\Gamma(\tilde{n}_{s,r}+b_{s,r}+1)}{\Gamma(\tilde{N}_s+B_s+1)} \cdot \frac{\prod_{i,j \neq s,r} \Gamma(\tilde{n}_{i,j}+b_{i,j})}{\prod_{i \neq s} \Gamma(\tilde{N}_i+B_i)} \tag{4.23}$$

Therefore Eqn (4.21) can be written as

$$p(e_r \mid e_s, m_c, \mathcal{D}_E)$$

$$= \mu \cdot \frac{\Gamma(\tilde{n}_{s,r}+b_{s,r}+1)}{\Gamma(\tilde{N}_s+B_s+1)} \cdot \frac{\prod_{i,j \neq s,r} \Gamma(\tilde{n}_{i,j}+b_{i,j})}{\prod_{i \neq s} \Gamma(\tilde{N}_i+B_i)} \cdot \frac{\prod_{i=1}^{K} \Gamma(\tilde{N}_i+B_i)}{\mu \cdot \prod_{i=1}^{K}\prod_{j=1}^{K} \Gamma(\tilde{n}_{i,j}+b_{i,j})}$$

$$= \frac{\tilde{n}_{s,r}+b_{s,r}}{\tilde{N}_s+B_s} \tag{4.24}$$

4.2.3   Issues in Estimating Dependence in Evidence

In a practical setting, it can be very difficult to estimate transition probabilities, especially if a large number of evidence sources are considered. For example, if one wishes to estimate a likelihood $p(E_t|M_t)$ for eight monitors, each having two values, it would require estimating a distribution with $2^8 = 256$ parameters; this is a considerable number of parameters to estimate, but in most cases it is possible to estimate such a distribution. However, if we wish to take into account evidence transition in our likelihood $p(E_t|E_{t-1}, M_t)$, the number of parameters balloons to $2^{16} = 65536$, which is a formidable number of parameters to estimate. In some cases, the benefit of the additional accuracy in this new model structure would strongly be outweighed by the difficulty in parameter estimation. Thus grouping of evidence space according to correlations between various evidences is beneficial to curb the curse of dimension problem. Chapter 13 will provide a procedure for reduction of dimension.

4.3 Temporally Dependent Modes

4.3.1 Mode Dependence

In practice, the status of a component in a control loop is very likely to be dependent on its previous status. For instance, a sticky valve will usually remain in the faulty mode until maintenance work is performed. In addition, the system mode may also change due to scheduled shifts in the operating point, which follow a specific pattern. By considering mode dependence, the extra mode information can be incorporated into the diagnosis framework.

The dynamic Bayesian model of Ghahramani (2002) is utilized in this work to represent mode dependence. At this point, we assume that the mode of the control loop is dependent only on its immediate previous status, hence a *Markov* process. By connecting lines between the consecutive modes to describe the temporal mode dependence, a hidden Markov model (Murphy 2002) is built (see Figure 4.5).

In order to pinpoint the probability of the current system mode, previous evidence also needs to be considered, as the current mode is dependent on the previous mode

$$p(M^t|E^1,\cdots,E^t)$$

$$= \frac{p(E^1,\cdots,E^t|M^t)p(M^t)}{p(E^1,\cdots,E^t)}$$

$$= \frac{\sum_{M^{t-1}}p(E^1,\cdots,E^t,M^{t-1}|M^t)p(M^t)}{p(E^1,\cdots,E^t)}$$

$$= \frac{\sum_{M^{t-1}}p(E^t|M^t,M^{t-1},E^1,\cdots,E^{t-1})p(E^1,\cdots,E^{t-1},M^{t-1}|M^t)p(M^t)}{p(E^1,\cdots,E^t)} \quad (4.25)$$

According to the hidden Markov model, if the current mode M^t is determined, the current evidence E^t is independent of the previous modes M^{t-i}, where $i \geq 1$, and previous evidence E^{t-j}, where $j \geq 1$, thus

$$p(M^t|E^1,\cdots,E^t)$$

$$= \frac{\sum_{M^{t-1}}p(E^t|M^t)p(E^1,\cdots,E^{t-1},M^{t-1}|M^t)p(M^t)}{p(E^1,\cdots,E^t)} \quad (4.26)$$

In this equation, the term $p(E^1,\cdots,E^{t-1},M^{t-1}|M^t)$ can be calculated according to Bayes' rule:

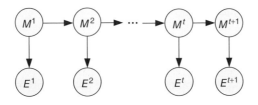

Figure 4.5 Bayesian model considering dependent mode

$$p(E^1, \cdots, E^{t-1}, M^{t-1}|M^t)$$
$$= \frac{p(M^t|E^1, \cdots, E^{t-1}, M^{t-1})p(E^1, \cdots, E^{t-1}, M^{t-1})}{p(M^t)} \qquad (4.27)$$

Combining Eqn (4.27) and Eqn (4.26) yields

$$p(M^t|E^1, \cdots, E^t)$$
$$= \frac{\sum_{M^{t-1}} p(E^t|M^t)p(M^t|E^1, \cdots, E^{t-1}, M^{t-1})p(E^1, \cdots, E^{t-1}, M^{t-1})}{p(E^1, \cdots, E^t)} \qquad (4.28)$$

M^t is independent of $E^1, \cdots, E^{t-1}$ with M^{t-1} known, so

$$p(M^t|E^1, \cdots, E^t)$$
$$= \frac{\sum_{M^{t-1}} p(E^t|M^t)p(M^t|M^{t-1})p(E^1, \cdots, E^{t-1}, M^{t-1})}{p(E^1, \cdots, E^t)}$$
$$= \frac{\sum_{M^{t-1}} p(E^t|M^t)p(M^t|M^{t-1})p(M^{t-1}|E^1, \cdots, E^{t-1})p(E^1, \cdots, E^{t-1})}{p(E^1, \cdots, E^t)}$$
$$= p(E^t|M^t)p(E^1, \cdots, E^{t-1})\frac{\sum_{M^{t-1}} p(M^{t-1}|E^1, \cdots, E^{t-1})p(M^t|M^{t-1})}{p(E^1, \cdots, E^t)} \qquad (4.29)$$

which means

$$p(M^t|E^1, \cdots, E^t) \propto p(E^t|M^t) \sum_{M^{t-1}} p(M^{t-1}|E^1, \cdots, E^{t-1})p(M^t|M^{t-1}) \qquad (4.30)$$

To calculate the probability of the current mode, the mode-transition probability $p(M^t|M^{t-1})$ is needed in addition to the current likelihood $p(E^t|M^t)$.

In addition to the ability to incorporate mode dependence, another advantage of using a hidden Markov model is that this method is less sensitive to inaccurate prior probabilities. In Eqn (4.30), the prior probability for the mode is not required. The necessary information to obtain the mode posterior probability only includes:

1. the mode transition probability, $p(M^t|M^{t-1})$
2. the evidence likelihood probability, $p(E^t|M^t)$
3. the mode posterior probability from the previous diagnosis, $p(M^{t-1}|E^1, \cdots, E^{t-1})$.

Note that items (1) and (2) can be obtained from the historical dataset or from experts. The prior probability is still needed to calculate the first mode probability $p(M^1|E^1)$ when the previous mode term does not exist,

$$p(M^1|E^1) \propto p(E^1|M^1)p(M^1) \qquad (4.31)$$

but according to Eqn (4.30), the prior probability will have a diminishing effect on the diagnosis result as more evidence becomes available. Recall the posterior probability equation for single evidence diagnosis:

$$p(M^t|E^t) \propto p(E^t|M^t)p(M^t) \qquad (4.32)$$

In Eqn (4.30) and Eqn (4.32), $p(E^t|M^t)$ appears in both, serving the function of updating probabilities when evidence emerges. The remaining terms in each equation exist before the new evidence E^t is available. As such, in the recursive solution, the term

$$\sum_{M^{t-1}} p(M^{t-1}|E^1, \cdots, E^{t-1}) p(M^t|M^{t-1})$$

is treated as the 'prior'. The difference is that this term is constantly updated when new evidence emerges. Thus, as more evidence becomes available and updates of the posterior, the impact of the initial prior probability will gradually diminish. With a sufficiently large number of new evidence observations, it can be said that the posterior is solely determined by data and is not affected by inaccurate initial prior probabilities.

4.3.2   Estimating Mode Transition Probabilities

Our goal is to calculate the likelihood probability of a mode M^t given previous mode M^{t-1} from the historical evidence dataset $\mathcal{D}$. Samples that are used to estimate the mode transition should include both the current and previous mode; such samples are called composite mode samples and are given by the following notation

$$d_M^{t-1} = \{M^{t-1}, M^t\} \tag{4.33}$$

Since our focus here is only on the mode transition, the evidence transition is not included. The new composite mode dataset $\mathcal{D}_M$, assembled from historical dataset $\mathcal{D}$, is defined as

$$\begin{aligned}\mathcal{D}_M &= \{d_M^1, \cdots, d_M^{\tilde{N}-1}\} \\ &= \{(M^1, M^2), \cdots, (M^{\tilde{N}-1}, M^{\tilde{N}})\}\end{aligned} \tag{4.34}$$

Figure 4.6 shows how composite mode samples are constructed from the historical dataset. In Figure 4.6, two composite mode samples are denoted by the white boxed areas, while a third sample is denoted by a cross-hatched area. Each of these three samples would constitute a separate composite mode sample defined in Eqn (4.33). Suppose that the mode-transition probability from $M^{t-1} = m_u$ to $M^t = m_v$, hence $p(m_v|m_u)$, is to be estimated:

$$p(M^t|M^{t-1}, \mathcal{D}_M) = p(m_v|m_u, \mathcal{D}_M) \tag{4.35}$$

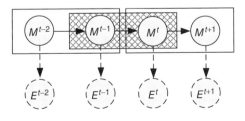

Figure 4.6   Historical composite mode dataset for mode transition probability estimation

where

$$m_v, m_u \in \mathcal{M} = \{m_1, \cdots, m_Q\} \tag{4.36}$$

The mode-transition probability can be calculated by marginalization over all possible mode-transition parameters,

$$p(m_v|m_u, \mathcal{D}_M)$$

$$= \int_{\Sigma_1, \cdots, \Sigma_Q} p(m_v|m_u, \mathcal{D}_M, \Upsilon_1, \cdots, \Upsilon_Q) f(\Upsilon_1, \cdots, \Upsilon_Q|m_u, \mathcal{D}_M) d\Upsilon_1 \cdots \Upsilon_Q \tag{4.37}$$

where $\Upsilon_i = \{\varphi_{i,1}, \cdots, \varphi_{i,Q}\}$ is the probability parameter set for mode transition from mode $M^{t-1} = m_i$; Q is the total number of possible modes – for example, $\varphi_{i,j} = p(m_j|m_i)$ – and Σ_i is the space of all possible parameter sets Υ_i, where

$$\sum_{j=1}^{Q} \varphi_{i,j} = 1$$

The transition probability from m_u to m_v is decided only by the parameter set Υ_u,

$$p(m_v|m_u, \mathcal{D}_M, \Upsilon_1, \cdots, \Upsilon_Q) = p(m_v|m_u, \Upsilon_u) \tag{4.38}$$

Eqn (4.37) can be re-written as

$$p(m_v|m_u, \mathcal{D}_M)$$

$$= \int_{\Sigma_1, \cdots, \Sigma_Q} p(m_v|m_u, \mathcal{D}_M, \Upsilon_1, \cdots, \Upsilon_Q) f(\Upsilon_1, \cdots, \Upsilon_Q|m_u, \mathcal{D}_M) d\Upsilon_1 \cdots \Upsilon_Q$$

$$= \int_{\Sigma_1, \cdots, \Sigma_Q} p(m_v|m_u, \Upsilon_u) f(\Upsilon_1, \cdots, \Upsilon_Q|m_u, \mathcal{D}_M) d\Upsilon_1 \cdots \Upsilon_Q$$

$$= \int_{\Sigma_1, \cdots, \Sigma_Q} \varphi_{u,v} f(\Upsilon_1, \cdots, \Upsilon_Q|m_u, \mathcal{D}_M) d\Upsilon_1 \cdots \Upsilon_Q \tag{4.39}$$

The second term in the integration can be calculated with Bayes' rule,

$$f(\Upsilon_1, \cdots, \Upsilon_Q|m_u, \mathcal{D}_M) = \frac{p(\mathcal{D}_M|\Upsilon_1, \cdots, \Upsilon_Q, m_u) f(\Upsilon_1, \cdots, \Upsilon_Q|m_u)}{p(\mathcal{D}_M|m_u)} \tag{4.40}$$

where $p(\mathcal{D}_M|m_u)$ is the scaling factor,

$$p(\mathcal{D}|m_u) = \int_{\Sigma_1, \cdots, \Sigma_Q} p(\mathcal{D}_M|\Upsilon_1, \cdots, \Upsilon_Q, m_u) f(\Upsilon_1, \cdots, \Upsilon_Q|m_u) d\Upsilon_1 \cdots d\Upsilon_Q \tag{4.41}$$

In Eqn (4.40), the first term in the numerator is the likelihood of historical composite mode data. It is only determined by the parameter set $\{\Upsilon_1, \cdots, \Upsilon_Q\}$, and is independent of m_u; in other words,

$$p(\mathcal{D}_M|\Upsilon_1, \cdots, \Upsilon_Q, m_u) = p(\mathcal{D}_M|\Upsilon_1, \cdots, \Upsilon_Q) = \prod_{i=1}^{Q} \prod_{j=1}^{Q} \varphi_{i,j}^{\hat{n}_{i,j}} \tag{4.42}$$

where $\hat{n}_{i,j}$ is the number of mode transitions from $M^{t-1} = m_i$ to $M^t = m_j$ in the historical composite mode dataset.

According to the common assumption that the priors of different parameter sets Υ_i and $\Upsilon_{j \neq i}$ are independent (Pernestal 2007),

$$f(\Upsilon_1, \cdots, \Upsilon_Q | m_u) = f(\Upsilon_1 | m_u) \cdots f(\Upsilon_Q | m_u) \qquad (4.43)$$

Here we use Dirichlet distribution to model the prior probability distribution of the mode-transition parameters Υ_i with parameters $c_{i,1}, \cdots, c_{i,Q}$,

$$f(\Upsilon_i | m_u) = \frac{\Gamma(\sum_{j=1}^{Q} c_{i,j})}{\prod_{j=1}^{Q} \Gamma(c_{i,j})} \prod_{j=1}^{Q} \varphi_{i,j}^{c_{i,j}-1} \qquad (4.44)$$

Therefore, we have

$$f(\Upsilon_1, \cdots, \Upsilon_Q | m_u) = \prod_{i=1}^{Q} \frac{\Gamma(\sum_{j=1}^{Q} c_{i,j})}{\prod_{j=1}^{Q} \Gamma(c_{i,j})} \prod_{j=1}^{Q} \varphi_{i,j}^{c_{i,j}-1} \qquad (4.45)$$

where $c_{i,j}$ can be considered as the number of prior samples for mode transition from m_i to m_j and $\Gamma(\cdot)$ is the gamma function,

$$\Gamma(x) = \int_0^\infty t^{x-1} e^{-t} dt \qquad (4.46)$$

In this application, all the x terms in the gamma functions are counts of mode transitions, and hence are positive integers so the factorial notation can be substituted:

$$\Gamma(x) = (x - 1)! \qquad (4.47)$$

Substituting Eqn (4.45) and (4.42) into Eqn (4.40) yields

$$f(\Upsilon_1, \cdots, \Upsilon_Q \mid m_u, \mathcal{D}_M)$$
$$= \frac{p(\mathcal{D}_M | \Upsilon_1, \cdots, \Upsilon_Q, m_u) f(\Upsilon_1, \cdots, \Upsilon_Q | m_u)}{p(\mathcal{D}_M | m_u)}$$
$$= \frac{1}{p(\mathcal{D}_M | m_u)} \cdot \prod_{i=1}^{Q} \frac{\Gamma(\sum_{j=1}^{Q} c_{i,j})}{\prod_{j=1}^{Q} \Gamma(c_{i,j})} \prod_{j=1}^{Q} \varphi_{i,j}^{c_{i,j}-1} \cdot \prod_{i=1}^{Q} \prod_{j=1}^{Q} \varphi_{i,j}^{\hat{n}_{i,j}} \qquad (4.48)$$

Let us define ρ as

$$\rho = \prod_{i=1}^{Q} \frac{\Gamma(\sum_{j=1}^{Q} c_{i,j})}{\prod_{j=1}^{Q} \Gamma(c_{i,j})} \qquad (4.49)$$

so that Eqn (4.48) can be written as

$$f(\Upsilon_1, \cdots, \Upsilon_K \mid m_u, \mathcal{D}_M)$$

$$= \frac{\rho}{p(\mathcal{D}_M \mid m_u)} \cdot \prod_{i=1}^{Q} \prod_{j=1}^{Q} \varphi_{i,j}^{c_{i,j}-1} \cdot \prod_{i=1}^{Q} \prod_{j=1}^{Q} \varphi_{i,j}^{\hat{n}_{i,j}}$$

$$= \frac{\rho}{p(\mathcal{D}_M \mid m_u)} \prod_{i=1}^{Q} \prod_{j=1}^{Q} \varphi_{i,j}^{\hat{n}_{i,j}+c_{i,j}-1} \tag{4.50}$$

From this, the transition probability from evidence m_u to m_v can be derived as

$$p(m_v \mid m_u, \mathcal{D}_M)$$

$$= \int_{\Sigma_1, \cdots, \Sigma_Q} \varphi_{u,v} f(\Upsilon_1, \cdots, \Upsilon_Q \mid m_u, \mathcal{D}_M) d\Upsilon_1 \cdots \Upsilon_Q$$

$$= \int_{\Sigma_1, \cdots, \Sigma_Q} \varphi_{u,v} \frac{\rho}{p(\mathcal{D}_M \mid m_u)} \prod_{i=1}^{Q} \prod_{j=1}^{Q} \varphi_{i,j}^{\hat{n}_{i,j}+c_{i,j}-1} d\Upsilon_1 \cdots \Upsilon_Q$$

$$= \frac{\rho}{p(\mathcal{D}_M \mid m_u)} \int_{\Sigma_1} \prod_{j=1}^{Q} \varphi_{1,j}^{\hat{n}_{1,j}+c_{1,j}-1} d\Upsilon_1 \cdots \int_{\Sigma_u} \varphi_{u,v}^{\hat{n}_{u,v}+c_{u,v}} \prod_{j \neq v} \varphi_{u,j}^{\hat{n}_{u,j}+c_{u,j}-1} d\Upsilon_u$$

$$\cdots \int_{\Sigma_Q} \prod_{j=1}^{Q} \varphi_{A,j}^{\hat{n}_{Q,j}+c_{K,j}-1} d\Upsilon_Q \tag{4.51}$$

In the equation above, $p(\mathcal{D}_M \mid m_u)$ is the scaling factor as defined in Eqn (4.40). Now, according to Eqn (4.41),

$$p(\mathcal{D}_M \mid m_u)$$

$$= \int_{\Sigma_1, \cdots, \Sigma_Q} p(\mathcal{D}_M \mid \Upsilon_1, \cdots, \Upsilon_Q, m_u) f(\Upsilon_1, \cdots, \Upsilon_Q \mid m_u) d\Upsilon_1 \cdots d\Upsilon_Q$$

$$= \int_{\Sigma_1, \cdots, \Sigma_Q} \prod_{i=1}^{Q} \frac{\Gamma(\sum_{j=1}^{Q} c_{i,j})}{\prod_{j=1}^{Q} \Gamma(c_{i,j})} \prod_{j=1}^{Q} \varphi_{i,j}^{c_{i,j}-1} \cdot \prod_{i=1}^{Q} \prod_{j=1}^{Q} \varphi_{i,j}^{\hat{n}_{i,j}} d\Upsilon_1 \cdots d\Upsilon_Q$$

$$= \rho \cdot \int_{\Sigma_1} \prod_{j=1}^{Q} \varphi_{1,j}^{\hat{n}_{1,j}+c_{1,j}-1} d\Upsilon_1 \cdots \int_{\Sigma_Q} \prod_{j=1}^{Q} \varphi_{Q,j}^{\hat{n}_{Q,j}+c_{Q,j}-1} d\Upsilon_K$$

$$= \rho \cdot \prod_{i=1}^{Q} \frac{\prod_{j=1}^{Q} \Gamma(\hat{n}_{i,j} + c_{i,j})}{\Gamma(\hat{N}_i + C_i)} \tag{4.52}$$

where $\hat{N}_i = \sum_j \hat{n}_{i,j}$ is the total number of historical composite mode samples, from mode m_i, and $C_i = \sum_j c_{i,j}$ is the corresponding total number of prior composite mode samples. Similarly, we can derive

$$
\int_{\Sigma_1} \prod_{j=1}^{Q} \varphi_{1,j}^{\hat{n}_{1,j}+c_{1,j}-1} d\Upsilon_1 \cdots \int_{\Sigma_u} \varphi_{u,v}^{\hat{n}_{u,v}+c_{u,v}} \prod_{j\neq v} \varphi_{u,j}^{\hat{n}_{u,j}+c_{u,j}-1} d\Upsilon_u
$$

$$
\cdots \int_{\Sigma_Q} \prod_{j=1}^{Q} \varphi_{Q,j}^{\hat{n}_{Q,j}+c_{K,j}-1} d\Upsilon_Q
$$

$$
= \frac{\Gamma(\hat{n}_{u,v}+c_{u,v}+1)}{\Gamma(\hat{N}_u+C_u+1)} \cdot \frac{\prod_{i,j\neq u,v}\Gamma(\hat{n}_{i,j}+c_{i,j})}{\prod_{i\neq u}\Gamma(\hat{N}_i+C_i)} \tag{4.53}
$$

Thus, Eqn (4.51) can be simplified as

$$
p(m_v|m_u,\mathcal{D}_E)
$$

$$
= \rho \cdot \frac{\Gamma(\hat{n}_{u,v}+c_{u,v}+1)}{\Gamma(\hat{N}_u+C_u+1)} \cdot \frac{\prod_{i,j\neq u,v}\Gamma(\hat{n}_{i,j}+c_{i,j})}{\prod_{i\neq u}\Gamma(\hat{N}_i+C_i)} \cdot \frac{\prod_{i=1}^{Q}\Gamma(\hat{N}_i+C_i)}{\rho \cdot \prod_{i=1}^{Q}\prod_{j=1}^{Q}\Gamma(\hat{n}_{i,j}+c_{i,j})}
$$

$$
= \frac{\hat{n}_{u,v}+c_{u,v}}{\hat{N}_u+C_u} \tag{4.54}
$$

where $\hat{n}_{u,v}$ is the number of composite mode data samples from m_u to m_v. Additionally $\hat{N}_u = \sum_j \hat{n}_{u,j}$ is the total number of composite mode data samples, from mode m_u to any other mode. Finally, $c_{u,v}$ is the number of prior samples for the mode transition from m_u to m_v, and $C_u = \sum_j c_{u,j}$ is the total number of prior mode transitions from m_u to any other mode.

4.4 Dependent Modes and Evidence

Up to this point, this chapter has discussed temporal dependence of modes and evidence in separate contexts. In any given Bayesian diagnosis problem either of these dependencies can exist, or even both exist simultaneously. Because of this, it is important to be able to merge both techniques in one unifying solution.

With both mode and evidence dependencies being considered, a dynamic Bayesian model (Ghahramani 2002) is established to describe the temporal dependence between data samples. In Figure 4.7, the directed lines, which represent dependencies, are added between consecutive modes and evidence. The model is also known as the autoregressive hidden Markov model (Murphy 2002).

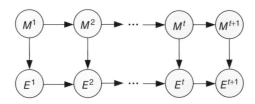

Figure 4.7   Dynamic Bayesian model that considers both mode and evidence dependence

With similar derivations as in the previous sections, a recursive solution based on the model structure in Figure 4.7 can be developed. In this autoregressive hidden Markov model, the previous evidence also needs to be taken into consideration to estimate current mode probability,

$$
\begin{aligned}
&p(M^t|E^1,\cdots,E^t)\\
&= \frac{p(E^1,\cdots,E^t|M^t)p(M^t)}{p(E^1,\cdots,E^t)}\\
&= \frac{\sum_{M^{t-1}}p(E^1,\cdots,E^t,M^{t-1}|M^t)p(M^t)}{p(E^1,\cdots,E^t)}\\
&= \frac{\sum_{M^{t-1}}p(E^t|M^t,M^{t-1},E^1,\cdots,E^{t-1})p(E^1,\cdots,E^{t-1},M^{t-1}|M^t)p(M^t)}{p(E^1,\cdots,E^t)}
\end{aligned}
\tag{4.55}
$$

Note that the current evidence is determined by both the current mode M^t and previous evidence E^{t-1},

$$
\begin{aligned}
&p(M^t|E^1,\cdots,E^t)\\
&= \frac{\sum_{M^{t-1}}p(E^t|M^t,E^{t-1})p(E^1,\cdots,E^{t-1},M^{t-1}|M^t)p(M^t)}{p(E^1,\cdots,E^t)}
\end{aligned}
\tag{4.56}
$$

By following a procedure similar to that in the mode-dependence section, a recursive solution can be presented as follows,

$$
p(M^t|E^1,\cdots,E^t) \propto p(E^t|M^t,E^{t-1}) \sum_{M^{t-1}} p(M^{t-1}|E^1,\cdots,E^{t-1})p(M^t|M^{t-1})
\tag{4.57}
$$

4.5 Notes and References

The main material of this chapter stems from Qi and Huang (2011) and Qi and Huang (2010). Readers can refer to Murphy (2002) for further readings on the hidden Markov model, and to Ghahramani (2002) for those on the dynamic Bayesian model.

References

Ghahramani Z 2002 An introduction to hidden Markov models and Bayesian networks. *International Journal of Pattern Recognition and Artifical Intelligence* **15**(1), 9–42.

Murphy K 2002 *Dynamic Bayesian networks: representation, inference, and learning.* PhD thesis University of California, Berkeley.

Pernestal A 2007 *A Bayesian approach to fault isolation with application to diesel engines* PhD thesis KTH School of Electrical Engineering, Sweden.

Qi F and Huang B 2010 Dynamic Bayesian approach for control loop diagnosis with underlying mode dependency. *Industrial & Engineering Chemistry Research* **46**, 8613–8623.

Qi F and Huang B 2011 Bayesian methods for control loop diagnosis in presence of temporal dependent evidences. *Automatica* **47**, 1349–1356.

5

Accounting for Incomplete Discrete Evidence

5.1 Introduction

A data-driven Bayesian procedure for control-loop diagnosis has been discussed in Chapter 3. However, the inability to handle incomplete evidence remains a barrier in its applicability. Due to instrument reliability issue, control-network traffic, or data storage problems, some process variable readings may not be always available in the historical record. Depending on the monitor calculation algorithms, the missing process variable will often lead to incomplete evidence, where some of the monitors readings cannot be calculated. In the Bayesian method introduced in Chapter 3, one requirement for using evidence is the availability of all monitor readings in an evidence sample. If any of the monitor readings are missing, the evidence for that sample is incomplete, and all the other monitor readings in that sample must be discarded. Consequently, requiring the availability of all monitors in a sample can significantly reduce the number of samples in the history that can be used to estimate likelihoods.

In this chapter, we will develop a strategy to handle incomplete evidence, based on the marginalization of the incomplete evidence likelihood. Missing monitor readings tend to happen for extended periods of time in certain data periods. The monitors that are missing thus form a pattern over a certain length of the dataset. This chapter proposes a method based on marginalization to solve the problem of a single missing pattern occuring in the history, and then proposes a generalization based on the EM algorithm that can handle multiple missing patterns.

5.2 The Incomplete Evidence Problem

When the data contains samples that have missed part of the evidence, the historical dataset can be segregated into two portions: the portion where the evidence is complete and the portion

Process Control System Fault Diagnosis: A Bayesian Approach, First Edition. Ruben Gonzalez, Fei Qi and Biao Huang.
© 2016 John Wiley & Sons, Ltd. Published 2016 by John Wiley & Sons, Ltd.

where it is incomplete:

$$\mathcal{D} = \{\mathcal{D}_c, \mathcal{D}_{ic}\}$$

where $\mathcal{D}_c$ is the dataset in which all evidence is complete, and the $\mathcal{D}_{ic}$ is the remaining dataset that only includes samples with incomplete evidence. The two datasets are called *the complete dataset* and *the incomplete dataset* respectively.

The concept of *missing patterns* needs to be introduced in order to properly classify the problem type. Suppose that a complete evidence sample consists of L monitor readings,

$$E = (\pi_1, \pi_2, \cdots, \pi_L)$$

and each single monitor reading π_i may take k_i different discrete values. Any one of these monitors could have missing values. If the same monitor set has missing values for multiple samples, each of these samples is said to belong to the same missing pattern. For example, data samples where monitors (π_1, π_3) are missing will all fall into the same missing pattern regardless of the values the available monitors take. As another example, consider a case where we have three binary monitors taking values of 0 and 1, but a value of $\times$ represents a missing value. The two evidence readings $(\times, 0, \times)$ and $(\times, 1, \times,)$ belong to the same missing pattern as the monitors with missing values are identical, but the two evidence readings $(1, 0, \times,)$ and $(1, \times, 1)$ fall into two different missing patterns as the missing monitor set is different. If only a single missing pattern exists in the data, it is classified as a *single missing pattern* problem. Otherwise, if more missing patterns exist, it is classified as a *multiple missing pattern* problem.

The unique underlying complete evidence matrix (UCEM) is a tool that can be used to visualize missing monitor patterns and the possible values the missing monitors can take. Without loss of generality, assume that the first q monitor readings are missing for a given missing pattern. Such a missing pattern can be represented as

$$E = (\times, \cdots, \times, \pi_{q+1}, \cdots, \pi_L)$$

Each of the available monitor readings $\pi_{q+1}, \cdots, \pi_L$ can take one of k_i possible discrete values. Therefore, there are in total

$$S = \prod_{i=q+1}^{L} k_i$$

different possible realizations of the available evidence values.

Each missing reading of the first q monitors could have been any one of its possible k_i output values. Thus each incomplete evidence may be one of the

$$R = \prod_{i=1}^{q_k}$$

possible complete evidence realizations.

A UCEM can be constructed for each missing pattern. This matrix is built by listing all unique incomplete evidence values (for a given pattern) in a column to the left of the matrix (which we will call the prefix column). Then each row of the matrix to the right lists all possible complete evidence values that can be taken by the prefix column. Consequently, the size of a UCEM is $S \times R$. Such a matrix, for instance, may resemble the following:

$$
\begin{array}{c}
(\times, \times, 0) \\
(\times, \times, 1)
\end{array}
\begin{bmatrix}
(0,0,0) & (0,1,0) & (1,0,0) & (1,1,0) \\
(0,0,1) & (0,1,1) & (1,0,1) & (1,1,1)
\end{bmatrix}
\tag{5.1}
$$

Each row of UCEM contains all possible complete evidence values that can be taken by the incomplete evidence in the left-hand column. All elements of this matrix are unique, so there should be no redundancy between any two different rows in a UCEM. In addition, all possible incomplete evidence values are listed. Therefore, all underlying complete evidence values can be located in the UCEM without repetition.

In the following discussion, let us consider the kth single missing pattern corresponding to a UCEM (denoted the kth UCEM) and let us consider the following notation set:

- The underlying complete evidence with location (i, j) in the kth UCEM is denoted $\epsilon_{i,j/k}$ and its likelihood parameter is denoted $\theta_{i,j/k}$.
- The number of historical samples with this underlying complete evidence is denoted $\eta_{i,j/k}$.
- The corresponding incomplete evidence with location (j) in the kth UCEM is denoted $\epsilon_{i/k}$; its likelihood is denoted $\lambda_{i/k}$.
- The number of historical samples matching this incomplete evidence pattern is denoted $\eta_{i/k}$.

For instance, in the UCEM in Eqn (14.6), which is assumed to be the first UCEM, the evidence $(0,0,0)$ is denoted $\epsilon_{1,1/1}$; the number of historical samples matching this evidence value is denoted $\eta_{1,1/1}$. The corresponding incomplete evidence, $(0, \times, \times)$, is denoted $\epsilon_{1/1}$; the number of times $(0, \times, \times,)$ occurs in the history is $\eta_{1/1}$, and its likelihood is $\lambda_{1/1}$.

5.3   Diagnosis with Incomplete Evidence

Recall the equation used for calculating posterior probability

$$
p(M|E, \mathcal{D}) \propto p(E|M, \mathcal{D})p(M)
\tag{5.2}
$$

As discussed in Chapter 3, the prior probability $p(M)$ is determined by a-priori information. Thus our interest is how to derive the likelihood of evidence $p(E|M, \mathcal{D})$ in the presence of incomplete evidence samples.

Despite some missing monitor readings, the available monitor readings still provide partial information of evidence. The likelihood probability of an incomplete historical data sample $d_{ic}^t = (\epsilon_{i/k}, M)$ is given as

$$
p(d_{ic}^t|\Theta, M) = p((\epsilon_{i/k}, M)|\theta, M) = p(\epsilon_{i/k}|\Theta, M) = \lambda_{i/k}
\tag{5.3}
$$

which is the summation of the likelihood of all the possible underlying complete evidence realizations. This term is essentially the marginalization of the likelihood over all possible complete evidence values in the ith row of the kth UCEM

$$p((\epsilon_{i/k}, M)|\theta, M) = \sum_{j=1}^{R} p(\epsilon_{i,j/k}|\Theta, M) = \sum_{j=1}^{R} \theta_{i,j/k} \tag{5.4}$$

As an example, let us take the UCEM in Eqn (5.1). The likelihood of incomplete evidence $(\times, \times, 0)$ is equal to the summation of likelihoods from $(0, 0, 0)$ to $(1, 1, 0)$ in the first row:

$$p((\epsilon_{1/1}, M)|\Theta, M) = \lambda_{1/1} = \sum_{j=1}^{4} \theta_{1,j/1} \tag{5.5}$$

5.3.1   Single Missing Pattern Problem

For a single missing pattern problem, where only one UCEM exists, the likelihood probability of the historical dataset (including both complete and incomplete samples) is

$$p(\mathcal{D}|\Theta, M) = p(\mathcal{D}_c|\Theta, M)p(\mathcal{D}_{ic}|\Theta, M)$$

$$= \prod_{i=1}^{S} \prod_{j=1}^{R} \theta_{i,j}^{\eta_{i,j}} \cdot \prod_{i=1}^{S} \lambda_i^{\eta_i} = \prod_{i=1}^{S} \left[\prod_{j=1}^{R} \theta_{i,j}^{\eta_{i,j}} \cdot \left(\sum_{k=1}^{R} \theta_{i,k} \right)^{\eta_i} \right] \tag{5.6}$$

Note that since only one UCEM exits for the single missing pattern problem, the subscript denoting the number of UCEMs can be omitted.

Here we again use Dicichlet distribution to model the distribution of likelihood parameters (Pernestal 2007), where parameters $a_{i,j}$ pertain to the prior sample numbers.

$$f(\Theta|M) = \frac{\Gamma(\sum_{i=1}^{S} \sum_{j=1}^{R} a_{i,j})}{\prod_{i=1}^{S} \prod_{j=1}^{R} \Gamma(a_{i,j})} \prod_{i=1}^{S} \prod_{j=1}^{R} \theta_{i,j}^{a_{i,j}-1} \tag{5.7}$$

By applying this distribution to our previous expression,

$$f(\Theta|M, \mathcal{D}) = \frac{p(\mathcal{D}|\Theta, M)f(\Theta|M)}{p(\mathcal{D}|M)}$$

$$= \frac{1}{p(\mathcal{D}|M)} \cdot \prod_{i=1}^{S} \left[\prod_{j=1}^{R} \theta_{i,j}^{\eta_{i,j}} \cdot \left(\sum_{k=1}^{R} \theta_{i,k} \right)^{\eta_i} \right]$$

$$\cdot \frac{\Gamma(\sum_{i=1}^{S} \sum_{j=1}^{R} a_{i,j})}{\prod_{i=1}^{S} \prod_{j=1}^{R} \Gamma(a_{i,j})} \prod_{i=1}^{S} \prod_{j=1}^{R} \theta_{i,j}^{a_{i,j}-1} \tag{5.8}$$

Now, let

$$c = \frac{\Gamma(\sum_{i=1}^{S} \sum_{j=1}^{R} a_{i,j})}{\prod_{i=1}^{S} \prod_{j=1}^{R} \Gamma(a_{i,j})}$$

then Eqn (5.8) can be written as

$$f(\Theta|M,\mathcal{D}) = \frac{c}{p(\mathcal{D}|M)} \cdot \prod_{i=1}^{S}\left[\prod_{j=1}^{R} \theta_{i,j}^{\eta_{i,j}} \cdot \left(\sum_{k=1}^{R}\theta_{i,k}\right)^{\eta_i}\right] \cdot \prod_{i=1}^{S}\prod_{j=1}^{R}\theta_{i,j}^{a_{i,j}-1}$$

$$= \frac{c}{p(\mathcal{D}|M)} \cdot \prod_{i=1}^{S}\left[\prod_{j=1}^{R} \theta_{i,j}^{\eta_{i,j}+a_{i,j}-1} \cdot \left(\sum_{k=1}^{R}\theta_{i,k}\right)^{\eta_i}\right]$$

$$= \frac{c}{\int_{\Omega} p(\mathcal{D}|\Theta,M)f(\Theta|M)d\Theta} \cdot \prod_{i=1}^{S}\left[\prod_{j=1}^{R} \theta_{i,j}^{\eta_{i,j}+a_{i,j}-1} \cdot \left(\sum_{k=1}^{R}\theta_{i,k}\right)^{\eta_i}\right] \quad (5.9)$$

Suppose that the likelihood of evidence $\epsilon_{s,r}$ is about to be estimated from historical data. It can be computed by marginalization over all possible likelihood parameter sets Θ,

$$p(\epsilon_{s,r}|M,\mathcal{D})$$

$$= \int_{\Omega} p(\epsilon_{s,r}|\Theta,M,\mathcal{D})f(\Theta|M,\mathcal{D})d\Theta$$

$$= \int_{\Omega} \theta_{s,r} \frac{c}{p(\mathcal{D}|M)} \prod_{i=1}^{S}\left[\prod_{j=1}^{R} \theta_{i,j}^{\eta_{i,j}+a_{i,j}-1} \cdot \left(\sum_{k=1}^{R}\Theta_{i,k}\right)^{\eta_i}\right]d\Theta$$

$$= \frac{c}{p(\mathcal{D}|M)} \int_{\Omega} \theta_{s,r} \prod_{i=1}^{S}\left[\prod_{j=1}^{R} \theta_{i,j}^{\eta_{i,j}+a_{i,j}-1} \cdot \left(\sum_{k=1}^{R}\theta_{i,k}\right)^{\eta_i}\right]d\Theta$$

$$= \frac{c}{\int_{\Omega} p(\mathcal{D}|\theta,M)f(\theta|M)} \cdot \int_{\Omega} \theta_{s,r} \prod_{i=1}^{S}\left[\prod_{j=1}^{R} \theta_{i,j}^{\eta_{i,j}+a_{i,j}-1} \cdot \left(\sum_{k=1}^{R}\theta_{i,k}\right)^{\eta_i}\right]d\Theta$$

$$= \left(\int_{\Omega} \theta_{s,r} \prod_{i=1}^{S}\left[\prod_{j=1}^{R} \theta_{i,j}^{\eta_{i,j}+a_{i,j}-1} \cdot \left(\sum_{k=1}^{R}\theta_{i,k}\right)^{\eta_i}\right]d\Theta\right)$$

$$\Big/ \left(\int_{\Omega} \prod_{i=1}^{S}\left[\prod_{j=1}^{R} \theta_{i,j}^{\eta_{i,j}+a_{i,j}-1} \cdot \left(\sum_{k=1}^{R}\theta_{i,k}\right)^{\eta_i}\right]d\Theta\right) \quad (5.10)$$

The denominator of Eqn (5.10) is

$$
\int_\Omega \prod_{i=1}^{S} \left[\prod_{j=1}^{R} \theta_{i,j}^{\eta_{i,j}+a_{i,j}-1} \cdot \left(\sum_{k=1}^{R} \theta_{i,k} \right)^{\eta_i} \right] d\Theta
$$

$$
= \int_\Omega \prod_{i=1}^{S} \prod_{j=1}^{R} \theta_{i,j}^{\eta_{i,j}+a_{i,j}-1} \cdot \sum_{t_{i,1},\cdots,t_{i,R}} \binom{\eta_i}{t_{i,1},\cdots,t_{i,R}} \theta_{i,1}^{t_{i,1}} \cdots \theta_{i,R}^{t_{i,R}} d\Theta
$$

$$
= \int_\Omega \prod_{i=1}^{S} \sum_{t_{i,1},\cdots,t_{i,R}} \binom{\eta_i}{t_{i,1},\cdots,t_{i,R}} \cdot \theta_{i,1}^{t_{i,1}+\eta_{i,1}+a_{i,1}-1} \cdots \theta_{i,R}^{t_{i,R}+\eta_{i,R}+a_{i,R}-1} d\Theta \tag{5.11}
$$

where

$$
\binom{\eta_i}{t_{i,1},\cdots,t_{i,R}} = \binom{t_{i,1}}{t_{i,1}} \binom{t_{i,1}+t_{i,2}}{t_{i,2}} \cdots \binom{t_{i,1}+t_{i,2}+\cdots+t_{i,R}}{t_{i,R}}
$$

$$
= \frac{\eta_i!}{t_{i,1}!\, t_{i,2}! \cdots t_{i,R}!} \tag{5.12}
$$

and $\sum_{t_{i,1},\cdots,t_{i,R}}$ is summation over all possible sequences of indices $t_{i,1}$ through $t_{i,R}$ such that $\sum_{j=1}^{R} t_{i,j} = \eta_i$. Similarly, the numerator in Eqn (5.10) is

$$
\int_\Omega \theta_{s,r} \prod_{i=1}^{S} \sum_{t_{i,1},\cdots,t_{i,R}} \binom{\eta_i}{t_{i,1},\cdots,t_{i,R}} \cdot \theta_{i,1}^{t_{i,1}+\eta_{i,1}+a_{i,1}-1} \cdots \theta_{i,R}^{t_{i,R}+\eta_{i,R}+a_{i,R}-1} d\Theta \tag{5.13}
$$

Now, let us denote

$$
\text{all sets of} \quad t_{i,1},\cdots,t_{i,R} \quad \text{as} \quad \mathscr{T}_i
$$

$$
\text{the combination of all} \quad \mathscr{T}_i \quad \text{as} \quad \mathscr{T}
$$

$$
\binom{\eta_i}{t_{i,1},\cdots,t_{i,R}} \quad \text{as} \quad \mathbf{C}_{\eta_i}^{\mathscr{T}_i}
$$

$$
\eta_{i,j} + a_{i,j} - 1 \quad \text{as} \quad q_{i,j}
$$

Then Eqn (5.11) can be expressed as

$$
\int_\Omega \prod_{i=1}^{S} \left[\sum_{\mathscr{T}_i} \mathbf{C}_{\eta_i}^{\mathscr{T}_i} \theta_{i,1}^{t_{i,1}+q_{i,1}} \cdots \theta_{i,R}^{t_{i,R}+q_{i,R}} \right] d\Theta
$$

$$
= \sum_{\mathscr{T}} \prod_{i=1}^{S} \mathbf{C}_{\eta_i}^{\mathscr{T}_i} \cdot \int_\Omega \prod_{i=1}^{S} \left[\theta_{i,1}^{t_{i,1}+q_{i,1}} \cdots \theta_{i,R}^{t_{i,R}+q_{i,R}} \right] d\Theta
$$

$$
= \sum_{\mathscr{T}} \frac{\prod_{i=1}^{S} \mathbf{C}_{\eta_i}^{\mathscr{T}_i} \prod_{j=1}^{R} \Gamma(t_{i,j}+q_{i,j}+1)}{\Gamma(N+A+1)} \tag{5.14}
$$

where $N = \sum_i \eta_i + \sum_i \sum_j \eta_{i,j}$ is the total number of historical data samples for mode M, including both complete and incomplete samples; $A = \sum_i \sum_j a_{i,j}$ is the total number of prior samples for complete evidence only.

With this new notation, Eqn (5.13) can be rewritten as

$$\int_\Omega \theta_{s,r} \prod_{i=1}^{S} \left[\sum_{\mathscr{T}_i} \mathbf{C}_{\eta_i}^{\mathscr{T}_i} \theta_{i,1}^{t_{i,1}+q_{i,1}} \cdots \theta_{i,R}^{t_{i,R}+q_{i,R}} \right] d\Theta$$

$$= \sum_{\mathscr{T}} \frac{\prod_{i=1}^{S} \mathbf{C}_{\eta_i}^{\mathscr{T}_i} \Gamma(t_{s,r} + q_{s,r} + 2) \prod_{j \neq r} \Gamma(t_{i,j} + q_{i,j} + 1)}{\Gamma(N + A + 1)} \tag{5.15}$$

Now, the likelihood of evidence $\epsilon_{s,r}$ can be computed as

$$p(\epsilon_{s,r} | M, \mathcal{D})$$

$$= \frac{1}{N + A} \cdot \frac{\sum_{\mathscr{T}} \prod_{i=1}^{S} \mathbf{C}_{\eta_i}^{\mathscr{T}_i} (t_{s,r} + q_{s,r} + 1)! \prod_{j \neq r} (t_{i,j} + q_{i,j})!}{\sum_{\mathscr{T}} \prod_{i=1}^{S} \mathbf{C}_{\eta_i}^{\mathscr{T}_i} \prod_{j=1}^{R} (t_{i,j} + q_{i,j})!} \tag{5.16}$$

As discussed previously, there is no repetition between different rows in a single UCEM. Thus the assignments of $t_{i,1}, \cdots, t_{i,R}$ are independent for different rows i, and Eqn (5.16) can be simplified as

$$p(\epsilon_{s,r} | M, \mathcal{D})$$

$$= \frac{1}{N + A} \cdot \frac{\sum_{\mathscr{T}_s} \mathbf{C}_{\eta_s}^{\mathscr{T}_s} (t_{s,r} + q_{s,r} + 1)! \prod_{j \neq r} (t_{s,j} + q_{s,j})!}{\sum_{\mathscr{T}_s} \mathbf{C}_{\eta_s}^{\mathscr{T}_s} \prod_{j=1}^{R} (t_{s,j} + q_{s,j})!}$$

$$\cdot \frac{\sum_{\mathscr{T} \backslash \mathscr{T}_s} \prod_{i \neq s} \mathbf{C}_{\eta_i}^{\mathscr{T}_i} \prod_{j=1}^{R} (t_{s,j} + q_{s,j})!}{\sum_{\mathscr{T} \backslash \mathscr{T}_s} \prod_{i \neq s} \mathbf{C}_{\eta_i}^{\mathscr{T}_i} \prod_{j=1}^{R} (t_{s,j} + q_{s,j})!}$$

$$= \frac{1}{N + A} \cdot \frac{\sum_{\mathscr{T}_s} \mathbf{C}_{\eta_s}^{\mathscr{T}_s} (t_{s,r} + q_{s,r} + 1)! \prod_{j \neq r} (t_{s,j} + q_{s,j})!}{\sum_{\mathscr{T}_s} \mathbf{C}_{\eta_s}^{\mathscr{T}_s} \prod_{j=1}^{R} (t_{s,j} + q_{s,j})!} \tag{5.17}$$

where $\mathscr{T} \backslash \mathscr{T}_s$ is the combination of all the possible set of $\mathscr{T}_i$ with $i \neq s$.

As a result, the likelihood equation is greatly simplified by completely removing other possible $\mathscr{T}_{i \neq s}$. Only the assignment of $t_{s,1}, \cdots, t_{s,R}$ to evidence samples $\epsilon_{s,1}, \cdots, \epsilon_{s,R}$ needs to be considered. However, the computational load is still heavy, and will grow exponentially in the same manner as R and η_s. Further simplification is required to make the likelihood computation practical.

Lemma 1

$$\sum_{\mathscr{T}_s} \binom{\eta_s}{t_1, \cdots, t_R} \prod_{j=1}^{R} (x_j + t_j)! = \prod_{k=1}^{R} x_k! \cdot \prod_{i=0}^{\eta_s - 1} \left(\sum_{j=1}^{R} x_j + i + R \right) \tag{5.18}$$

Here we use mathematical induction to prove this Lemma. First, let $n_s = 1$, $R = 2$, which is the simplest missing data case, where only one incomplete data sample with two possible evidence values exists.

$$RHS = \prod_{j=1}^{2} x_j! \prod_{i=0}^{0} \left(\sum_{j=1}^{2} x_j + i + 2 \right) = x_1! x_2! (x_1 + x_2 + 2) \tag{5.19}$$

$$LHS = \sum_{\mathscr{T}} \binom{1}{t_1, t_2} \prod_{j=1}^{2} (x_j + t_j)! = (x_1 + 1)! x_2! + x_1! (x_2 + 1)!$$

$$= x_1! x_2! (x_1 + x_2 + 2) = RHS \tag{5.20}$$

There are two parameters that need to be inducted: n_s and R. Suppose the Lemma holds for n_s and R.

1. *Induction on R*

   Note that Eqn (5.18) can be written as

$$\sum_{\mathscr{T}_s} \binom{n_s}{t_1, \cdots, t_R} \prod_{j=1}^{R} \prod_{r=1}^{t_j} (x_j + r)! = \prod_{i=0}^{n_s-1} \left(\sum_{j=1}^{R} x_j + i + R \right) \tag{5.21}$$

Thus

$$\prod_{i=0}^{n_s-1} \left(\sum_{j=1}^{R+1} x_j + i + R + 1 \right)$$

$$= \prod_{i=0}^{n_s-1} \left(\sum_{j=1}^{R-1} (x_j + 1) + [(x_R + x_{R+1} + 1) + 1] + i \right)$$

$$= \sum_{\mathscr{T}_s} \binom{n_s}{t_1, \cdots, t_{R-1}, T} \prod_{j=1}^{R-1} \prod_{r=1}^{t_j} (x_j + r)$$

$$\cdot (x_R + x_{R+1} + 1 + 1) \cdots (x_R + x_{R+1} + 1 + T)$$

$$= \sum_{\mathscr{T}_s} \binom{n_s}{t_1, \cdots, t_{R-1}, T} \prod_{j=1}^{R-1} \prod_{r=1}^{t_j} (x_j + r)$$

$$\cdot (x_R + 1 + x_{R+1} + 1) \cdots (x_R + 1 + x_{R+1} + 1 + T - 1)$$

$$= \sum_{\mathscr{T}_s} \binom{n_s}{t_1, \cdots, t_{R-1}, T} \prod_{j=1}^{R-1} \prod_{r=1}^{t_j} (x_j + r) \cdot \binom{T}{t_R, t_{R+1}} \prod_{j=R}^{R+1} \prod_{r=1}^{t_j} (x_R + r) \tag{5.22}$$

In Eqn (5.22),

$$\left(\begin{array}{c} \eta_s \\ t_1, \cdots, t_{R-1}, T \end{array} \right) \cdot \left(\begin{array}{c} T \\ t_R, t_{R+1} \end{array} \right)$$

$$= \frac{\eta_s}{t_1! \cdots t_{R-1}! T!} \cdot \frac{T!}{t_R! t_{R+1}!} = \left(\begin{array}{c} \eta_s \\ t_1, \cdots, t_{R+1} \end{array} \right) \qquad (5.23)$$

so

$$\prod_{i=0}^{n_s-1} \left(\sum_{j=1}^{R+1} x_j + i \mid R \mid 1 \right) - \sum_{\mathcal{T}_s} \left(\begin{array}{c} \eta_s \\ t_1, \cdots, t_{R+1} \end{array} \right) \prod_{j=1}^{R+1} \prod_{r=1}^{t_j} (x_j + r) \qquad (5.24)$$

Multiplying Eqn (5.24) with $\prod_{j=1}^{R+1} x_i!$ generates

$$\sum_{\mathcal{T}_s} \left(\begin{array}{c} \eta_s \\ t_1, \cdots, t_{R+1} \end{array} \right) \prod_{j=1}^{R+1} (x_j + t_j)! = \prod_{j=1}^{R+1} x_j! \prod_{i=0}^{n_s-1} \left(\sum_{j=1}^{R+1} x_j + i + R + 1 \right) \qquad (5.25)$$

2. *Induction on n_s*
   The right-hand side of Eqn 5.21 is

$$\prod_{i=0}^{n_s} \left(\sum_{j=1}^{R} x_j + i + R \right)$$

$$= \prod_{i=0}^{n_s-1} \left(\sum_{j=1}^{R} x_j + i + R \right) \cdot \left(\sum_{j=1}^{R} x_j + n_s + R \right)$$

$$= \sum_{\mathcal{T}_s} \left(\begin{array}{c} \eta_s \\ t_1, \cdots, t_R \end{array} \right) \prod_{j=1}^{R} (x_j + t_j)! \cdot \left(\sum_{j=1}^{R} x_j + n_s + R \right)$$

$$= \sum_{\mathcal{T}_s} \left(\begin{array}{c} \eta_s \\ t_1, \cdots, t_R \end{array} \right) \prod_{j=1}^{R} (x_j + t_j)! \sum_{k=1}^{R} (x_k + 1 + t_k)$$

$$= \sum_{k=1}^{R} \sum_{\mathcal{T}_s} \left(\begin{array}{c} \eta_s \\ t_1, \cdots, t_R \end{array} \right) \prod_{j=1}^{R} (x_j + t_j)! (x_k + t_k + 1) \qquad (5.26)$$

Following the multinomial theorem, which indicates that

$$(x_1 + x_2 + \cdots + x_m)^{n+1} = \sum_{\mathcal{T}_s} \left(\begin{array}{c} n \\ t_1, \cdots, t_m \end{array} \right) x_1^{t_1} \cdots x_m^{t_m} \cdot (x_1 + \cdots + x_m)$$

$$= \sum_{j=1}^{m} \sum_{\mathcal{T}_s} \left(\begin{array}{c} n \\ t_1, \cdots, t_m \end{array} \right) x_1^{t_1} \cdots x_m^{t_m} x_j$$

$$= \sum_{\mathcal{T}_s} \left(\begin{array}{c} n \\ t_1, \cdots, t_m \end{array} \right) x_1^{t_1} \cdots x_m^{t_m} \qquad (5.27)$$

Eqn (5.26) can be written as

$$
\sum_{k=1}^{R}\sum_{\mathcal{T}_s}\binom{\eta_s}{t_1,\cdots,t_R}\prod_{j=1}^{R}(x_j+t_j)!(x_k+t_k+1)=\sum_{\mathcal{T}_s}\binom{\eta_s}{t_1,\cdots,t_R}\prod_{j=1}^{R}(x_j+t_j)!
$$

$$(5.28)$$

Multiplying Eqn (5.24) with $\prod_{k=1}^{R+1}x_k!$ yields

$$
\sum_{\mathcal{T}_s}\binom{\eta_s+1}{t_1,\cdots,t_R}\prod_{j=1}^{R}(x_j+t_j)!=\prod_{k=1}^{R}x_k!\cdot\prod_{i=0}^{\eta_s}\left(\sum_{j=1}^{R+1}x_j+i+R+1\right)
$$

$$(5.29)$$

which completes the induction and the proof.

With Lemma 1, Eqn (5.17) can be further simplified. The denominator is

$$
(N+A)\sum_{\mathcal{T}_s}\binom{\eta_s}{t_1,\cdots,t_R}\prod_{j=1}^{R}(t_j+\eta_{s,j}+a_{s,j}-1)!
$$

$$
=(N+A)\prod_{k=1}^{R}(\eta_{s,k}+a_{s,k}-1)!\cdot\prod_{i=0}^{\eta_s-1}\left(\sum_{j=1}^{R}[\eta_{s,j}+a_{s,j}]+i\right)
$$

$$(5.30)$$

and the numerator is

$$
\sum_{\mathcal{T}_s}\binom{\eta_s}{t_1,\cdots,t_R}(t_r+\eta_{s,r}+a_{s,r})!\cdot\prod_{j\neq r}(t_j+\eta_{s,j}+a_{s,j}-1)!
$$

$$
=(\eta_{s,r}+a_{s,r})!\prod_{k\neq r}(\eta_{s,k}+a_{s,k}-1)!\cdot\prod_{i=0}^{\eta_s-1}\left(\sum_{j=1}^{R}[\eta_{s,j}+a_{s,j}]+i+1\right)
$$

$$(5.31)$$

By substituting Eqn (5.30) and Eqn (5.31) in Eqn (5.17), the following result can be obtained for the likelihood of evidence $\epsilon_{s,r}$,

$$
p(\epsilon_{s,r}|M,\mathcal{D})=\frac{\eta_{s,r}+a_{s,r}}{N+A}\cdot\frac{\sum_{j=1}^{R}(\eta_{s,j}+a_{s,j})+\eta_s}{\sum_{j=1}^{R}(\eta_{s,j}+a_{s,j})}
$$

$$
=\frac{\eta_{s,r}+a_{s,r}}{N+A}\cdot\left(1+\frac{\eta_s}{\sum_{j=1}^{R}(\eta_{s,j}+a_{s,j})}\right)
$$

$$(5.32)$$

where $N=\sum_i\eta_i+\sum_i\sum_j\eta_{i,j}$ is the total number of historical data samples for mode M, including both complete and incomplete samples; $A=\sum_i\sum_j a_{i,j}$ is the total number of prior samples of complete evidence samples only.

5.3.2   Multiple Missing Pattern Problem

The solution for single missing patterns had a convenient solution and was easily derived from the Bayesian framework, but taking the same Bayesian marginalization procedure for multiple missing patterns is more difficult. However, an alternative approach, the EM algorithm, is commonly used when estimating probability distributions with missing data. Fortunately, the EM algorithm can provide a solution that is consistent with the Bayesian single missing pattern solution, and can also be applied to the multiple missing pattern problem.

Using the EM algorithm to solve the problem of missing data has been illustrated in Section 2.3 of Chapter 2. Using the Q-function and applying it to the discrete missing data problem, a convenient formulation was obtained that included an analytical solution for each maximization step. By applying this analytical maximization to the EM algorithm, one arrives at the following solution:

1. Estimate the likelihood using complete data

$$\hat{p}(E|M,\mathcal{D}) = p(E|M,\mathcal{D}_c)$$

2. Obtain the expected number of data samples for each missing data pattern (or UCEM) and sum them

$$n'_{s,r} = \sum_{e=\text{UCEM}_k} n_s \cdot \frac{n_{s,r} + a_{s,r}}{\sum_{j=1}^{R}(n_{s,j} + a_{s,j})}$$

$$= \sum_{e=\text{UCEM}_k} n_s \cdot \frac{\hat{p}(e_{s,r}|M,\mathcal{D})}{\sum_{j=1}^{R}\hat{p}(e_{s,r}|M,\mathcal{D})}$$

3. Re-estimate the likelihood using these new data samples

$$\hat{p}(e_{s,r}|M,\mathcal{D}) = \frac{n_{s,r} + a_{s,r} + n'_{s,r}}{N + A}$$

4. Iterate between steps 2 and 3 until $\hat{p}(E|M,\mathcal{D})$ converges.

As one can see, this solution is a generalization of the Bayesian solution to a single missing pattern. In the case of a single missing pattern, the solution converges after one iteration.

5.3.3 Limitations of the Single and Multiple Missing Pattern Solutions

Using the Bayesian or EM algorithm to estimate distributions with missing data is not without its limitations. The first limitation is dimensionality; when dimensionality is high, the possible number of likelihood parameters is very large and can be strongly affected by allocating missing values in the incomplete evidence samples. This problem can be worsened if the monitors are independent. This is because the EM algorithm assumes that information from the available monitors provides information about the monitor values that are missing. If there is no link between the available and missing monitors, and if the existing data does not sufficiently demonstrate this (either due to high dimensionality or data scarcity), then the EM algorithm may spuriously make use of nonexistent relationships when calculating likelihood parameters.

Dimensionality

Estimating joint distributions from discrete data suffers from the *curse of dimensionality*. As the number of monitors increases, the number of parameters in the joint distribution grows very rapidly. If each monitor takes k_i different values, the total number of parameters in $\Theta(n(\Theta))$ is

$$n(\Theta) = \prod_{i=1}^{n} k_i$$

Thus, if all the monitors could take the same number of values, the growth in parameters Θ is exponential with respect to the number of monitors, dramatically increasing the amount of data required for a reliable estimate. If the evidence dimension is high, estimating Θ can be a challenge even if there is no incomplete evidence in the data. Both the Bayesian and EM algorithm approaches use estimates of Θ from the reduced complete dataset, which may be unreliable if the dimension is too high (or if data is too sparse), casting doubt on the results obtained from either approach. Thus existence of independence between various evidences could help alleviate the curse of dimension problem. Chapter 13 provides a detailed procedure for testing independence of various evidences.

Independence

Both the Bayesian and EM algorithm approaches make a 'soft classification' of the missing elements within the incomplete evidence. If the missing element is strongly dependent on other elements that are present, then the classification will be reliable. However, if that missing element is independent of all other elements that are not missing, then classification will be unreliable.

While independent evidence is a challenge for the incomplete evidence problem, assuming independence allows us to split the data and estimate separate likelihoods altogether. For example, if E_1 is independent of E_2 given M, we can construct likelihoods $p(E_1|M)$ and $p(E_2|M)$ separately. If there is a missing value in E_1 at time t we can discard it without affecting E_2 at time t. Detailed discussion on assuming independence (and checking whether independence can be assumed) can be found in Chapter 8.

5.4   Notes and References

The main content of this chapter stems from Qi et al. (2010). Some other typical missing data handling algorithms include maximum-likelihood estimation (Cox and Hinkley 1974), EM algorithms (Little and Rubin 1987; Schafer 1997), multi-imputation (Rubin 1987; Sinharay et al. 2001), generalized estimating equations (Zeger et al. 1988), selection models (Little 1995; Verbeke and Molenberghs 2000), Gibbs sampling (Geman and Geman 1984; Korb and Nicholson 2004), and so on. A comparison between the approach presented in this chapter and the Gibbs sampling method is available in Qi et al. (2010).

References

Cox D and Hinkley D 1974 *Theoretical Statistics*. Chapman & Hall.

Geman S and Geman D 1984 Stochastic relaxation, gibbs distributions, and the bayesian restoration of images. *IEEE Transactions on Pattern Analysis and Machine Intelligence* **6**, 721–741.

Korb K and Nicholson A 2004 *Bayesian artificial intelligence*. Chapman & Hall/CRC.

Little RJA 1995 Modeling the dropout mechanism in repeated-measures studies. *Journal of the American Statistical Association* **90**, 1112–1121.

Little RJA and Rubin DB 1987 *Statistical Analysis with Missing Data*. Wiley.

Pernestal A 2007 *A Bayesian approach to fault isolation with application to diesel engines*. PhD thesis KTH School of Electrical Engineering, Sweden.

Qi F, Huang B and Tamayo E 2010 A Bayesian approach for control loop diagnosis with missing data. *AIChE Journal* **56**, 179–195.

Rubin DB 1987 *Multiple Imputation for Noresponse in Surveys*. Wiley.

Schafer JL 1997 *Analysis of Incomplete Multivariate Data*. Chapman & Hall.

Sinharay S, Stern HS and Russell D 2001 The use of multiple imputation for the analysis of missing data. *Psychological Methods* **6**, 317–329.

Verbeke G and Molenberghs G 2000 *Linear Mixed Models for Longitudinal Data*. Springer-Verlag.

Zeger SL, Liang KY and Albert PS 1988 Models for longitudinal data: A generalized estimating equation approach. *Biometrics* **44**, 1049–1060.

6

Accounting for Ambiguous Modes: A Bayesian Approach

6.1 Introduction

One of the challenges to be addressed in this book is the problem of missing information about the mode. A similar challenge of missing information about the evidence has been addressed in Qi and Huang (2010a) and in the previous chapter as well.

As indicated in Chapter 2, a mode comprises a set of states for individual components. For example, consider a system with two components c_1, c_2; let c_1 represent a sensor and c_2 represent a valve. A mode must have information for both components $[c_1, c_2]$. If, for this system, the sensor has three states (positive bias, no bias, and negative bias), while the valve has three states (no stiction, moderate stiction, and severe stiction), there is a total of nine possible modes.

When information about any of the system components is missing, the mode is said to be ambiguous. For example, let us say that in a certain segment of historical data we are unsure of the state of the valve, but we know there is moderate stiction. The corresponding mode is $[c_1 = \times, c_2 = 2]$. The mode is ambiguous as it could be one of three specific modes $[c_1 = 1, c_2 = 2], [c_1 = 2, c_2 = 2], [c_1 = 3, c_2 = 2]$. The aim of this chapter is to determine how to deal with ambiguous modes, such as this, when they appear in the data.

6.2 Parametrization of Likelihood Given Ambiguous Modes

6.2.1 Interpretation of Proportion Parameters

When data in the history is taken from an ambiguous mode, a proportion of this data may belong to any of the specific modes within the ambiguous mode, for example the specific mode $[c_1 = 1, c_2 = 2]$ within the ambiguous mode $[c_1 = \times, c_2 = 2]$. In order to deal with ambiguity, we consider a set of unknown parameters $\theta\{\frac{M}{m}\}$ that indicate the proportion of data under the potentially ambiguous mode m to any of the specific modes $M \subseteq m$. For example, we might have ten data points for mode $[c_1 = \times, c_2 = 2]$ and know that three of them belonged to mode

Process Control System Fault Diagnosis: A Bayesian Approach, First Edition. Ruben Gonzalez, Fei Qi and Biao Huang.
© 2016 John Wiley & Sons, Ltd. Published 2016 by John Wiley & Sons, Ltd.

$[c_1 = 1, c_2 = 2]$, $\theta\{\frac{[1,2]}{[\times,2]}\} = 3/10$. Note that in this book, the boldface M can indicate any observed mode that has occurred in the data (including an ambiguous one) but M (or m, if we are talking about observations) can only represent a specific (unambiguous) mode.

When expressed as a probability, $\theta\{\frac{M}{m}\}$ equates to

$$\theta\{\tfrac{M}{m}\} = p(M|m)$$

which is the probability that the specific mode M occurs given the potentially ambiguous mode m.

As a practical example of defining θ, let us consider the same system with the valve and sensor each being able to take on three states. The resulting modes are as follows:

$$\begin{bmatrix} m_1 \\ m_2 \\ m_3 \\ m_4 \\ m_5 \\ m_6 \\ m_7 \\ m_8 \\ m_9 \end{bmatrix} = \begin{bmatrix} c_1 = 1, \ c_2 = 1 \\ c_1 = 1, \ c_2 = 2 \\ c_1 = 1, \ c_2 = 3 \\ c_1 = 2, \ c_2 = 1 \\ c_1 = 2, \ c_2 = 2 \\ c_1 = 2, \ c_2 = 3 \\ c_1 = 3, \ c_2 = 1 \\ c_1 = 3, \ c_2 = 2 \\ c_1 = 3, \ c_2 = 3 \end{bmatrix}$$

If we consider data coming from the ambiguous mode $[c_1 = \times, c_2 = 2]$ or, equivalently, the set of modes $\{m_2, m_5, m_8\}$, a certain proportion $\theta\{\frac{m_2}{m_2, m_5, m_8}\}$ of that data actually belongs to mode m_2, another proportion $\theta\{\frac{m_5}{m_2, m_5, m_8}\}$ belongs to m_5, and a final proportion of the data $\theta\{\frac{m_8}{m_2, m_5, m_8}\}$ belongs to m_8. Each member θ in Θ (which contains all θ values) is an unknown quantity unless the following conditions apply:

$$\theta\{\tfrac{m_i \not\subseteq m_k}{m_k}\} = 0 \qquad \text{e.g.} \quad \theta\{\tfrac{m_2}{m_1, m_4, m_7}\} = 0 \tag{6.1}$$

$$\theta\{\tfrac{m_i = m_k}{m_k}\} = 1 \qquad \text{e.g.} \quad \theta\{\tfrac{m_2}{m_2}\} = 1 \tag{6.2}$$

For the first condition, the mode m_2 is not possible given the ambiguous mode $\{m_1, m_4, m_7\}$, thus none of the data from this ambiguous mode can belong to mode m_2. For the second exception, given the mode m_2, all of the data in m_2 will belong to m_2; it cannot belong anywhere else. In a probabilistic sense, it would be equivalent to stating $p(M = m_2|m_2) = 1$, or $p(m_2|m_2) = 1$ for shorthand.

6.2.2   Parametrizing Likelihoods

The principle data-driven component of Bayesian inference is the likelihood. When combined with prior probabilities, the posterior is calculated as

$$p(M|E) = \frac{p(E|M)p(M)}{p(E)} \tag{6.3}$$

$$p(E) = \sum_M p(E|M)p(M) \tag{6.4}$$

When ambiguous modes are present, obtaining likelihoods $p(E|M)$ can be quite challenging, as the conditioning variable M is unknown. In the discrete case, when no prior samples are taken, the likelihood is obtained as

$$p(E|M) = \frac{n(E, M)}{n(M)}$$

However, if we are calculating $p(E|m_1)$ and there is an ambiguous mode $\{m_1, m_2\}$ in the data, then we need to take into account the data in $\{m_1, m_2\}$ that belongs to mode m_1. Thus, the likelihood is calculated as

$$p(E|m_1) = \frac{n(E, m_1) + \theta\{\frac{m_1}{m_1, m_2}\}n(E, \{m_1, m_2\})}{n(m_1) + \theta\{\frac{m_1}{m_1, m_2}\}n\{m_1, m_2\}}$$

Let us now revisit the term, *support*, as given in Shafer (1976), which is calculated in the same manner as probability, except that now ambiguous modes can be used as the conditioning variable:

$$S(E|\boldsymbol{M}) = \frac{n(E, \boldsymbol{M})}{n(\boldsymbol{M})}$$

where, again, the boldface $\boldsymbol{M}$ indicates that the mode can be ambiguous. The previous likelihood expression can then be given as

$$p(E|m_1) = \frac{n(E, m_1) + \theta\{\frac{m_1}{m_1, m_2}\}S(E|\{m_1, m_2\})n\{m_1, m_2\}}{n(m_1) + \theta\{\frac{m_1}{m_1, m_2}\}n\{m_1, m_2\}}$$

In this chapter, it is assumed that $S(E|\boldsymbol{M})$ is derived from discrete evidence, but continuous evidence can be used in its stead as well; note that the utilization of continuous evidence will be addressed in Chapter 8.

The expression for the likelihood $p(E|M, \Theta)$ accounted for only one ambiguous mode, but in general the likelihood can be expressed to take into account multiple ambiguous modes:

$$p(E|M, \Theta) = \frac{\sum\limits_{\boldsymbol{M} \supseteq M} \theta\{\frac{M}{\boldsymbol{M}}\}S(E|\boldsymbol{M})n(\boldsymbol{M})}{\sum\limits_{\boldsymbol{M} \supseteq M} \theta\{\frac{M}{\boldsymbol{M}}\}n(\boldsymbol{M})} \tag{6.5}$$

where the summation limit $\boldsymbol{M} \supseteq M$ cycles through all historical modes and includes every historical mode m_k that can support the mode in question, M. Note that the term Θ includes the variables in Θ that can support the mode M; other variables in Θ do not play any role in calculating $p(E|M)$.

6.2.3   Informed Estimates of Likelihoods

As with Dempster–Shafer theory, the presence of ambiguity in the historical data will result in probability ranges, as the likelihoods depend on variables in Θ. The lower-bound probability is called the *belief* while the upper-bound probability is called the *plausibility*

$$Bel(E|M) = \min_{\Theta} p(E|M, \Theta) \tag{6.6}$$

$$Pl(E|M) = \max_{\Theta} p(E|M, \Theta) \tag{6.7}$$

In addition to the belief and plausibility, there is also a probability between these two boundaries that represents the best estimate of the likelihood. This best estimate can be obtained by using prior probabilities. For example, let us consider an ambiguous mode $\{m_1, m_2\}$. If there is complete ignorance as to whether any of the data belongs to mode m_1 or mode m_2, then the data should be divided evenly according to the principle of insufficient reason. However, if there is prior knowledge that mode m_1 happens three times as frequently as mode m_2, in other words $p(m_1) = 3p(m_2)$, then it stands to reason that 75% of the data in $\{m_1, m_2\}$ belongs to m_1 and 25% belongs to m_2. This would equate to the following parameters:

$$\hat{\theta}\{\tfrac{m_1}{m_1,m_2}\} = p(m_1|\{m_1, m_2\}) = 0.75$$

$$\hat{\theta}\{\tfrac{m_2}{m_1,m_2}\} = p(m_2|\{m_1, m_2\}) = 0.25$$

These estimates of $\theta\{\tfrac{m_1}{m_1,m_2}\}$ and $\theta\{\tfrac{m_2}{m_1,m_2}\}$ are called *informed* estimates, as they make use of prior information. In general, the informed estimate of a proportion parameter is given as

$$\hat{\theta}\{\tfrac{M}{m}\} = \frac{p(M)}{\sum\limits_{M \subseteq m} p(M)} \tag{6.8}$$

where the summation limit $M \subseteq m$ cycles through all unambiguous modes M that are contained within the historical (and potentially ambiguous) mode m. Note that logically, informed estimates in $\hat{\Theta}$ are also subject to the conditions in Eqns (6.1) and (6.2). The informed likelihood can serve as an educated guess as to what the likelihood value should be and is important in the techniques mentioned later in this chapter.

6.3   Fagin–Halpern Combination

While a parametrized expression now exists for the likelihood in Eqn (6.5), applying Bayes' rule in Eqn (6.3) is quite difficult, and successive combinations will yield more complex results. Fagin and Halpern (1991) proposed a conditioning rule that can be used to combine likelihood ranges with prior probabilities.

$$Bel(M|E) = \frac{Bel(M, E)}{Bel(M, E) + Pl(\bar{M}, E)}$$

$$Pl(M|E) = \frac{Pl(M, E)}{Pl(M, E) + Bel(\bar{M}, E)}$$

where $\bar{M}$ denotes all specific modes that are not M. When applied to Bayes' rule of combination, the Fagin–Halpern combination rule can be given as

$$Bel(M|E) = \frac{Bel(E|M)Bel(M)}{Bel(E|M)Bel(M) + \sum\limits_{\bar{M}} Pl(E|\bar{M})Pl(\bar{M})} \tag{6.9}$$

$$Pl(M|E) = \frac{Pl(E|M)Pl(M)}{Pl(E|M)Pl(M) + \sum\limits_{\bar{M}} Bel(E|\bar{M})Bel(\bar{M})} \tag{6.10}$$

This rule gives the largest possible boundary, and is suitable for combining a single likelihood term with a prior probability. The difficulty is that when successive combinations are used (which is of importance for assuming independence or applying this method dynamically), the Fagin–Halpern rule yields boundaries that quickly grow to a point where the end result is uninformative.

For example, if we consider evidence E where certain elements E_1 are independent of the other elements E_2, ($E = [E_1, E_2]$, $E_1 \perp E_2$), then

$$p(E|M) = p(E_1|M)p(E_2|M)$$

When we consider ambiguous modes, $p(E|M, \Theta)$, the same applies

$$p(E|M, \Theta) = p(E_1|M, \Theta)p(E_2|M, \Theta)$$

If we use the Fagin–Halpern combination to combine the two results

$$Bel(E|M) = \frac{Bel(E_1|M)Bel(E_2|M)}{Bel(E_1|M)Bel(E_2|M) + \sum_M Pl(E_1|\bar{M})Pl(E_1|\bar{M})}$$

$$Pl(E|M) = \frac{Pl(E_1|M)Pl(E_2|M)}{Pl(E_1|M)Pl(E_2|M) + \sum_M Bel(E_1|\bar{M})Bel(E_2|\bar{M})}$$

the end result would yield probability boundaries larger than the original boundaries by maximizing and minimizing the original expression $p(E|M, \Theta)$. As a result, Fagin–Halpern boundaries are too conservative for separating likelihoods, and are also too conservative for sequential combination as done in dynamic applications.

6.4   Second-order Approximation

Up to this point, we have considered two solutions.

1. Directly applying the Bayesian solution in Eqn (6.3) to the parametrized likelihood in Eqn (6.5). Here it was found that the expression grew successively more complicated with each combination step.
2. Applying the Fagin–Halpern method in Eqns (6.9) and (6.10). This was found to be suitable for combining a single likelihood with a belief, but since probability boundaries grew with each successive combination it was found to be unsuitable for separating evidence into independent groups and for dynamic application.

In order to retain some of the properties of the first (direct) solution while avoiding the increasingly complicated expressions of the second (Fagin–Halpern) solution, we propose a second-order approximation. We can first express Eqn (6.5) as a second-order function, and perform combination, ignoring higher-order terms. Thus the parametrized likelihood will continue to be a second-order expression with respect to Θ. While this is an approximate solution, due to the fact that the domain of Θ is restricted to values between 0 and 1, there is generally not enough room in the domain to deviate strongly from second-order behaviour. Hence, the second-order approximation is reasonable throughout the entire domain of $p(E|M, \Theta)$. Finally, the second-order approximation is exact at its reference point. In this case, we set the reference point for Θ to be the *informed* probability, which is our best guess at the value for Θ. Thus the second-order approximation inherently makes use of a best-guess probability that can be easily obtained from this method.

6.4.1   Consistency of Θ Parameters

One important advantage of the second-order combination rule over the Fagin – Halpern combination method is the ability to assume consistent Θ parameters. Let us consider a case where we would like to separate $p(E|M, \Theta)$ into two independent distributions so that

$$p(E|M, \Theta) = p(E_1|M, \Theta)p(E_2|M, \Theta) \tag{6.11}$$

For the Fagin – Halpern method of combination, it is assumed that Θ can take on different values for E_1 than E_2. In making this assumption, the probability boundaries are grown in such a way as to encompass all possible probability results from all values of Θ given that they are allowed to independently vary for E_1 and E_2. In reality, the values of Θ used for E_1 must be the same values used for E_2 if the independent combination in Eqn (6.11) is to hold true.

Whether directly combining independent likelihoods $p(E_1|M, \Theta)$ and $p(E_2|M, \Theta)$ or combining their approximation, because all values of Θ must be the same between the two sources of evidence, the terms in Θ can be collected. Thus, when applying the combination rule to second-order approximations, zeroth, first, and second-order terms of Θ are collected, and higher-order terms are ignored.

When evaluating evidence at different time intervals $p(E^t|M^t, \Theta)$, the values of Θ do not change with time because same historical data with the same values of Θ are used to evaluate the likelihood at each different time step t. Thus, not only is collecting Θ terms valid when combining likelihoods from independent evidence, it is also valid when applying the second-order rule in a dynamic fashion. Dynamic application of the second-order combination rule will be covered later on in this chapter.

6.4.2   Obtaining a Second-order Approximation

The second-order approximation of a function $f(x)$ is given by the Taylor series

$$f(x) \approx f(\hat{x}) + J(x - \hat{x}) + \frac{1}{2}(x - \hat{x})^T H(x - \hat{x})$$

where $\hat{x}$ is a reference point around which the approximation is taken (the approximation is exact at $\hat{x}$ but becomes worse when x is further away from $\hat{x}$), J is the Jacobian matrix (first-order derivatives evaluated at $\hat{x}$)

$$J_i = \frac{\partial f(x)}{\partial x_i}\bigg|_{\hat{x}}$$

and H is the Hessian matrix (second-order derivatives also evaluated at $\hat{x}$)

$$H_{i,j} = \left.\frac{\partial^2 f(x)}{\partial x_i \partial x_j}\right|_{\hat{x}}$$

When applied to our problem, the second-order approximation of $p(E|M, \Theta)$ is calculated with respect to Θ. A convenient reference point for Θ is the informed estimate $\hat{\Theta}$. For the Jacobian, the expression is given as

$$J_M[i] = \left.\frac{\partial p(E|M, \Theta)}{\partial \theta\{\frac{M}{m_i}\}}\right|_{\hat{\Theta}}$$

Note that because values of Θ are not variable in the conditions stated by Eqns (6.1) and (6.2) – these nonvarible conditions exist whenever $m_i \not\supset m$ – the following derivatives have zero value

$$\left.\frac{\partial p(E|M, \Theta)}{\partial \theta\{\frac{M}{m_i}\}}\right|_{\hat{\Theta}} = 0 \quad \forall \quad m_i \not\supset m$$

The expressions for the partial derivatives with respect to $p(E|M, \Theta)$ are obtained by differentiating Eqn (6.5). For compactness of notation, we introduce S, n, and $\hat{\theta}$ as vectors.

$$S = [S(E|m_1), S(E|m_2), \ldots, S(E|m_n)]$$
$$n = [n(m_1), n(m_2), \ldots, n(m_n)]$$
$$\hat{\theta} = \left[\hat{\theta}\{\tfrac{M}{m_1}\}, \hat{\theta}\{\tfrac{M}{m_2}\}, \ldots, \hat{\theta}\{\tfrac{M}{m_n}\}\right]$$

For nonzero conditions, the partial differentials for the Jacobian are then given as

$$\left.\frac{\partial p(E|M, \Theta)}{\partial \theta\{\frac{M}{m_i}\}}\right|_{\hat{\Theta}} = \frac{n_i S_i}{\sum_k n_k \hat{\theta}} - \frac{n_i \sum_k S_k n_k \hat{\theta}_k}{\left(\sum_k n_k \hat{\theta}_k\right)^2}$$

Terms for the Hessian can be written as

$$H_M[i,j] = \left.\frac{\partial^2 p(E|M, \Theta)}{\partial \theta\{\frac{M}{m_i}\} \partial \theta\{\frac{M}{m_j}\}}\right|_{\hat{\Theta}}$$

with similar zero-derivative conditions

$$\left.\frac{\partial^2 p(E|M, \Theta)}{\partial \theta\{\frac{M}{m_i}\} \partial \theta\{\frac{M}{m_j}\}}\right|_{\hat{\Theta}} = 0 \quad \forall \quad \begin{matrix} m_i \not\supset M \\ m_j \not\supset M \end{matrix}$$

For nonzero conditions, the second-order partial differentials for the Hessian can be written as

$$\left.\frac{\partial^2 p(E|M, \Theta)}{\partial \theta\{\frac{M}{m_i}\} \partial \theta\{\frac{M}{m_j}\}}\right|_{\hat{\Theta}} = -\frac{n_i S_j + n_j S_i}{\left(\sum_k n_k \hat{\theta}_k\right)^2} + \frac{n_i n_j \sum_k S_k n_k \hat{\theta}_k}{\left(\sum_k n_k \hat{\theta}_k\right)^3}$$

With terms for the Jacobian and Hessian already defined, the resulting second-order expression is

$$p(E|M,\Theta) = \hat{p}(E|M) + \boldsymbol{J}_M(\Theta - \hat{\Theta}) + \frac{1}{2}(\Theta - \hat{\Theta})^T \boldsymbol{H}_M(\Theta - \hat{\Theta}) \qquad (6.12)$$

where $\hat{p}(E|M)$ is the informed likelihood estimate

$$\hat{p}(E|M) = p(E|M,\hat{\Theta})$$

6.4.3 The Second-order Bayesian Combination Rule

After the second-order approximation has been taken for the likelihood, it can be combined with priors. Priors often do not have any ambiguity, but in case they do, we consider a general case where both the priors and the likelihoods are represented by second-order expressions.

$$p(E|M,\Theta) = \hat{p}(E|M) + \boldsymbol{J}_{(E|M)}(\hat{\Theta} - \Theta) + \frac{1}{2}(\hat{\Theta} - \Theta)^T \boldsymbol{H}_{(E|M)}(\hat{\Theta} - \Theta)$$

$$p(M|\Theta) = \hat{p}(M) + \boldsymbol{J}_{(M)}(\hat{\Theta} - \Theta) + \frac{1}{2}(\hat{\Theta} - \Theta)^T \boldsymbol{H}_{(M)}(\hat{\Theta} - \Theta)$$

$$p(E) = \sum_m \hat{p}(E|M)\hat{p}(M)$$

Bayesian combination is performed by taking the following product:

$$p(M|E,\Theta) = \frac{1}{p(E)} p(E|M,\Theta) p(M|\Theta)$$

By collecting terms with respect to $(\hat{\Theta} - \Theta)$, the posterior probability is expressed as

$$p(M|E,\Theta) = \hat{p}(M|E) + \boldsymbol{J}_{(M|E)}(\hat{\Theta} - \Theta) + \frac{1}{2}(\hat{\Theta} - \Theta)^T \boldsymbol{H}_{(M|E)}(\hat{\Theta} - \Theta) \qquad (6.13)$$

where the terms $\hat{p}(M|E)$, $\boldsymbol{J}_{(M|E)}$ and $\boldsymbol{H}_{(M|E)}$ are calculated as

$$\hat{p}(M|E) = \frac{1}{p(E)} \hat{p}(E|M)\hat{p}(M) \qquad (6.14)$$

$$\boldsymbol{J}_{(M|E)} = \frac{1}{p(E)} [\boldsymbol{J}_{(M)}\hat{p}(E|M) + \boldsymbol{J}_{(E|M)}\hat{p}(M)] \qquad (6.15)$$

$$\boldsymbol{H}_{(M|E)} = \frac{1}{p(E)} [\boldsymbol{H}_{(M)}\hat{p}(E|M) + \boldsymbol{H}_{(E|M)}\hat{p}(M) + \qquad (6.16)$$

$$\boldsymbol{J}_{(M)}^T \boldsymbol{J}_{(E|M)} + \boldsymbol{J}_{(E|M)}^T \boldsymbol{J}_{(M)}]$$

These expressions form the second-order update rules. In addition, these rules can be used to combine independent likelihoods

$$p(E|M,\Theta) = p(E_1|M,\Theta) p(E_2|M,\Theta)$$

However, there is no normalization constant $p(E)$, and thus the second-order rules for combining independent evidence are

$$\hat{p}(E|M) = \hat{p}(E_1|M)\hat{p}(E_2|M) \tag{6.17}$$

$$\boldsymbol{J}_{(E|M)} = [\boldsymbol{J}_{(E_1|M)}\hat{p}(E_2|M) + \boldsymbol{J}_{(M|E_1)}\hat{p}(M|E_2)] \tag{6.18}$$

$$\boldsymbol{H}_{(E|M)} = [\boldsymbol{H}_{(E_1|M)}\hat{p}(E_2|M) + \boldsymbol{H}_{(M|E_1)}\hat{p}(M|E_2) + \tag{6.19}$$

$$\boldsymbol{J}_{(M|E_1)}^T \boldsymbol{J}_{(M|E_2)} + \boldsymbol{J}_{(M|E_2)}^T \boldsymbol{J}_{(M|E_1)}]$$

6.5   Brief Comparison of Combination Methods

An example comparing the Fagin–Halpern and second-order boundaries is borrowed from Gonzalez and Huang (2013). In this example, evidence is available from six independent sources in a seven-mode system. When a new piece of evidence was available, the sources of evidence yielded the support as given in Table 6.1. These six sources of evidence are combined with the following prior:

$$[p(m_1), \ldots, p(m_7)] = [0.27,\ 0.09,\ 0.09,\ 0.19,\ 0.11,\ 0.15,\ 0.11]$$

Combination was done using the exact method (which resulted in a very complicated function), the second-order method, and the Fagin–Halpern method. The posterior probability results

Table 6.1   Support from example scenario

Mode	Support $s(E\|M)$						Number of observations
	e_1	e_2	e_3	e_4	e_5	e_6	
{1}	0.29	0.15	0.17	0.17	0.20	0.40	164
{2}	0.14	0.25	0.15	0.15	0.15	0.25	82
{3}	0.30	0.20	0.32	0.22	0.10	0.30	79
{4}	0.20	0.20	0.23	0.59	0.22	0.22	89
{5}	0.23	0.19	0.17	0.20	0.36	0.21	64
{6}	0.17	0.24	0.14	0.14	0.13	0.23	62
{7}	0.09	0.08	0.11	0.11	0.14	0.34	103
{1, 2}	0.18	0.21	0.20	0.16	0.16	0.34	31
{1, 3}	0.29	0.21	0.39	0.18	0.15	0.40	31
{1, 4}	0.29	0.24	0.19	0.39	0.22	0.31	32
{1, 5}	0.27	0.16	0.21	0.30	0.34	0.44	29
{1, 6}	0.20	0.22	0.15	0.14	0.22	0.27	29
{1, 7}	0.16	0.31	0.28	0.14	0.16	0.36	34
{2, 6}	0.22	0.21	0.19	0.22	0.26	0.24	18
{3, 4}	0.25	0.23	0.27	0.34	0.16	0.27	21
{4, 5}	0.21	0.19	0.25	0.43	0.27	0.31	20
{5, 6}	0.29	0.20	0.15	0.23	0.23	0.24	16
{4, 5, 6}	0.19	0.19	0.26	0.32	0.30	0.22	18

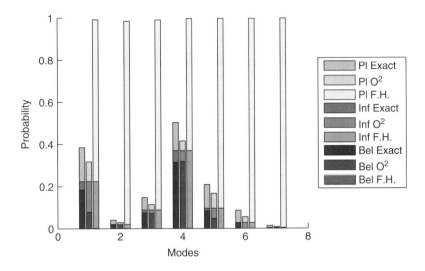

Figure 6.1   Diagnosis result for support in Table 6.1

are shown in Figure 6.1. Three probabilities are shown, the plausibility (indicated by lightest bar), the informed probability (mid-coloured bar, the same height for all three methods), and the belief (the darkest bar). From what can be seen from six combinations, the Fagin–Halpern probability ranges are extremely large, as was expected, with a range of practically 0–100% for all modes. Such a result is definitively inconclusive for diagnosis. By contrast, the second-order method yields probability boundaries that are much closer to the true result.

6.6   Applying the Second-order Rule Dynamically

6.6.1   *Unambiguous Dynamic Solution*

One of the strongest motivations for using the second-order combination rule is its ability to express ambiguity, even after successive combination. Successive combination is heavily used when solutions are applied in a dynamic manner to take into account autodependent modes. The dynamic solution to the autodependent mode problem has been discussed in Chapters 2 and 4, but it was only applicable to the case where no modes had ambiguity. However, the second-order probability expression in Eqn (6.13) is formulated in a manner that enables easy application in a dynamic setting.

In Chapter 2, it was noted that the probability transition solution was

$$p(M^t|E^{t-1}) = \sum_{i=1}^{n} p(M^t|m_i^{t-1})p(m_i^{t-1}|E^{t-1}) \qquad (6.20)$$

where $p(m_i^{t-1}|E^{t-1})$ is the posterior probability of mode m_i at time $t-1$. The probability transition rule calculates a prior probability at time t ($p(M^t|E^{t-1})$) using the posterior probability from $t-1$ ($p(m_i^{t-1}|E^{t-1})$). This resultant prior $p(M^t|E^{t-1})$ is used as the prior probability, to calculate the posterior at time t using Bayes' theorem:

$$p(m_i^t|E^t) = \frac{p(E^t|M)p(M^t|E^{t-1})}{\sum_M p(E^t|M)p(M^t|E^{t-1})} \tag{6.21}$$

6.6.2   The Second-order Dynamic Solution

In this chapter, the second-order version of Bayes' theorem has already been derived. The remaining task is to define the *second-order probability transition rule*. Consider a second-order posterior probability that was obtained at time $t - 1$

$$p(M^{t-1}|E^{t-1}, \Theta) = \hat{p}(M^{t-1}|E^{t-1}) + \boldsymbol{J}_{(M^{t-1}|E^{t-1})}(\hat{\Theta} - \Theta)+ \tag{6.22}$$

$$\frac{1}{2}(\hat{\Theta} - \Theta)^T \boldsymbol{H}_{(M^{t-1}|E^{t-1})}(\hat{\Theta} - \Theta)$$

The second-order prior probability at time t can be obtained by directly applying Eqn (6.22) to Eqn (6.20)

$$p(M^t|E^{t-1}, \Theta) = \sum_{i=1}^{n} p(M^t|m_i^{t-1})p(m_i^{t-1}|E^{t-1}, \Theta)$$

If we consider each second-order term for $p(M^{t-1}|E^{t-1}, \Theta)$ in Eqn (6.22), a second-order transition rule can be made for each term when transitioning to M^t.

$$\hat{p}(M^t|E^{t-1}) = \sum_{i=1}^{n} p(M^t|m_i^{t-1})\hat{p}(M^{t-1}|E^{t-1}) \tag{6.23}$$

$$\boldsymbol{J}_{(M^t|E^{t-1})} = \sum_{i=1}^{n} p(M^t|m_i^{t-1})\boldsymbol{J}_{(M^{t-1}|E^{t-1})} \tag{6.24}$$

$$\boldsymbol{H}_{(M^t|E^{t-1})} = \sum_{i=1}^{n} p(M^t|m_i^{t-1})\boldsymbol{H}_{(M^{t-1}|E^{t-1})} \tag{6.25}$$

This resulting probability is used as a prior probability at time t, which can be updated to a posterior using the previously proposed second-order Bayesian combination rule:

$$\hat{p}(E^t) = \sum_M \hat{p}(E^t|M)\hat{p}(M^t|E^{t-1})$$

$$\hat{p}(M^t|E^t) = \frac{1}{\hat{p}(E^t)}\hat{p}(E^t|M)\hat{p}(M^t|E^{t-1})$$

$$\boldsymbol{J}_{(M|E)} = \frac{1}{\hat{p}(E^t)}[\boldsymbol{J}_{(M^t|E^{t-1})}\hat{p}(E^t|M) + \boldsymbol{J}_{(E^t|M)}\hat{p}(M^t|E^{t-1})]$$

$$\boldsymbol{H}_{(M|E)} = \frac{1}{\hat{p}(E^t)}[\boldsymbol{H}_{(M^t|E^{t-1})}\hat{p}(E^t|M) + \boldsymbol{H}_{(E^t|M)}\hat{p}(M^t|E^{t-1})+$$

$$\boldsymbol{J}_{(M^t|E^{t-1})}^T \boldsymbol{J}_{(E^t|M)} + \boldsymbol{J}_{(E^t|M)}^T \boldsymbol{J}_{(M^t|E^{t-1})}]$$

6.7   Making a Diagnosis

After using the second-order combination rule to merge prior probabilities and likelihoods, the results can be used for diagnosis. The second-order approximation is convenient as it can be used to define four quantities useful for diagnosis:

1. The informed probability $\hat{p}(M|E)$
2. The belief $Bel(M|E)$
3. The plausibility $Pl(M|E)$
4. The expected probability $E_\Theta[p(M|E,\Theta)]$

6.7.1   Simple Diagnosis

The first quantity $\hat{p}(M|E)$ can be used as a simple diagnosis reference. This is a convenient quantity to use because it is explicitly available in the second-order result. In fact, if one simply wishes to obtain a simple diagnosis, $\hat{p}(M|E)$ is the only term that needs to be calculated; $\boldsymbol{J}_{(M|E)}$ and $\boldsymbol{H}_{(M|E)}$ are not needed. One simply chooses the mode that has the largest informed posterior $\hat{p}(M|E)$.

6.7.2   Ranged Diagnosis

It may be desirable to also convey information about the ambiguity associated with a diagnosis. For example, if evidence is located in a region where historical data tends to be ambiguous, the range of possible probability will be large. Conversely, in regions where historical data tends to be unambiguous, the probability ranges will be small. Probability ranges can be calculated according to

$$Bel(M|E) = \min_\Theta p(M|E,\Theta)$$

$$Pl(M|E) = \max_\Theta p(M|E,\Theta)$$

Because the expression is quadratic, the resulting minimization and maximization problems can be solved using quadratic programming methods; hence converting the expression to a second-order approximation greatly simplifies the minimization and maximization procedures. Note that the following constraints must be applied:

$$0 \le \theta\{\tfrac{M}{m_k}\} \le 1$$

While the probability boundaries are approximate, the function $p(M|E,\Theta)$ is generally well-approximated by the second-order approximation over its domain (limited between 0 and 1 for all Θ). The probability boundaries serve to give an estimate of how reliable the diagnosis is, and how adversely it is affected by ambiguity in the historical data.

6.7.3   Expected Value Diagnosis

One can also use the second-order approximation to obtain an expected value $E_\Theta[p(M|E,\Theta)]$. However, in doing this, one must treat Θ as a random variable and construct a probability distribution for it. Due to the fact that elements in Θ can be seen as probabilities themselves

$$\theta\{\tfrac{M}{m_k}\} = p(M|\boldsymbol{m}_k)$$

an appropriate distribution for Θ is the Dirichlet distribution, a probability distribution often used to define the distribution of probability estimates.

The Dirichlet Distribution for Expected Values of $p(M|E, \Theta)$

Let us consider $\Theta\{\tfrac{\bullet}{m_k}\}$, the set of elements in Θ that pertain to the ambiguous mode $\boldsymbol{m}_k$

$$\Theta\{\tfrac{\bullet}{m_k}\} = \left[\theta\{\tfrac{m_1}{m_k}\},\dots,\theta\{\tfrac{m_n}{m_k}\}\right]$$

These elements behave like a complete set of discrete probabilities, and thus follow the Dirichlet distribution

$$f(\Theta\{\tfrac{\bullet}{m_k}\} \mid \alpha\{\tfrac{\bullet}{m_k}\}) = \frac{\Gamma\left(\sum_i \alpha\{\tfrac{m_i}{m_k}\}\right)}{\prod_i \Gamma(\alpha\{\tfrac{m_i}{m_k}\})} \prod_i \theta\{\tfrac{m_i}{m_k}\}^{\alpha\{\tfrac{m_i}{m_k}\}-1}$$

$$= \frac{1}{B(\alpha)} \prod_i \theta\{\tfrac{m_i}{m_k}\}^{\alpha\{\tfrac{m_i}{m_k}\}-1}$$

$$\alpha\{\tfrac{\bullet}{m_k}\} = \left[\alpha\{\tfrac{m_1}{m_k}\},\dots,\alpha\{\tfrac{m_n}{m_k}\}\right]$$

where $B(\alpha)$ is a normalization constant. As previously mentioned, the values $\Theta\{\tfrac{\bullet}{m_k}\}$ can be seen as probabilities and must be on the interval between 0 and 1; furthermore, they must sum to one. The terms $\alpha\{\tfrac{\bullet}{m_k}\}$ are shape parameters that can be interpreted as the number of prior samples. Thus the Dirichlet distribution is a probability distribution of probability estimates given the samples. The expected value of $\theta\{\tfrac{m_i}{m_k}\}$ is given as

$$E(\theta\{\tfrac{m_i}{m_k}\}) = \frac{\alpha\{\tfrac{m_i}{m_k}\}}{\sum \alpha\{\tfrac{\bullet}{m_k}\}} \tag{6.26}$$

As an example to aid in the interpretation of a Dirichlet distribution, let us assign the following values to the parameters:

$$\theta\{\tfrac{m_i}{m_k}\} = p(m_i|\boldsymbol{m}_k)$$

$$\alpha\{\tfrac{m_i}{m_k}\} = n(m_i|\boldsymbol{m}_k)$$

where $p(m_i|\boldsymbol{m}_k)$ is the probability of mode m_i given ambiguous mode $\boldsymbol{m}_k$ and $n(m_i|\boldsymbol{m}_k)$ is the frequency of mode m_i given ambiguous mode $\boldsymbol{m}_k$. The Dirichlet distribution is given as

$$f(p(M|\boldsymbol{m}_k) \mid n(M|\boldsymbol{m}_k)) = \frac{\Gamma\left(\sum_i n(m_i|\boldsymbol{m}_k)\right)}{\prod_i \Gamma(n(m_i|\boldsymbol{m}_k))} \prod_i p(m_i|\boldsymbol{m}_k)^{n(m_i|\boldsymbol{m}_k)-1}$$

Using this Dirichlet distribution, the expected value of $\theta\{\frac{m_i}{m_k}\}$ can be calculated as

$$E[p(m_i|\boldsymbol{m}_k)] = \frac{n(m_i|\boldsymbol{m}_k)}{\sum_j n(m_j|\boldsymbol{m}_k)} = \frac{n(m_i|\boldsymbol{m}_k)}{n(\boldsymbol{m}_k)}$$

This is the probability one would expect to obtain given the samples $n(m_j|\boldsymbol{m}_k)$. Note that the expected values are always in the interval of 0 and 1, and sum to one (given $\boldsymbol{m}_k$) as they must.

The previous Dirichlet distribution was used to denote the probability density function (PDF) $f(\Theta\{\frac{\bullet}{m_k}\})$ for Θ parameters pertaining to a single ambiguous mode. Now we would like to express a PDF $f(\Theta)$ pertaining to all ambiguous modes. Note that parameter sets for different ambiguous modes are independent of each other, resulting in the following expression for $f(\Theta)$

$$f(\Theta) = \prod_k f(\Theta\{\tfrac{\bullet}{m_k}\})$$

This probability distribution defines the possible values of Θ and will be used for calculating the expected value.

Calculating the expected value of $p(M|E, \Theta)$

The posterior probability $p(M|E, \Theta)$ was previously given in Eqn (6.13). By making use of the probability distribution over Θ in $f(\Theta)$, the expected value is given as

$$E[p(M|E)] = \int_\Theta \left[\hat{p}(M|E) + \boldsymbol{J}_{(M|E)}\Delta\Theta + \frac{1}{2}\Delta\Theta^T \boldsymbol{H}_{(M|E)}\Delta\Theta\right] \prod_k f(\Theta\{\tfrac{\bullet}{m_k}\})d\Theta$$

$$= \text{Const} + \int_\Theta \left[\boldsymbol{J}^*_{(M|E)}\Theta + \frac{1}{2}\Theta^T \boldsymbol{H}_{(M|E)}\Theta\right] \prod_k f(\Theta\{\tfrac{\bullet}{m_k}\})d\Theta \qquad (6.27)$$

where

$$\text{Const} = \hat{p}(M|E) - \boldsymbol{J}_{(M|E)}\hat{\Theta} + \frac{1}{2}\hat{\Theta}^T \boldsymbol{H}_{(M|E)}\hat{\Theta}$$

$$\boldsymbol{J}^*_{(M|E)} = (\boldsymbol{J}_{(M|E)} - \hat{\Theta}^T \boldsymbol{H}_{(M|E)})$$

The second-order expression of $p(M|E, \Theta)$ is linear with respect to Θ and contains terms no higher than second order. Because of this, the expected value can be expressed in terms of means and variances, which have well-known solutions for the Dirichlet distribution.

$$\int_\Theta \theta\{\tfrac{M}{m_i}\} \prod_k f(\Theta\{\tfrac{\bullet}{m_k}\})d\Theta = E(\theta\{\tfrac{M}{m_i}\})$$

$$\int_\Theta \theta\{\tfrac{M}{m_i}\}\theta\{\tfrac{M}{m_j}\} \prod_k f(\Theta\{\tfrac{\bullet}{m_k}\})d\Theta = E(\theta\{\tfrac{M}{m_i}\})E(\theta\{\tfrac{M}{m_j}\})$$

$$\int_\Theta \theta^2\{\tfrac{M}{m_i}\} \prod_k f(\Theta\{\tfrac{\bullet}{m_k}\})d\Theta = \left[E(\theta\{\tfrac{M}{m_i}\})\right]^2 + \text{Var}(\theta\{\tfrac{M}{m_i}\})$$

For the Dirichlet distribution, the means and variances are given by

$$A(\boldsymbol{m}_i) = \sum_{\boldsymbol{m}_k \subset \boldsymbol{m}_i} \alpha(m_k | \boldsymbol{m}_i)$$

$$E(\theta\{\tfrac{M}{\boldsymbol{m}_i}\}) = \frac{\alpha(M)}{A(\boldsymbol{m}_i)}$$

$$\text{Var}(\theta\{\tfrac{M}{\boldsymbol{m}_i}\}) = \frac{\alpha(M)[A(\boldsymbol{m}_i) - \alpha(M)]}{A(\boldsymbol{m}_i)^3 + A(\boldsymbol{m}_i)^2}$$

where $\alpha(M)$ represents the prior sample of the unambiguous mode M. If one uses the frequency of mode occurrences $n(M)$, the shape parameters $\alpha(M)$ will be quite large and the expected value of $p(M|E, \Theta)$ will be nearly identical to the informed estimate. This is because large values in shape parameters $\alpha(M)$ yield a very sharp distribution centred at $\hat{\Theta}$. If one is unsure about the accuracy of $\hat{\Theta}$ it is best to divide all $\alpha(M)$ by a common factor so that the largest $\alpha(M)$ is no larger than 10. Values larger than 10 can result in fairly narrow distributions.

The means and variances can be applied to Eqn (6.27). First, one has to separate the squared terms of Θ associated with variance

$$E[p(E|M)] = \text{Const} +$$

$$\int_\Theta \left[\boldsymbol{J}^*_{(M|E)} \Theta + \frac{1}{2} [\Theta^T (\boldsymbol{H}_{(M|E)} - \boldsymbol{H}_D) \Theta + [\Theta^2]^T (\boldsymbol{H}_D \boldsymbol{1})] \right] f(\Theta) d\Theta$$

where $\boldsymbol{H}_D$ is a diagonal matrix containing the diagonal elements of $\boldsymbol{H}_{(M|E)}$ and $\boldsymbol{1}$ is a column vector of ones. Furthermore, Θ is a column vector of parameters and Θ^2 is a column vector containing squared values of Θ. By expressing the integrals as expected values and variances, the solution is reduced to

$$E[p(E|M)] = \text{Const} + \boldsymbol{J}^*_{(M|E)} E(\Theta) + \frac{1}{2} E(\Theta)^T (\boldsymbol{H}_{(M|E)} - \boldsymbol{H}_D) E(\Theta)$$

$$+ \frac{1}{2} [\text{Var}(\Theta) + E(\Theta)^2]^T \boldsymbol{H}_D \boldsymbol{1} \tag{6.28}$$

This can be solved in order to obtain the expected value of the posterior.

When $E_\Theta[p(M|E, \Theta)]$ approaches $\hat{p}(M|E)$

As was mentioned earlier, if the number of prior samples $n(M)$ is large, one can justify using the informed estimate $\hat{p}(E|M)$ by assuming strong prior knowledge. In such a case, the prior samples α are large so that $\alpha \to \infty$. The expected values and variances will then take the following values:

$$E(\theta\{\tfrac{M}{\boldsymbol{m}_i}\}) = \hat{\theta}\{\tfrac{M}{\boldsymbol{m}_i}\}$$

$$\text{Var}(\theta\{\tfrac{M}{\boldsymbol{m}_i}\}) = 0$$

If this occurs, it can be shown that Eqn (6.27) will revert to the informed estimate $\hat{p}(M|E)$.

$$
\begin{aligned}
E[p(M|E)] &= \text{Const} + \boldsymbol{J}^*_{(M|E)}\hat{\Theta} + \frac{1}{2}\left[\hat{\Theta}^T(\boldsymbol{H}_{(M|E)} - \boldsymbol{H_D})\hat{\Theta} + [\boldsymbol{0} + \hat{\Theta}^2]^T\boldsymbol{H_D}\boldsymbol{1}\right] \\
&= \text{Const} + \left[\boldsymbol{J}_{(M|E)} - \hat{\Theta}^T\boldsymbol{H}_{(M|E)}\right]\hat{\Theta} \\
&\quad + \frac{1}{2}\left[\hat{\Theta}^T(\boldsymbol{H}_{(M|E)} - \boldsymbol{H_D})\hat{\Theta} + \hat{\Theta}^T\boldsymbol{H_D}\hat{\Theta}\right] \\
&= \text{Const} + [\boldsymbol{J}_{(M|E)}\hat{\Theta} - \hat{\Theta}^T\boldsymbol{H}_{(M|E)}\hat{\Theta}] + \frac{1}{2}\hat{\Theta}^T\boldsymbol{H}_{(M|E)}\hat{\Theta} \\
&= \left[\hat{p}(M|E) - \boldsymbol{J}_{(M|E)}\hat{\Theta} + \frac{1}{2}\hat{\Theta}^T\boldsymbol{H}_{(M|E)}\hat{\Theta}\right] \\
&\quad + \boldsymbol{J}_{(M|E)}\hat{\Theta} - \frac{1}{2}\hat{\Theta}^T\boldsymbol{H}_{(M|E)}\hat{\Theta} \\
&= \hat{p}(M|E)
\end{aligned}
$$

This result justifies using $\hat{p}(M|E)$ when the prior sample size is large or, equivalently, if one has high confidence in the priors. However, if one is not confident in the value of $\hat{\Theta}$ (or, equivalently, if one is not confident that prior probabilities adequately represent proportions in the ambiguous modes), one should use small sample sizes of α and calculate the expected value for diagnosis.

6.8   Notes and References

The majority of the material in this chapter stems from work presented in Gonzalez and Huang (2013). The use of ambiguity originated in Dempster (1968) while the definition of the support function was stated in the related work Shafer (1976). In addition to the material in Gonzalez and Huang (2013), this chapter also includes information on dynamic inference, which was not previously covered; the dynamic solution was obtained by combining techniques from Gonzalez and Huang (2013) with material that originated in Qi and Huang (2010b).

References

Dempster A 1968 A generalization of Bayesian inference. *Journal of the Royal Statistical Society. Series B* **30**(2), 205–247.

Fagin R and Halpern J 1991 A new approach to updating beliefs. *Uncertainty in Artificial Intelligence, 6*. North-Holland, pp. 347–374.

Gonzalez R and Huang B 2013 Control loop diagnosis from historical data containing ambiguous operating modes: Part 2. Information synthesis based on proportional parameterization. *Journal of Process Control* **23**(4), 1441–1454.

Qi F and Huang B 2010a A Bayesian approach for control loop diagnosis with missing data. *AIChE Journal* **56**(1), 179–195.

Qi F and Huang B 2010b Dynamic Bayesian approach for control loop diagnosis with underlying mode dependency. *Industrial and Engineering Chemistry Research* **49**, 8613–8623.

Shafer G 1976 *A Mathematical Theory of Evidence*. Princeton University Press.

7

Accounting for Ambiguous Modes: A Dempster–Shafer Approach

7.1 Introduction

Inference with ambiguous hypotheses has existed since the late 1960s, with Dempster (1968) and Shafer (1976) being the first to contribute major publications in this field, which became known as Dempster–Shafer theory. Being proposed as a generalization to Bayesian inference, Dempster–Shafer theory was shown to be able to account for both probabilistic uncertainty and ignorance, making the claim being that Bayesian inference cannot adequately express ignorance. Since its inception, there has been a vast amount of literature published on Dempster–Shafer theory, which includes a wide array of combination rules, interpretation results, and criticisms.[1]

In Chapter 6, a solution to the diagnosis problem with ambiguous modes was proposed using the parametrized Bayesian method. However, Dempster–Shafer theory provides an alternative solution. It may seem that Demspter–Shafer theory can be readily applied to the ambiguous mode problem, but further investigation reveals some difficult challenges that ultimately requires restructuring and generalization of some of the basic concepts of the theory. Nevertheless, because the intended scope of Dempster–Shafer theory is quite broad, the method in this chapter does not require the assumption of Θ being identical for different evidence sources, as in Chapter 6.

7.2 Dempster–Shafer Theory

7.2.1 Basic Belief Assignments

The principal difference between Dempster–Shafer theory and Bayesian inference is the interpretation of probability. In the Bayesian sense, all of the supported hypotheses must be mutually exclusive. For example, let us borrow the problem in Chapter 6 with a valve and a

---

[1] Criticisms are mainly due to the subjective nature of Dempster–Shafer theory.

Process Control System Fault Diagnosis: A Bayesian Approach, First Edition. Ruben Gonzalez, Fei Qi and Biao Huang.
© 2016 John Wiley & Sons, Ltd. Published 2016 by John Wiley & Sons, Ltd.

sensor, with each of the two components having three states.

$$
\begin{bmatrix} m_1 \\ m_2 \\ m_3 \\ m_4 \\ m_5 \\ m_6 \\ m_7 \\ m_8 \\ m_9 \end{bmatrix} = \begin{bmatrix} c_1 = 1\,, c_2 = 1 \\ c_1 = 1\,, c_2 = 2 \\ c_1 = 1\,, c_2 = 3 \\ c_1 = 2\,, c_2 = 1 \\ c_1 = 2\,, c_2 = 2 \\ c_1 = 2\,, c_2 = 3 \\ c_1 = 3\,, c_2 = 1 \\ c_1 = 3\,, c_2 = 2 \\ c_1 = 3\,, c_2 = 3 \end{bmatrix}
$$

All modes in this case are exclusive hypotheses. When there is ambiguity in any of the modes, overlap between the hypotheses will exist. Probability can only be applied to a set of exclusive hypotheses, which in our case is an unambiguous mode m.

$$
p(M) = \frac{n(M)}{n}
$$

Dempster–Shafer theory, however, makes use of a basic belief assignment (BBA), which is equivalent to the support function defined Chapter 6.

$$
S(M) = \frac{n(M)}{n}
$$

$$
1 = \sum_{M} S(M)
$$

where M can contain ambiguous modes (e.g., $m_k = \{m_1, m_2, m_3\}$ which occurs when $c_1 = 1$ is observed and c_2 is missing); when ambiguous modes are in the data, overlapping hypotheses can occur. Dempster–Shafer theory aims to express the probability in terms of the BBA. When invoking the Θ parameter notation used in Chapter 6, the probability is expressed as

$$
p(M|\Theta) = \sum_{m_k \cap M \neq \emptyset} \theta\{\tfrac{M}{m_k}\} S(m_k)
$$

where M is the mode of interest and $\theta\{\tfrac{M}{m_k}\}$ is an unknown proportion parameter that represents the probability of M given m_k.

$$
\theta\{\tfrac{M}{m_k}\} = p(M|m_k)
$$

The parameters $\theta\{\tfrac{M}{m_k}\}$ have a set of constraints. The first constraint

$$
0 \leq \theta\{\tfrac{M}{m_k}\} \leq 1
$$

states that, because $\theta\{\tfrac{M}{m_k}\}$ is a probability, it cannot be larger than 1 or smaller than 0. In two special cases, $\theta\{\tfrac{M}{m_k}\}$ is not random at all, but must take on specific values based on logical constraints. In the first case, if the mode of interest M completely contains the mode m_k, then all the support to m_k must apply to M.

$$
\theta\{\tfrac{M}{m_k}\} = 1 \qquad \forall \ M \supseteq m_k \tag{7.1}
$$

For example, if the mode of interest M is the ambiguous mode $\{m_1, m_2, m_3\}$ and the supported mode is $\{m_1, m_3\}$ then all support given to $\{m_1, m_3\}$ in the history must also be given to $\{m_1, m_2, m_3\}$.

In the second case, if the mode of interest M has nothing in common with the mode m_k, none of the support given to m_k can apply to M.

$$\theta\{\tfrac{M}{m_k}\} = 0 \qquad \forall \; M \cap m_k = \emptyset \tag{7.2}$$

Outside of these conditions, $\theta\{\tfrac{M}{m_k}\}$ is a flexible (or unknown) value between 0 and 1.

$$0 \le \theta\{\tfrac{M}{m_k}\} \le 1 \qquad \forall \; M \not\supseteq m_k, \quad M \cap m_k = \emptyset \tag{7.3}$$

7.2.2 Probability Boundaries

Dempster–Shafer theory concerns itself with boundaries on the probability. The plausibility and belief can be obtained by maximizing and minimizing $p(M|\Theta)$ over the unknown parameters Θ. The optimization in this problem is linear with respect to Θ and is constrained by the previous conditions. Because the BBA $S(m_k)$ values (which serve as coefficients on Θ) are nonnegative, the belief can be obtained by setting all flexible values of Θ to zero.

$$Bel(M) = Ex(M) = \sum_{m_k \subseteq M} S(m_k)$$

Because the condition $m_k \supseteq M$ excludes support from all flexible values of Θ, it is called the *exclusive condition* $Ex(M)$. For Dempster–Shafer theory, the exclusive probability is the solution to the belief or lower-bound probability.

In a similar manner, the plausibility can be obtained by setting all flexible values of Θ to 1.

$$Pl(M) = In(M) = \sum_{m_k \cap M \neq \emptyset} S(m_k)$$

The condition $m_k \cap M \neq \emptyset$ includes support from all flexible values of Θ and so is called the *inclusive condition* $In(M)$. In this way, the inclusive probability is the solution to the plausibility or upper-bound probability.

7.2.3 Dempster's Rule of Combination

Dempster's rule of combination is made to combine two probability assessments of M, and is said to be a generalization of Bayeisan combination. Dempster's rule can be expressed as

$$S_{1,2}(M) = \frac{1}{1-K} \sum_{M = m_i \cap m_j \neq \emptyset} S_1(m_i) S_2(m_j)$$

$$K = \sum_{\emptyset = m_i \cap m_j} S_1(m_i) S_2(m_j)$$

Here, to find out the combined support $S_{1,2}(M)$ of the mode M we search for all modes in S_1 and S_2 that intersect to yield M (expressed as $M = m_i \cap m_j \neq \emptyset$). However, there is no such thing as support to the empty set $\emptyset$, which denotes conflict. Support to conflict is denoted as K, and is normalized out (as $1 - K$), because BBAs are not allowed to support conflict.

In Dempster–Shafer theory, a BBA is called Bayesian if it contains no support for ambiguous hypotheses. In our application, the BBA is Bayesian if support is only given to unambiguous modes. If BBAs are Bayesian, then Dempster's rule will revert to Bayes' theorem.

$$S_{1,2}(M) = \frac{1}{1 - K} \sum_{m = m_i \cap m_j \neq \emptyset} S_1(m_i)S_2(m_j) = \frac{1}{1 - K}S_1(M)S_2(M)$$

$$K = \sum_{\emptyset = m_i \cap m_j} S_1(\boldsymbol{m}_i)S_2(\boldsymbol{m}_j) = 1 - \sum_m S_1(M)S_2(M)$$

so that the end result is

$$S_{1,2}(M) = \frac{1}{\sum_m S_1(M)S_2(M)} S_1(M)S_2(M)$$

which indeed resembles Bayes' theorem.

Dempster's rule will always yield support to intersections between $S_1(\boldsymbol{m}_i)$ and $S_2(\boldsymbol{m}_j)$. Because of this, ambiguity is reduced after every combination; the idea is that information from $S_1(\boldsymbol{m}_i)$ will be applied to ambiguity in $S_2(\boldsymbol{m}_j)$ and information from $S_2(\boldsymbol{m}_j)$ will be applied to ambiguity in $S_1(\boldsymbol{m}_i)$. Applying information to each others' ambiguity will reduce the uncertainty between the two BBAs.

Because each combination results in a reduction in ambiguity, the more combinations that are performed, the more the resulting BBA will resemble a Bayesian BBA. In fact, if any BBA is combined with a Bayesian BBA, the resulting BBA will be Bayesian. In such a case, precise information from the Bayesian BBA will be applied to the uncertainties in the other BBAs, resulting in zero uncertainty after combination.

7.2.4   Short-cut Combination for Unambiguous Priors

If the prior probability is Bayesian, Dempster–Shafer combination of any BBA will yield a Bayesian result. In such a case, a short-cut solution is available to calculate the posterior, which is much less computationally intensive than applying Dempster's rule directly. In addition, when implementing a dynamic solution, successive combinations yield a Bayesian result, thus it makes sense to use a Bayesian prior in order to cut computational loads. Choosing Bayesian priors not only reduces computational loads, it also yields a dynamic application that is fully compatible with the dynamic Bayesian method; the posteriors are always Bayesian, thus the probability transition technique will be the same as in the dynamic Bayesian solution.

Let us consider a BBA $S_1(M)$ that is a Bayesian prior $p_1(M)$, and another BBA $S_2(M)$ that contains ambiguity. Using Dempster's rule for combination

$$S_{12}(M) = \frac{1}{1 - K} \sum_{M = m_i \cap \boldsymbol{m}_j \neq \emptyset} p_1(m_i)S_2(\boldsymbol{m}_j)$$

Now the condition

$$M = m_i \cap \boldsymbol{m}_j \neq \emptyset$$

is only true when $m_i = M$ and when $m_j \supseteq M$. Because of this, we can factor out $p_1(M)$ and replace the condition with $m_j \supseteq m$.

$$S_{12}(M) = \frac{1}{1-K} p_1(M) \sum_{m_j \supseteq} S_2(m_j)$$

Now because M is unambiguous, the condition $m_j \supseteq M$ is equivalent to $m_j \cap M \neq \emptyset$ so that

$$S_{12}(M) = \frac{1}{1-K} p_1(M) \sum_{m_j \cap M \neq \emptyset} S_2(m_j)$$

From earlier results, we can see that the S_2 term amounts to the *inclusive probability* of an unambiguous mode M (or equivalently, the Dempster–Shafer plausibility).

$$S_{12}(M) = \frac{1}{1-K} p_1(M) In_2(M) \tag{7.4}$$

Because $1-K$ is a normalization constant, so that $\sum_M S_{12}(M) = 1$, we can define $1-K$ as

$$1 - K = \sum_M p_1(M) In_2(M) \tag{7.5}$$

Eqns (7.4) and (7.5) together define the short-cut evaluation of Dempster's rule with a Bayesian prior.

Dynamic Application

When using the short-cut rule, the posterior result is always Bayesian, thus the transition rule is still the same

$$S^t(M) = \sum_k p(M^t | m_k^{t-1}) S^{t-1}(m_k)$$

After Dempster's rule was applied at a time step $t-1$, the resulting probability $S^{t-1}(m_k)$ can be converted to the prior $S^t(M)$ at time t using the transition rule. Dempster's combination rule is then used to update $S^t(M)$ with more evidence.

7.3 Generalizing Dempster–Shafer Theory

In the Fagin–Halpern combination rule, discussed in Section 6.3, boundaries grow after each combination; in the second-order Bayesian method boundaries tend to stay relatively constant. However, in the Dempster–Shafer method, boundaries tend to shrink after successive combinations. The reason for this is that Dempster's rule does not make the assumption that Θ values are identical but that they are independent and that information from one BBA can make up for the ambiguity in another. As a rule for application, if reference data from different information sources is taken from the same time window, it is best to use the second-order Bayesian method because Θ values will be identical. However, if the evidence data for each source comes from different time intervals, it is better to apply generalized Dempster–Shafer theory as Θ values will be independent.

Applying Dempster–Shafer theory to our problem does not come without difficulties, as will be seen in Section 7.3.1; the BBA does not adequately describe how ambiguity affects the likelihood. Because of this, the BBA and Dempster's rule need to be generalized in order to better fit the problem in question.

Previously, when discussing Dempster–Shafer theory, we concerned ourselves with the probabilities of all modes in the history $p(\boldsymbol{M})$. However, we are only interested in diagnosing unambiguous modes $p(M)$. From this point forward, we will be only considering the problem of diagnosing unambiguous modes $p(M)$ with potentially ambiguous modes $\boldsymbol{M}$ in the history.

7.3.1   Motivation: Difficulties with BBAs

The difficulty with using Dempster–Shafer theory is representing the likelihood as a BBA. Dempster–Shafer theory can be used to describe direct probabilities with ambiguity.

$$p(M|E, \Theta) = \sum_{\boldsymbol{m}_k \cap m \neq \emptyset} \theta\{\tfrac{M}{\boldsymbol{m}_k}\} S(\boldsymbol{m}_k|E) \tag{7.6}$$

where the BBA terms in S are given as

$$S(\boldsymbol{M}|E) = \frac{n(\boldsymbol{M}, E)}{n(E)}$$

This is the case where we sample data at random from the entire evidence history $\mathcal{D}$, and we assume that the mode frequencies in $\mathcal{D}$ represent the mode probabilities for the population. When evaluating the probability directly, we consider the frequency of the mode and evidence occurring $n(\boldsymbol{M}, E)$ and divide by the total number of times the evidence occurs $n(E)$.

The disadvantage for direct evaluation is that in most cases some modes occur quite rarely, thus we cannot trust that $\mathcal{D}$ is representative of the mode frequency. Instead, it may be better to obtain the priors using process knowledge (from both this process, and possibly other similar processes) and use Bayes' theorem to combine likelihoods from the data with these priors. The result of this approach is that we can be much more flexible with the data we use for $\mathcal{D}$ to estimate likelihoods.

Evaluating likelihoods, however, poses a problem for Dempster–Shafer theory, as the likelihood with respect to S and Θ is given in Chapter 6 as

$$p(E|M, \Theta) = \frac{\sum\limits_{\boldsymbol{m}_k \geq M} \theta\{\tfrac{M}{\boldsymbol{m}_k}\} S(E|\boldsymbol{m}_k) n(\boldsymbol{m}_k)}{\sum\limits_{\boldsymbol{m}_k \geq m} \theta\{\tfrac{M}{\boldsymbol{m}_k}\} n(\boldsymbol{m}_k)} \tag{7.7}$$

which is very different from Eqn (7.6). There are in fact, two main functional differences between the Dempster–Shafer problem in Eqn (7.6) and our problem in Eqn (7.7).

Difference 1

In the Dempster–Shafer problem, the term $S(\boldsymbol{m}_k|E)$ functions as a *nonnegative* coefficient on $\theta\{\tfrac{M}{\boldsymbol{m}_k}\}$, which means that increasing $\theta\{\tfrac{M}{\boldsymbol{m}_k}\}$ never decreases $p(M|E, \Theta)$, that is

$$\frac{\partial\, p(M|E, \Theta)}{\partial\, \theta\{\tfrac{M}{\boldsymbol{m}_k}\}} \geq 0 \tag{7.8}$$

In our problem, because we have a fractional expression in Eqn (7.7), it is possible for an increase in $\theta\{\frac{M}{m_k}\}$ to result in a decrease in $p(E|M, \boldsymbol{\theta}\{m\})$ (an example of this is shown later)

$$\frac{\partial \, p(E|M, \Theta)}{\partial \, \theta\{\frac{M}{m_k}\}} \not\geq 0 \tag{7.9}$$

Since derivatives are no longer nonnegative, $Bel(E|M)$ is not necessarily solved by using the exclusive probability, and $Pl(E|M)$ is no longer solved by using the inclusive probability.

Difference 2

In the Dempster–Shafer problem, the term $S(\boldsymbol{m}_k|E)$ functions as a *constant* coefficient on $\theta\{\frac{M}{m_k}\}$ with respect to m. This means that as long as m_i and m_j are both in $\boldsymbol{m}_k$, the partial derivatives of $p(m_i|E, \Theta), p(m_j|E, \Theta)$ are the same with respect to their θ parameters on m.

$$\frac{\partial \, p(m_i|E, \Theta)}{\partial \, \theta\{\frac{m_i}{m_k}\}} = \frac{\partial \, p(m_j|E, \Theta)}{\partial \, \theta\{\frac{m_j}{m_k}\}} \qquad m_i, m_j \subset \boldsymbol{m}_k \tag{7.10}$$

In our problem, because of the fractional expression, the normalization constant on the denominator can change with respect to M, thus the partial derivatives of $p(E|m_i, \Theta), p(E|m_j, \Theta)$ are not necessarily the same with respect to their θ parameters on m.

$$\frac{\partial \, p(E|m_i, \Theta)}{\partial \, \theta\{\frac{m_i}{m_k}\}} \neq \frac{\partial \, p(E|m_j, \Theta)}{\partial \, \theta\{\frac{m_j}{m_k}\}} \qquad m_i, m_j \subset \boldsymbol{m}_k \tag{7.11}$$

Example of differences

Let us consider a simple two-mode system with one ambiguous mode so that

- $\boldsymbol{m}_1 = m_1$
- $\boldsymbol{m}_2 = m_2$
- $\boldsymbol{m}_3 = \{m_1, m_2\}$

The historical data for this system is presented in Table 7.1.

Table 7.1 Frequency counts from example

	m_1	m_2	$\{m_1, m_2\}$	All M
e_1	12	4	7	23
e_2	5	9	6	20
All e	17	13	13	43

Now let us consider a case where e_1 is observed. We can see that directly evaluating the probability of m yields

$$p(m_1|e_1, \Theta) = \frac{12}{23} + \theta\{\tfrac{m_1}{m_1,m_2}\}\frac{7}{23}$$

$$p(m_2|e_1, \Theta) = \frac{4}{23} + \theta\{\tfrac{m_2}{m_1,m_2}\}\frac{7}{23}$$

One can observe that the derivative of these expressions with respect to $\theta\{\tfrac{M}{m_1,m_2}\}$ is $7/23$. This result is identical for both m_1, m_2 and is also nonnegative. Conversely, when evaluating the likelihoods of E given the modes m_1 and m_2 we obtain

$$p(e_1|m_1, \Theta) = \frac{12 + \theta\{\tfrac{m_1}{m_1,m_2}\}7}{17 + \theta\{\tfrac{m_1}{m_1,m_2}\}13}$$

$$p(e_1|m_2, \Theta) = \frac{4 + \theta\{\tfrac{m_2}{m_1,m_2}\}7}{9 + \theta\{\tfrac{m_2}{m_1,m_2}\}13}$$

We can further see that the derivatives can be obtained as

$$\frac{\partial\, p(e_1|m_1, \Theta)}{\partial\, \theta\{\tfrac{m_1}{m_1,m_2}\}} = \frac{-37}{(17 + \theta\{\tfrac{m_1}{m_1,m_2}\}13)^2}$$

$$\frac{\partial\, p(e_1|m_2, \Theta)}{\partial\, \theta\{\tfrac{m_2}{m_1,m_2}\}} = \frac{11}{(9 + \theta\{\tfrac{m_2}{m_1,m_2}\}13)^2}$$

We can see that the derivative of $p(e_1|m_1, \Theta)$ is always negative, while the derivative of $p(e_1|m_2, \Theta)$ is always positive. Thus the derivatives for likelihoods are not identical, nor are they nonnegative.

Implications of the differences

The implication of these results is that we cannot adequately express the likelihood expression $p(E|M, \Theta)$ in Eqn (7.7) in the Dempster–Shafer BBA format presented in Eqn (7.6). A more general form of the BBA is required to express $p(E|M, \Theta)$, and a new combination rule must also be constructed.

7.3.2   Generalizing the BBA

The objective of the generalized BBA is to express $p(E|m_i, \Theta)$ as a first-order approximation of Θ

$$p(E|M, \Theta) = G[:, m]^T \Theta[:, m] \qquad (7.12)$$

with G and Θ taking structures that allow us to easily define a generalized Dempster's rule of combination. In this book $G[:, m]$ denotes the mth column of G, while $G[m, :]$ denotes the mth row of G.

In this book, Θ is the matrix form of Θ, with each row representing the potentially ambiguous mode m_k, and each column representing an unambiguous mode m_i so that

$$\Theta[k, i] = \theta\{\tfrac{m_i}{m_k}\}$$

The matrix G has the same dimensions as Θ where elements can be calculated as

$$G[k, i] = \begin{cases} 0 & m_i \cap m_k = \emptyset \\ \tilde{p}(E|m_i) & m_i = m_k \\ \frac{\partial p(E|m_i)}{\partial \Theta[k,i]} & m_i \subset m_k \end{cases} \qquad (7.13)$$

where

$$\tilde{p}(E|m_i) = p(E|m_i, \hat{\Theta}) - \sum_{m_k \supset m_i} \hat{\theta}\{\tfrac{m_i}{m_k}\} \frac{\partial \, p(E|m_i, \Theta)}{\partial \, \theta\{\tfrac{m_i}{m_k}\}} \bigg|_{\hat{\Theta}}$$

$$\frac{\partial p(E|m_i)}{\partial \Theta[k,i]} = \frac{\partial \, p(E|m_i, \Theta)}{\partial \, \theta\{\tfrac{m_i}{m_k}\}} \bigg|_{\hat{\Theta}}$$

Note that $\hat{\Theta}$ is the reference value of Θ, around which the approximation is centered. In Chapter 6, $\hat{\Theta}$ was defined as the *informed probability*. In this chapter, because of the properties of Dempster's rule, it is best to use the *inclusive value* of Θ.

$$\hat{\Theta} = \Theta^*$$

where

- Θ^* is the *inclusive value*, which sets all *flexible* values to 1
- Θ_* is the *exclusive value*, which sets all *flexible* values to 0.

Note that the mathematical conditions for *flexible* values were given in Eqn (7.3).

With G and Θ defined in this manner, Eqn (7.12) is able to express the likelihood as a first-order approximation. While first-order approximations are generally not as accurate as second-order approximations (which were performed in Chapter 6), due to the fact that generalized Dempster–Shafer combinations converge to an unambiguous result, the order of the approximation with respect to Θ is not as much of an issue as the reference value chosen for G.

One important feature of G is that the mode can be extracted by the values taken on by the row of G. For example, let us consider a three-mode system, with possible ambiguous modes $\{m_1, m_2\}$, $\{m_1, m_3\}$ and $\{m_2, m_3\}$. The structure of G is defined as

$$G = \begin{bmatrix} & m_1 & m_2 & m_3 \\ m_1 & G\{\tfrac{m_1}{m_1}\} & 0 & 0 \\ m_2 & 0 & G\{\tfrac{m_2}{m_2}\} & 0 \\ m_3 & 0 & 0 & G\{\tfrac{m_3}{m_3}\} \\ \{m_1, m_2\} & G\{\tfrac{m_1}{m_1,m_2}\} & G\{\tfrac{m_2}{m_1,m_2}\} & 0 \\ \{m_1, m_3\} & G\{\tfrac{m_1}{m_1,m_3}\} & 0 & G\{\tfrac{m_3}{m_1,m_3}\} \\ \{m_2, m_3\} & 0 & G\{\tfrac{m_2}{m_2,m_3}\} & G\{\tfrac{m_3}{m_2,m_3}\} \end{bmatrix}$$

We can therefore recover the mode from the appropriate row of G by analysing the zero elements. For example, the fourth row of G is

$$G[4,:] = \left[G\{\tfrac{m_1}{m_1,m_2}\} \quad G\{\tfrac{m_2}{m_1,m_2}\} \quad 0 \right] \to \{m_1, m_2\}$$

where each column in this row pertains to modes m_1, m_2, m_3. Because the third element is zero, m_3 is not supported and hence $\{m_1, m_2\}$ *is* supported. Thus for every row of G, zeros indicate that the corresponding modes are not supported. In a similar manner, we can see that the row

$$G[2,:] = \left[0 \quad G\{\tfrac{m_2}{m_2}\} \quad 0 \right] \to m_2$$

only supports m_2. In this way, we can see that the mode can be recovered by determining which elements in $G[k,:]$ are equal to zero.

This generalized form of the BBA, or GBBA, because it is a generalization, can be used to express Dempster–Shafer BBAs as well. If we consider the conditions set forth in Eqns (7.8) and (7.10), and the GBBA construction method in Eqn (7.13) we can set two conditions that must be met to allow a GBBA to be classified as a BBA.

1. Every nonzero element in G must be nonnegative.
2. Every nonzero element in a given row $G[k,:]$ must have identical values.

Taking the GBBA structure from our previous three-mode system, G would be a BBA if it took the form

$$
G = \begin{array}{c|ccc}
 & m_1 & m_2 & m_3 \\
\hline
m_1 & S(m_1) & 0 & 0 \\
m_2 & 0 & S(m_2) & 0 \\
m_3 & 0 & 0 & S(m_3) \\
\{m_1, m_2\} & S\{m_1, m_2\} & S\{m_1, m_2\} & 0 \\
\{m_1, m_3\} & S\{m_1, m_3\} & 0 & S\{m_1, m_3\} \\
\{m_2, m_3\} & 0 & S\{m_2, m_3\} & S\{m_2, m_3\}
\end{array}
$$

Probability Boundaries on GBBAs

When GBBAs are applied, the inclusive and exclusive probabilities have similar definitions to BBAs

$$Ex(M) = \sum_{m_k \subseteq m} G[m_k, m] = G[:,m]^T \Theta_*[:,m] \tag{7.14}$$

$$In(M) = \sum_{m_k \supseteq m} G[m_k, m] = G[:,m]^T \Theta^*[:,m] \tag{7.15}$$

where $\Theta^*[:,m]$ sets all flexible θ values to 1, while $\Theta_*[:,m]$ sets all flexible θ values to 0. While the inclusive and exclusive probability definitions are similar, they are not the solutions to the belief and plausibility.

$$Bel(M) = \min_{\Theta} \boldsymbol{G}[:, m]^T \boldsymbol{\Theta}[:, m] \neq Ex(M)$$

$$Pl(M) = \max_{\Theta} \boldsymbol{G}[:, m]^T \boldsymbol{\Theta}[:, m] \neq In(M)$$

7.3.3 Generalizing Dempster's Rule

As described above (see Section 7.2.3), Dempster's rule of combination is given for BBAs in the following form:

$$S(\boldsymbol{m}_k) = \frac{1}{1-K} \sum_{\boldsymbol{m}_k = \boldsymbol{m}_i \cap \boldsymbol{m}_j \neq \emptyset} S(\boldsymbol{m}_i) S(\boldsymbol{m}_j) \tag{7.16}$$

$$1 - K = \sum_{\boldsymbol{m}_k = \boldsymbol{m}_i \cap \boldsymbol{m}_j \neq \emptyset} S(\boldsymbol{m}_i) S(\boldsymbol{m}_j) \tag{7.17}$$

In a similar manner, the generalized Dempster's rule of combination is applied to the rows of $\boldsymbol{G}$ that pertain to $\boldsymbol{m}_k$ (or equivalently, $G[\boldsymbol{m}_k, :]$)

$$G_{12}[\boldsymbol{m}_k, :] = \frac{1}{1-K} \sum_{\boldsymbol{m}_k = \boldsymbol{m}_i \cap \boldsymbol{m}_j \neq \emptyset} G_1[\boldsymbol{m}_i, :] \circ G_2[\boldsymbol{m}_j, :] \tag{7.18}$$

$$1 - K = \sum_{\boldsymbol{m}_k = \boldsymbol{m}_i \cap \boldsymbol{m}_j \neq \emptyset} \text{mean}_{x \neq 0}(G[\boldsymbol{m}_i, :] \circ G[\boldsymbol{m}_j, :]) \tag{7.19}$$

where $X \circ Y$ denotes the Hadamard (or element-wise) product between X and Y, while $\text{mean}_{x \neq 0}(X)$ is the mean of the nonzero values of X.

The interesting property about the Hadamard product is that it conserves properties of the intersection $x \cap y$ operation. For example, we can see that

$$[\, 0 \ X_2 \ 0 \,] \circ [\, 0 \ Y_2 \ Y_3 \,] = [\, 0 \ X_2 Y_2 \ 0 \,]$$

Analogously, when obtaining sets from these row vectors, we can see that

$$m_2 \cap \{m_2, m_3\} = m_2$$

When taking this into account, we can see that the generalized Dempster's rule of combination truly generalizes Dempster's rule. For example, let us consider two BBA values for $\{m_1, m_2\}$ and $\{m_2, m_3\}$. In Dempster's rule, we could allocate the product

$$S\{m_1, m_2\} S\{m_2, m_3\}$$

to mode m_2. In the generalized Dempster's rule, the Hadamard product

$$[S\{m_1, m_2\}, S\{m_1, m_2\}, \ 0] \circ [0, S\{m_2, m_3\}, S\{m_2, m_3\}]$$
$$= [0, S\{m_1, m_2\} S\{m_2, m_3\}, 0]$$

is allocated to mode m_2 (which one can see from the product result). When the GBBA is consistent with a Dempster–Shafer BBA, the Hadamard product serves no purpose as all nonzero elements in the ambiguous mode m_k have the same support, regardless of the unambiguous modes $M \subset m_k$ supported. The same final result can be obtained without Hadamard products as long as the sets and the BBA are known. However, when an ambiguous mode m_k in a GBBA allocates different support to each unambiguous mode $M \subset m_k$, the Hadamard product allows us keep track of how M supports each $M \subset m_k$.

7.3.4 Short-cut Combination for Unambiguous Priors

The generalized Dempster's rule shares another property with Dempster's rule: repeated combination will result in shrinking ambiguity; consequently, combination with a Bayesian prior will result in an unambiguous posterior. In fact, the Bayesian short-cut for the generalized Dempster's rule is the same as the Bayesian short-cut for Dempster's rule. If $p_1(M)$ is a Bayesian prior, which is combined with a GBBA (G_2), the resulting GBBA (G_{12}) can be expressed as

$$G_{12}(m_i, m_i) = \frac{1}{1 - K} p_1(m_i) In_2(m_i) \tag{7.20}$$

$$G_{12}(m_i, m_{j \neq i}) = 0$$

$$1 - K = \sum_M p_1(M) In_2(M) \tag{7.21}$$

This can be shown by analysing the generalized rule of combination. However, we first need to define $p_1(M)$ as a GBBA

$$G_1 = \begin{bmatrix} p_1(m_1) & 0 & 0 \\ 0 & \ddots & 0 \\ 0 & 0 & p_1(m_n) \end{bmatrix}$$

From the generalized rule,

$$G_{12}[m, :] = \frac{1}{1 - K} \sum_{m = m_i \cap m_j \neq \emptyset} G_1[m_i, :] \circ G_2[m_j, :]$$

From this we can see that the condition $m = m_i \cap m_j \neq \emptyset$ is only satisfied when $m_i = m$ and when $m_j \supseteq m$. This allows us to factor out $G_1[m, :]$, which yields

$$G_{12}[m, :] = \frac{1}{1 - K} G_1[m, :] \sum_{m_j \supseteq m} G_2[m_j, :]$$

We can see that for row 1 of G_1, the form is $[X, 0, \ldots, 0]$, which means that the first row of G_{12} will also take the form $[X, 0, \ldots, 0]$, where all elements other than the first one are zero.

In the same manner, the second row of G_1 and G_{12} will take the form of $[0, X, 0, \ldots, 0]$, so on and so forth. From this we can see that

$$G_{12}[m, m] = \frac{1}{1 - K} G_1[m, m] \sum_{m_j \supseteq m} G_2[\boldsymbol{m}_j, m]$$

$$G_{12}[m_i, m_{j \neq i}] = 0$$

Now one can observe that the summation term is identical to that of inclusive probability described in Eqn (7.14), and that $G_1[m, m] = p_1(M)$. Thus,

$$G_{12}[m, m] = \frac{1}{1 - K} p_1(M) In_2(M)$$

$$G_{12}[m_i, m_{j \neq i}] = 0$$

where the normalization constant ensures that the diagonal sum of G_{12} is 1.

$$1 - K = \sum_M p_1(M) In_2(M)$$

This result provides a quick method to combine GBBAs with a Bayesian prior. In addition, the result validates the consistency of the generalized Dempster's rule with Dempster's original rule.

Dynamic application

When using the short-cut rule, the posterior result is always Bayesian, thus the transition rule is still the same

$$G^t(M, M) = \sum_k p(M^t | m_k^{t-1}) G^{t-1}(m_k, m_k)$$

After Dempster's rule was applied at a time step $t - 1$, the resulting probability $G^{t-1}(M, M)$ can be converted to the prior $G^t(M, M)$ at time t using the transition rule. Dempster's combination rule is then used to update $G^t(M, M)$ with more evidence.

7.4   Notes and References

The majority of the content in this chapter stems from Gonzalez and Huang (2013), which gives a brief overview on the motivation and methodology of the generalized Dempster–Shafer method. The combination rule in this chapter was based on Dempster's rule, with its well-known format originally occurring in Shafer (1976). In addition, other combination rules are available, but all are based on the traditional BBA. Stentz and Ferson (2002) gives an overview of many popular combination rules.

References

Dempster A 1968 A generalization of Bayesian inference. *Journal of the Royal Statistical Society. Series B* **30**(2), 205–247.

Gonzalez R and Huang B 2013 Data-driven diagnosis with ambiguous hypotheses in historical data: a generalized Dempster–Shafer approach. *Proceedings from the 16th International Conference on Information Fusion.*

Shafer G 1976 *A Mathematical Theory of Evidence*. Princeton University Press.

Stentz K and Ferson S 2002 *Combination of evidence in Dempster Shafer theory*. Technical report. Sandia National Laboratories, Albuquerque, New Mexico 87185 and Livermore, California 94550.

8

Making use of Continuous Evidence Through Kernel Density Estimation

8.1 Introduction

When discussing material in Chapters 6 and 7, it was assumed that evidence was discretized in order to construct alarms. However, in the case of process monitors, the raw evidence is often continuous and discretization can result in the loss of valuable information. It is shown in this chapter that, for a single monitor, discretization can be optimized to yield exactly the same result as continuous methods, but that in higher dimensions optimal discretization can be a challenge as the optimal regions can take on strange shapes.

Often times, parametric methods are used to estimate continuous distributions. However, parametric methods make assumptions about distribution shape. In the case of process monitors, the distributions can take very unusual shapes. For example, in certain situations the control performance monitor results from the filtering and correlation (FCOR) algorithm by Huang and Shah (1999) will have bimodal behaviour with peaks near 0 and 1. The Gaussian mixture model approach is parametric but uses multiple Gaussian distributions to approximate the data distribution. The challenge for the Gaussian mixture model is that one is required to know beforehand how many Gaussian distributions are required, and estimation algorithms (such as EM) are not guaranteed to converge to the globally optimal solution. Furthermore, while Gaussian mixture distributions model multi-modal distributions quite well, there is still difficulty with distributions exhibiting nonlinear behaviour between variables.

This chapter discusses kernel density estimation, a density estimation technique for continuous variables, which is nonparametric. The main advantage of kernel density estimation is that it naturally follows the shape of the data and can adequately model the distributions regardless of the shape taken. Furthermore, the process of obtaining a kernel density estimate is not iterative, but is obtained in a single step; because of this, kernel density estimates yield the same consistent result for a single dataset. However, much like the discrete method, the kernel density method suffers from the curse of dimensionality: performance will degrade in

Process Control System Fault Diagnosis: A Bayesian Approach, First Edition. Ruben Gonzalez, Fei Qi and Biao Huang.
© 2016 John Wiley & Sons, Ltd. Published 2016 by John Wiley & Sons, Ltd.

higher dimensions. Nevertheless, the performance degrades at a slower rate for kernel density methods.

This chapter discusses many important aspects of kernel density estimation, including performance relative to discrete methods, how to obtain the critically important bandwidth parameter, and how to reduce dimensionality.

8.2 Performance: Continuous vs. Discrete Methods

In previous chapters, discrete evidence was used mainly because of its ease of interpretation and the fact that any distribution can be discretized to estimate the probability distribution. Discretization is a nonparametric method in the sense that it does not require knowledge of the distribution shape in order to approximate it, although discrete distributions are categorical and have parametric properties such as the ability to perform Bayesian parameter updating. The drawback of discretization, however, is that it results in a loss of information. By contrast, kernel density estimation is a method that is applicable to continuous distributions and suffers much less loss of information. It is also nonparametric, able to fit any type of distribution, which is the one key advantage of discrete methods. A comparison between discrete and kernel density methods is shown in Table 8.1.

The advantages of the kernel density approach over the discrete approach are intuitive, as there is much less loss of information. Mathematically, it has been shown that kernel density estimation converges to the true probability density function faster than discretization (Wand and Jones 1995). However, most readers may not be concerned so much with the accuracy of the density function estimate as with the false diagnosis rate. The material that follows in this section will help readers understand the diagnosis performance characteristics of both methods.

8.2.1 Average False Negative Diagnosis Criterion

In order to assess diagnostic performance, we must first choose a valid performance criterion. In this chapter, we calculate the *average false negative diagnosis rate* F_N as our performance criterion.

Our criterion takes note of every instance of a false diagnosis. In this chapter, the event that mode M diagnosed (or *chosen*) is denoted $C(M)$. From this, the events of true and false diagnosis can be expressed as

Table 8.1 Comparison between discrete and kernel methods

Discrete	Kernel density
Computationally light (when implemented intelligently)	Computationally heavy (proportional to data)
Suffers from information loss	Does not suffer from information loss
Performance suffers exponentially with increased dimensionality	Performance suffers exponentially with increased dimensionality (but slower than discrete case)

- $\mathcal{C}(M)|M$ represents a (true) diagnosis of M, when M is the true underlying mode.
- $\mathcal{C}(\bar{M})|M$ represents a (false) diagnosis of $\bar{M}$ when M is the true underlying mode (where $\bar{M}$ represents some mode other than M).

The average false diagnosis rate can be calculated as

$$F_N = \sum_{k=1}^{n_m} p(\mathcal{C}(\bar{m}_k)|m_k)p(m_k)$$

where $p(m_k)$ is the prior probability of m_k; a flat prior could be used if the prior probabilities are not known. When using Bayesian methods, which account for prior probability, the diagnosis is based on the maximum posterior probability.

$$\mathcal{C}(m_k) \quad \text{if} \quad m_k = \arg\max_M [p(E|M)p(M)]$$

The posterior probability density function is a likelihood function of E – in other words $p(E|M)$ – weighted by the prior probability $p(M)$. For certain regions of E, the mode M will be diagnosed. Ideally, this is when E closely resembles data in M. Our notation of this region of E is given as

$$E|\mathcal{C}(M) = \text{The region of } E \text{ where } M \text{ is diagnosed}$$

Since E is continuous, we obtain $p(\mathcal{C}(\bar{M})|M)$ by integration over the regions of E where M *is not diagnosed*.

$$p(\mathcal{C}(\bar{M})|M) = \int_{E|\mathcal{C}(\bar{M})} p(E|M)p(M) \ dE$$

This formal definition of F_N requires integration, which can be quite difficult to perform. In practice, the value of F_N is much easier to estimate using Monte-Carlo simulations. This approach can be performed by taking the following steps:

1. Use training data to construct likelihoods $p(E|M)$.
2. Obtain validation data in proportion to the prior probabilities, for example if the probability of m_1 is 40%, then 40% of the data must come from m_1.
3. Go through each of the data points E^t, taking note of its true mode M^t, and evaluate the posterior probability

$$p(M|E^t) = \frac{p(E^t|M)p(M)}{p(E^t)}$$

4. Find the mode with the maximum posterior probability, and set it as the diagnosed mode $\mathcal{C}(M)$.
5. In this step, we tally the diagnosis results. So if E^t is correctly diagnosed we add one to the tally of correct results n_{cor}

$$n_{cor} = n_{cor} + 1 \quad \text{if} \quad \mathcal{C}(M) = M^t$$

Otherwise, if E^t is incorrectly diagnosed, we add one to the tally of incorrect results n_{inc}

$$n_{inc} = n_{inc} + 1 \quad \text{if} \quad \mathcal{C}(M) \neq M^t$$

6. After tallying results for all E^t, the false negative rate F_N can be obtained using a simple quotient

$$F_N = \frac{n_{inc}}{n_{inc} + n_{cor}}$$

8.2.2 *Performance of Discrete and Continuous Methods*

When evaluating performance of methods between continuous and discretized evidence, one will arrive at two conclusions:

1. Continuous methods perform better than discrete methods, unless discrete methods are optimized in terms of boundary selection, at which point their performance is equal.
2. In a one-dimensional system, the optimal discretization boundary is easy to obtain, but in higher dimensions, defining the optimal discretization boundary can be difficult, if not infeasible.

Note that neither of these conclusions suggest that kernel density estimation provides a more accurate measure of the probability distribution than discrete methods, as proved in Wand and Jones (1995). In this case we are more concerned about diagnostic performance, where, theoretically, performance of discrete and continuous methods can be the same given an optimal discretization scheme. However, practically speaking, achieving an optimal discretization scheme can be extremely difficult. Since kernel density estimates yield better probability distribution measures than discrete methods, one does not need to worry about constructing optimal boundaries; performance is close to optimal by merit of a better density estimate. Furthermore, in practice kernel density estimation is used when knowledge of the distribution shape is unavailable. Since knowledge of the distribution shape is required for optimal discretization, kernel density estimation is much more widely applicable.

When Discrete Methods Perform as Well as Continuous

The discrete method performs as well as the continuous method when the discretization regions coincide with $E|\mathcal{C}(M)$

$$E|\mathcal{C}(M) = \text{The region of } E \text{ where } M \text{ is diagnosed}$$

As an example of defining optimal cutoff boundaries, let us consider a two-mode system in which each mode is Gaussian, having the distributions given in Figure 8.1(a). When continuous methods are used, we diagnose the mode that has the highest probability: if Mode 1 has a higher density function than Mode 2, Mode 1 is diagnosed. From this example, we can see that the diagnosis region for $E|\mathcal{C}(m_1)$ is where $E < 0$ and the region for $E|\mathcal{C}(m_2)$ is where $E \geq 0$. Let us consider a case where $E < 0$. Even if Mode 1 is more probable, it may not be the true mode because Mode 2 has nonzero probability in this region. In Figure 8.1(b), the overlapped regions are shaded in a darker color, and these regions represent the false negative rates F_N. If this region is small, we have a low probability of misdiagnosis.

If we consider the discrete case, we must create discretization boundaries to analyse the probability distribution. The best boundaries for discretization are the ones set by the

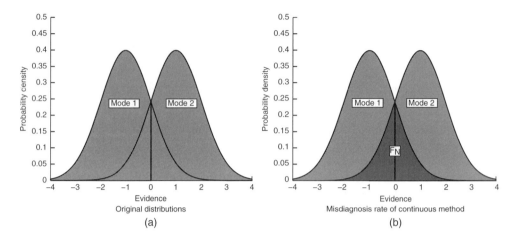

Figure 8.1   Grouping approaches for kernel density method

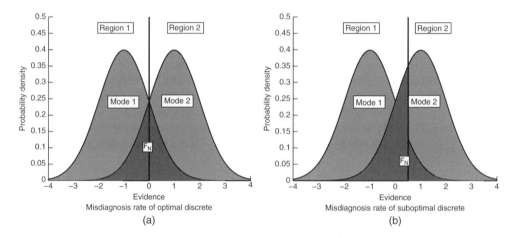

Figure 8.2   Discrete method performance

continuous distribution: in this example, we have one discrete region where $E < 0$ and one discrete region where $E \geq 0$. The resulting false negative rate is shown in Figure 8.2(a), which is identical to the continuous false diagnosis rate.

Now let us consider the case in which we have no knowledge of the continuous distributions. In this case we set the discretization regions to a new arbitrary location, where one region occurs at $E < 0.5$ and another region occurs at $E \geq 0.5$, as shown in Figure 8.2(b). Values of E to the left of this boundary will diagnose m_1 and values of E to the right of this boundary will diagnose m_2. When the boundary is shifted to this location, we can see that the shaded region representing F_N has somewhat grown. This is because when shifting this boundary from 0 to 0.5 the probability of falsely diagnosing Mode 2 (the right-hand part of the dark region) decreased, but at a rate slower than the probability of falsely diagnosing Mode 1 (the left-hand part of the dark region) grew.

From this example we can see that the optimal discretization region is given by the continuous distributions, and that shifting this region in any way will result in an increase in false diagnosis rates. Discrete methods can perform as well as continuous methods when optimally discretized, even when their estimate of the probability density is inferior to the real density. However, optimal discretization requires knowledge of the continuous distributions, which begs the question as to why one would discretize the evidence in the first place. A short answer to this question is that the motivation for discretization is not for performance but for computational simplicity.

In addition to comparing performance between discrete and continuous cases, this example illustrates a useful procedure for optimal discretization; that is, to analyse the continuous distributions – either defined parametrically or through kernel density estimation – and note the regions where the continuous density for a target mode M is highest. This approach, however, is only straightforward in one-dimensional cases because discretization boundaries can be more complex in higher dimensions.

The Feasibility of Optimal Discrete Methods in Higher Dimensions

In cases of dependent evidence E in dimensions two or higher, optimal discretization becomes a much more complicated endeavour than in our previous example, as discretization regions become more difficult to define, as they can be nonlinear.

As an example of the difficulty of discretization, consider the two-dimensional three-mode system in Figure 8.3. In this case, we have three Gaussian distributions, and two of them are quite correlated. This type of behaviour can occur quite easily in practice.

The challenge with this system is that it is easiest to define discretization boundaries one at a time for each piece of evidence; this is called the element-wise approach. In this way, the discretization boundaries will be linear and follow the direction of the axes as seen in Figure 8.4(a). However, as one can see, this discretization scheme is quite suboptimal. When boundaries are drawn in this manner, Region 1 will diagnose Mode 1, Regions 2 and 4 will

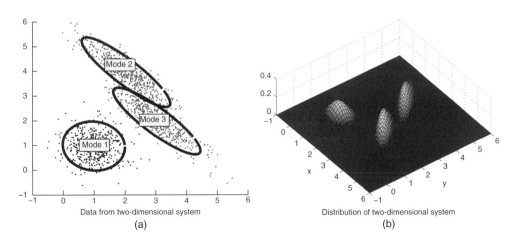

Figure 8.3 Two-dimensional system with dependent evidence

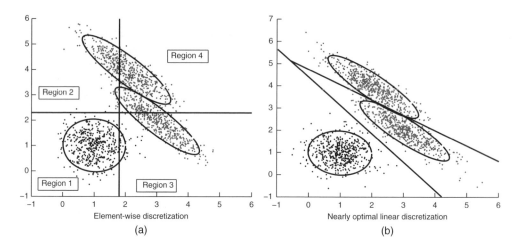

Figure 8.4   Two-dimensional discretization schemes

diagnose Mode 2, and Region 3 will diagnose Mode 3. This scheme will yield a false negative rate of $F_N \approx 0.1233$. However, one can see that in this figure, the modes are actually quite well separated; the kernel density method has a false negative rate of $F_N = 0.0020$.

If one allows a more complex method to define the boundary one can use visual inspection to draw linear boundaries, as done in Figure 8.4(b). In such a case, the false diagnosis rate will be much closer to optimal $F_N = 0.0025$. However, optimally defining linear boundaries in higher dimensions becomes a very difficult task, especially when data takes on nonlinear shapes.

8.3   Kernel Density Estimation

8.3.1   From Histograms to Kernel Density Estimates

Kernel density estimation is a nonparametric method used to estimate continuous probability density functions. The process of arriving at a kernel density estimate from discrete frequency data is an intuitive one. First, let us consider Figure 8.5, the histogram visualization of a distribution, which is synonymous to discretization. In this case, we divide the x-axis into discrete segments and we count the frequency of data points residing in each bin.

In the histogram, we divide x into discrete segments, but what if we allowed the segments to be centred around each data point? This would mean that, centred around each data point, we would place a rectangular function with area $1/(n \times \text{bin width})$, so that the distribution integrated to one. After summing these rectangular functions, the end result – the centred histogram – would be slightly smoother than the histogram in Figure 8.6.

The centred histogram is actually a kernel density estimation. The kernel in this case is a block with area $1/(n \times \text{width})$. Instead of using this kernel, we could choose a smoother one, for example a Gaussian density function divided by n. This would yield the result in Figure 8.7.

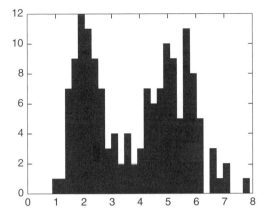

Figure 8.5　Histogram of distribution

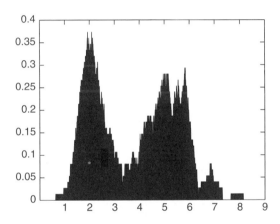

Figure 8.6　Centered histogram of distribution

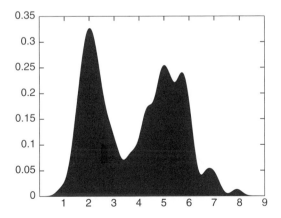

Figure 8.7　Gaussian kernel density estimate

8.3.2 Defining a Kernel Density Estimate

From the example above, a kernel density estimate takes a set of sampled data points and places a kernel function over it, centred around each data point. As shown in Figure 8.8, the kernel functions (denoted by the blue dotted line), centred around five data points (denoted by the black dots on the x-axis) will sum to yield a smooth function (denoted by the solid black curve).

Mathematically, the kernel density estimation procedure can be defined as

$$f(x) \approx \frac{1}{n} \sum_{i=1}^{n} \frac{1}{|H|^{1/2}} K(H^{1/2}(x - D_i)) \tag{8.1}$$

where D denotes a multivariate dataset with n entries, $K(x)$ represents the kernel function, and H represents the bandwidth, which will be discussed in more detail in Section 8.3.4. The kernel function itself can take on many forms so long as it is nonnegative and the integral over the entire domain is equal to one.

$$\int_{-\infty}^{\infty} K(x) \, dx = 1$$

The two most popular kernels are

1. the Epanechnikov kernel, due to its asymptotic efficiency
2. the standard multivariate normal kernel, due to its excellent differentiation properties and its ease of application in higher dimensions.

In our applications, we will use the standard multivariate normal kernel.

$$K(z) = \frac{1}{\sqrt{(2\pi)^d}} \exp\left(z^T z\right)$$

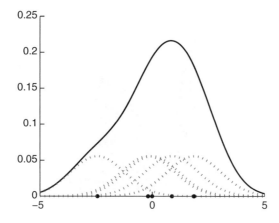

Figure 8.8   Kernels summing to a kernel density estimate

where d is the dimensionality of the data. Using this kernel, the kernel density estimate takes the following form:

$$f(x) \approx \frac{1}{n} \sum_{i=1}^{n} \frac{1}{\sqrt{(2\pi)^d |H|}} \exp\left([x - D_i]^T H^{-1} [x - D_i]\right) \tag{8.2}$$

This estimate is nonparametric, but its smoothness hinges on the bandwidth parameter H. In the same way that the width of bins in a histogram affects its smoothness, the kernel bandwidth H will affect the smoothness of the kernel density estimate.

When performing Bayesian diagnosis, the kernel density estimate is used to express the likelihood term

$$p(E|M) = f(E|M)$$

$$p(M|E) = \frac{p(E|M)p(M)}{\sum_M p(E|M)p(M)}$$

8.3.3 Bandwidth Selection Criterion

Selecting the bandwidth is not a trivial problem. The goal is to select a bandwidth that minimizes the asymptotic mean integrated square error (AMISE). One might recall the popular mean squared error (MSE) criterion

$$MSE = E\left[(\hat{f}(x) - f(x))^2\right]$$

where in this case $\hat{f}(x)$ is the kernel density estimate and $f(x)$ is the real density estimate. The MISE integrates this error over all values of x

$$MISE = \int E\left[(\hat{f}(x) - f(x))^2\right] dx$$

The MISE is generally intractable, hence the asymptotic approximation is used instead. In the univariate case, the AMISE is given by Wand and Jones (1995) as

$$AMISE = \frac{1}{4}h^4 R(f'')\mu_2(K)^2 + (nh)^{-1}R(K) \tag{8.3}$$

where n is the number of data points and h is the bandwidth – a scalar for the univariate case – and where

$$R(K) \equiv \int K^2(z) \, dz$$

$$\mu_2(K) \equiv \int z^2 K(z) \, dz$$

Wand and Jones (1995) also defined a multivariate version of the AMISE criterion, which is given as

$$AMISE\left[\hat{f}(x; h)\right] = \frac{1}{4}\mu_2(K)^2 \left[\text{vech}^T(H)\right] \Psi \left[\text{vech}(H)\right] + \frac{R(K)}{n\sqrt{|H|}} \tag{8.4}$$

where

$$\mu_2(K) = \int_{\mathbb{R}^d} z_i^{\ 2} K(z) \ dz \qquad \forall i$$

$$R(K) = \int_{\mathbb{R}^d} K(z)^2 \ dz < \infty$$

$$\Psi = \int_{\mathbb{R}^d} \mathrm{vech}(M) \mathrm{vech}^T(M) \ dz$$

$$M = 2\mathrm{D}^2[f(z)] - \mathrm{dg}\left(\mathrm{D}^2[f(z)]\right)$$

In the multivariate AMISE criterion expression, $\mathrm{vech}(H)$ takes the lower diagonal of H and strings it out column-wise into a vector. In addition, $\mathrm{D}^2[f(z)]$ is the Hessian of $f(z)$, and the operator $\mathrm{dg}(A)$ sets all off-diagonal elements of A to zero, the equivalent of the `diag(diag(A))` command in MATLAB.

8.3.4 Bandwidth Selection Techniques

The bandwidth selection criterion makes the assumption that we know the real density function $f(x)$. Obviously, if we knew the real density function $f(x)$, we would not need a kernel density estimate. The main idea of using the AMISE criterion, however, is to select an optimal bandwidth for a distribution we know; this bandwidth will be close to optimal for similar distributions. In practice, the concern with selecting appropriate bandwidths has more to do with the amount of data and its general spread than it has to do with the specifics of the distribution.

Optimal Bandwidths for Multivariate Normal Distributions

The most popular method to select bandwidths is to use the optimal kernel density estimate for multivariate normal distributions. This bandwidth is defined as

$$H_N = \left(\frac{4}{n(d+2)}\right)^{\frac{2}{d+4}} \Sigma$$

where Σ is the covariance matrix of the target multivariate normal distribution, n is the number of data points and d is the dimension. Now in practice Σ is not available to us, but the sample covariance matrix S can be easily obtained, resulting in the following bandwidth selection:

$$H_N = \left(\frac{4}{n(d+2)}\right)^{\frac{2}{d+4}} S$$

If the relationships between the variables is linear, it is best to use the full covariance matrix estimate S. However, if the variables exhibit nonlinear relationships, one may wish to set the off-diagonal elements of S to zero so that the diagonal elements are all that remain.

Adaptive Bandwidth Estimation Techniques

One problem with choosing a single bandwidth is that the distribution tends to be over-smoothed near the peaks of the distribution but under-smoothed near the tails. This is analogous to the problem of histogram bins being well-estimated in regions of high probability but very sparse and disjointed in regions of low probability.

In order to improve performance in these extreme cases, the bandwidth is modified so that the kernel function's height is proportional to the probability at that point. The reasoning behind this is that if the kernel is uniform, each kernel is expected to have a similar number of data points within its domain.

The bandwidth height is modified by setting a localized scalar λ_i in front of every individual bandwidth

$$H_i = \lambda_i^{-2/p} H_N \tag{8.5}$$

The scaling parameter λ_i can be calculated using a 'pilot' density estimate $\hat{f}_{H_N}^p(x)$, which is obtained using the optimal normal bandwidth H_N:

$$\lambda_i = \left(\frac{\hat{f}_{H_N}^p(D[i])}{g} \right)^\alpha$$

where α is a user-defined parameter and g is the geometric mean of $\hat{f}_H^p(D[i])$, such that

$$\log(g) = \frac{1}{n} \sum_i \log\left[\hat{f}_H^p(D[i]) \right]$$

If one wishes the kernel height to be proportional to the probability at that point, the user-defined parameter α should be set to 1; however Abramson (1982) found that $\alpha = 1$ was too aggressive for univariate cases and suggested that in these cases one should set $\alpha = 0.5$, which is a half-way point between the aggressively adaptive $\alpha = 1$ and the nonadaptive $\alpha = 0$. Terrel and Scott (1992) also commented on this phenomenon after rigorous theoretical analysis, suggesting that $\alpha = 0.5$ should also work well in higher dimensions. However, Terrel and Scott (1992) also mentioned that while adaptive kernel density estimation is useful for smaller sample sizes, it converges to the true density more slowly as the sample size increases. Thus for larger sample sizes, smaller values of α should be used.

One can see that setting $H_i = \lambda_i^{-2/p} H_N$ results in a kernel density estimate given as

$$\hat{f}(x) = \frac{1}{n} \sum_{i=1}^n \frac{1}{|H_i|^{1/2}} K\left(H_i^{1/2}(x - D_i) \right)$$

$$= \frac{1}{n} \sum_{i=1}^n \frac{1}{|\lambda_i^{-2/p} H_N|^{1/2}} K\left(H_i^{1/2}(x - D_i) \right)$$

$$= \frac{1}{n} \sum_{i=1}^n \frac{\lambda_i}{|H_N|^{1/2}} K\left(H_i^{1/2}(x - D_i) \right)$$

so that the height of each kernel is modified by λ_i.

8.4 Dimension Reduction

One point of difficulty for kernel density estimation is the problem of dimensionality. Much like its discrete counterpart, the difficulty of estimating kernel densities increases exponentially

Table 8.2   The curse of dimensionality

Dimension	Required data points
1	40
2	84
3	175
4	366
5	765
6	1600

with respect to its dimension; this is referred to as the *curse of dimensionality*. The rate at which the difficulty increases is of the order $O[n^{-4/(d+4)}]$.

As an example, let us consider a one-dimensional system. In order to adequately estimate a kernel density, around 40 data points will be needed; the number of data points required to achieve the same quality of estimate is shown in Table 8.2. As one might observe, for every additional dimension, the quantity of data required is increased by $40^{1/5}$.

Because of this problem, kernel density applications for large systems must always consider a dimension-reduction scheme. In this chapter, we will consider two main schemes: independence assumptions and independent component analysis (ICA).

8.4.1   Independence Assumptions

One method to reduce dimensionality is to introduce independence assumptions. For example, let us consider a six-dimensional system $E = [E_1, E_2, E_3, E_4, E_5, E_6]$. If, for mode M, the first three pieces of evidence can be considered to be independent of the second three, $[E_1, E_2, E_3] \perp [E_4, E_5, E_6]$, then we can calculate the joint probability as a product

$$\text{if } [E_1, E_2, E_3]|M \perp [E_4, E_5, E_6]|M$$

$$p(E|M) = p(E_1, E_2, E_3|M)p(E_4, E_5, E_6|M)$$

Even though the joint probability is six-dimensional, we can break the problem down into two kernel density estimates: $p(E_1, E_2, E_3|M)$ and $p(E_4, E_5, E_6|M)$, which are both three-dimensional distributions. In this way, a six-dimensional problem is reduced to a three-dimensional one.

If one suspects some evidence to be independent, there is a test that can be used to verify whether this assumption can be made. The *mutual information criterion* (MIC) can be used to check for independence.

$$MIC(E_1, E_2) = \int_{E_1} \int_{E_2} p(E_1, E_2|M) \log \left(\frac{p(E_1, E_2|M)}{p(E_1|M)p(E_2|M)} \right) dE_1 \ dE_2$$

The MIC can be calculated numerically using kernel density estimates for $p(E_1, E_2|M)$, $p(E_1|M)$ and $p(E_2|M)$, and then numerically integrating the result over a suitable range of E_1, E_2.

It may not always be possible to break down the evidence into purely independent groups. However, a much more lenient conditional dependence assumption can also result in dimension reduction. Consider a case in which all evidence $E = [E_1, E_2, E_3, E_4]$ is caused by a single underlying factor, which is best observed by E_1. If this is the case, then it is reasonable to break down the dimensions by conditioning with respect to E_1:

$$p(E_1, E_2, E_3, E_4|M) = p(E_1|M)p(E_2|E_1, M)p(E_3|E_1, M)p(E_4|E_1, M)$$

$$= p(E_1|M)\frac{p(E_1, E_2|M)}{p(E_1|M)}\frac{p(E_1, E_3|M)}{p(E_1|M)}\frac{p(E_1, E_4|M)}{p(E_1|M)}$$

By using conditional probability, the highest dimension of this problem has been reduced from four to two. A variation of the MIC (the conditional MIC or CMIC) can be used to test for conditional independence

$$CMIC(E_1, E_2)$$

$$= \int_{E_1}\int_{E_2} p(E_1, E_2|E_r, M) \log\left(\frac{p(E_1, E_2|E_r, M)}{p(E_1|E_r, M)p(E_2|E_r, M)}\right) dE_1\ dE_2\ dE_r$$

$$= \int_{E_1}\int_{E_2} \frac{p(E_1, E_2, E_r|M)}{p(E_r|M)} \log\left(\frac{p(E_1, E_2, E_r|M)p(E_r|M)}{p(E_1|E_r, M)p(E_2|E_r, M)}\right) dE_1\ dE_2\ dE_r$$

where E_r is the reference evidence.

8.4.2 Principal and Independent Component Analysis

In addition to making independence assumptions, one can also attempt to explain the data with respect to a set of independent components. Independent component analysis (ICA) is a generalization of principal component analysis (PCA) and is a useful tool for dimension reduction. Both techniques assume a model in which the data observations y are assumed to be linear combinations of latent variables f

$$y - \mu = At$$

The difference between PCA and ICA is that in PCA the independent latent variables t are assumed to be Gaussian, while in ICA they can follow any distribution, hence the generalization. Both PCA and ICA aim to define the loading matrix A. The procedure for PCA is standard and will not be discussed. For ICA, a variety of algorithms exist, some of which have been discussed in Hyvarinen and Oja (2000); Hyvarinen (2000) provided a MATLAB package entitled *Fast ICA*, with an algorithm that is both computationally efficient and reasonably accurate.

8.5 Missing Values

As in the discrete evidence case, it is possible for some values to be missing from the data. For the discrete case, Bayesian marginalization and the EM algorithm are typically the most popular solutions. For missing values in kernel density estimates, kernel density regression for the missing values is an effective and popular solution.

8.5.1   Kernel Density Regression

The Zeroth-order (Nadaraya–Watson) Method

Kernel density regression was first proposed by Nadaraya (1964), shortly after the widely cited paper on kernel density estimation by Parsen (1962). It can be derived by first noting the following regression function:

$$\hat{y} = g(x) = \frac{\int y f(y, x) dy}{f(x)}$$

We let the joint probability estimate for $f(y, x)$ be

$$\hat{f}(y, x) = \frac{1}{n |H_x|^{1/2} |H_y|^{1/2}} \sum_{i=1}^{n} K\left(H_x^{-1/2}[X_i - x]\right) K\left(H_y^{-1/2}[Y_i - y]\right)$$

$$\hat{f}(x) = \frac{1}{n |H_x|^{1/2}} \sum_{i=1}^{n} K\left(H_x^{-1/2}[X_i - x]\right)$$

where $K(x)$ is a kernel function, such as the standard multivariate normal distribution. The kernel functions are set to be independent between x and y to facilitate the integration over y. Independent kernels do not suggest that x and y are independent, but that the bandwidth matrix H is simply diagonal.

$$\int y \hat{f}(y, x) dy$$

$$= \int y \frac{1}{n |H_x|^{1/2} |H_y|^{1/2}} \sum_{i=1}^{n} K\left(H_x^{-1/2}[X_i - x]\right) K\left(H_y^{-1/2}[Y_i - y]\right) dy$$

$$= \frac{1}{n |H_x|^{1/2}} \sum_{i=1}^{n} K\left(H_x^{-1/2}[X_i - x]\right) \int \frac{y}{|H_y|^{1/2}} K\left(H_y^{-1/2}[Y_i - y]\right) dy$$

$$= \frac{1}{n |H_x|^{1/2}} \sum_{i=1}^{n} K\left(H_x^{-1/2}[X_i - x]\right) Y_i$$

This results in the following estimator for y

$$g(x) = \frac{\frac{1}{n |H_x|^{1/2}} \sum_{i=1}^{n} K\left(H_x^{-1/2}[X_i - x]\right) Y_i}{\frac{1}{n |H_x|^{1/2}} \sum_{i=1}^{n} K\left(H_x^{-1/2}[X_i - x]\right)}$$

$$= \frac{\sum_{i=1}^{n} K\left(H_x^{-1/2}[X_i - x]\right) Y_i}{\sum_{i=1}^{n} K\left(H_x^{-1/2}[X_i - x]\right)} \tag{8.6}$$

This result amounts to a weighted average of historical Y values based on the proximity of the corresponding X values to the query value x; hence it is a *locally weighted average*.

The First-order Method

The Nadaraya–Watson method is a fairly popular method of nonparametric, or kernel density, regression. However, it has been shown to be biased toward flat functions of y with respect to x. In order to reduce this bias, the first-order method was proposed. Here, instead of using a locally-weighted average, we wish to use a locally weighted linear model

$$y_i = \alpha + \beta^T (X_i - x) + \epsilon_i$$

We can arrive at the ordinary least squares solution by setting

$$Z_i = \begin{bmatrix} 1 \\ X_i - x \end{bmatrix}$$

and then performing the following operation to obtain a linear model:

$$\begin{bmatrix} \hat{\alpha}(x) \\ \hat{\beta}(x) \end{bmatrix} = \left[\sum_{i=1}^{n} Z_i Z_i^T \right]^{-1} \left[\sum_{i=1}^{n} Z_i Y_i \right]$$

Due to the way this problem was posed (being centered around x when we defined Z) $\hat{\alpha}(x)$ serves as our estimate of y.

$$\hat{y} = \hat{g}(x) = \hat{\alpha}(x)$$

However, for the kernel density variation, we use the kernel as a weighting function to create a locally weighted linear solution

$$\begin{bmatrix} \hat{g}(x) \\ \hat{\beta}(x) \end{bmatrix} = \left[\sum_{i=1}^{n} K \left(H_x^{-1/2}[X_i - x] \right) Z_i Z_i^T \right]^{-1} \left[\sum_{i=1}^{n} K \left(H_x^{-1/2}[X_i - x] \right) Z_i Y_i \right] \quad (8.7)$$

The locally weighted linear solution does not suffer as much from bias toward flat regression estimates. Higher-order solutions also exist, but their improvement over the first-order method tends to be quite minimal.

8.5.2 *Applying Kernel Density Regression for a Solution*

Let us consider a dataset X wherein some data entries are incomplete

$$X = \begin{bmatrix} X_c \\ X_{ic} \end{bmatrix}$$

Within each incomplete data entry, there are values that are present, denoted z, and values that are missing, denoted y. For each incomplete data entry, $X_{ic}[i]$, we use kernel density regression $\hat{g}(z|X_c)$ on z, based on the complete dataset X_c, to estimate the missing values y.

$$\hat{y} = \hat{g}(z|X_c)$$
$$\hat{X}_{ic}[i] = [z, \hat{y}]$$

We now have a complete data estimate

$$\hat{X}_c = \begin{bmatrix} X_c \\ \hat{X}_{ic} \end{bmatrix}$$

This complete data estimate can either be used as the kernel density estimate or we could perform another iteration of the estimation procedure.

$$\hat{y} = \hat{g}\left(z | \left\{ \hat{X}_c \setminus \hat{X}_{ic}[i] \right\} \right)$$

$$\hat{X}_{ic}[i] = [z, \hat{y}]$$

where $\left\{ \hat{X}_c \setminus \hat{X}_{ic}[i] \right\}$ denotes a set difference (or simply that we remove $\hat{X}_{ic}[i]$ from the dataset $\hat{X}_c$. This iterative scheme converges fairly quickly; it seldom requires more than ten iterations. For most intents and purposes, a single iteration will yield an adequate result.

8.6   Dynamic Evidence

Previously, it was shown that the likelihood for autodependent discrete evidence can be obtained as

$$p(E^t | E^{t-1}, m_k) = \frac{n(E^t, E^{t-1}, m_k)}{n(E^{t-1}, m_k)} \tag{8.8}$$

where $n(E^t, E^{t-1}, m_k)$ is the number of times that E^t, E^{t-1} and m_k are jointly observed in history, while $n(E^{t-1}, m_k)$ is the number of times that $n(E^{t-1})$ and m_k are jointly observed in the history. This solution was modified in the previous chapters to include prior samples in order to prevent possible division by zero.

A very similar solution can be arrived at by using the kernel density method to estimate the likelihood. In order to condition on both E^t and E^{t-1}, the rule of conditioning is applied

$$p(Y | X) = \frac{p(X \cap Y)}{p(X)}$$

By applying the rule of conditioning to the solution given by kernel density estimation

$$p(E^t | E^{t-1}, M^t) = \frac{p(E^t, E^{t-1} | M^t)}{p(E^{t-1} | M^t)} \tag{8.9}$$

This result requires two kernel density estimates to be evaluated:

1. $p(E^t, E^{t-1} | M^t)$, for the likelihood of joint present and past evidence given the mode M^t
2. $p(E^{t-1} | M)$, for the likelihood of past evidence given the mode M^t.

The resulting ratio of likelihoods can be used in the same manner as a likelihood. For example, when used in Bayesian diagnosis, the dynamic solution is applied as follows:

$$p(E^t | E^{t-1}, M^t) = \frac{p(E^t, E^{t-1} | M^t)}{p(E^{t-1} | M^t)}$$

$$p(M^t | E^t, E^{t-1}) = \frac{p(E^t | E^{t-1}, M^t) p(M^t)}{\sum_M p(E^t | E^{t-1}, M^t) p(M^t)}$$

Dimensionality Reduction

The use of dynamic evidence, however, has the problem of adding dimensionality to the data. Thus, when applying a dynamic evidence solution, dimensionality reduction techniques such as ICA and dependence analysis are even more important. Because of the curse of dimensionality, it is desirable to test whether it is necessary to include past evidence in the likelihood. Again, this test can be performed using the MIC, but with a focus on past and present evidence:

$$MIC(E_k^t, E_k^{t-1})$$

$$= \int_{E_k^t} \int_{E_k^{t-1}} p(E_k^t, E_k^{t-1}|M) \log \left(\frac{p(E_k^t, E_k^{t-1}|M)}{p(E_k^t|M)p(E_k^{t-1}|M)} \right) dE_k^t \ dE_k^{t-1}$$

If $MIC(E_k^t, E_k^{t-1})$ is a small number (generally less than 0.2), then we do not need to include past evidence for this particular evidence source E_k.

8.7   Notes and References

This chapter is primarily based on Gonzalez and Huang (2014), which provides a continuous nonparametric version of the techniques described in Qi and Huang (2010) and Pernestal (2007). Material on kernel density estimation was based on work in Wand and Jones (1995) and Duong and Hazelton (2005), including bandwidth selection and the AMISE criterion. The MIC technique was obtained from Tourassi et al. (2001) and Peng et al. (2005). Further reading on ICA is available in Hyvarinen and Oja (2000).

References

Abramson I 1982 On bandwidth variation in kernel estimates – a square-root law. *Annals of Statistics* **10**(4), 513–525.

Duong T and Hazelton ML 2005 Cross-validation bandwidth matrices for multivariate kernel density estimation. *Scandinavian Journal of Statistics* **32**, 485–506.

Gonzalez R and Huang B 2014 Control-loop diagnosis using continuous evidence through kernel density estimation. *Journal of Process Control* **24**(5), 640–651.

Huang B and Shah S 1999 *Performance Assessment of Control Loops*. Springer.

Hyvarinen A, Karhunen, J and Oja, E 2000 *Fast ICA package for MATLAB and Octave*. Helsinki Univ. of Technology, Espoo, Finland.

Hyvarinen A and Oja E 2000 Independent component analysis: algorithms and applications. *Neural Networks* **13**(4–5), 411–430.

Nadaraya EA 1964 On estimating regression. *Theory of Probability and Its Applications* **9**(1), 141–142.

Parsen E 1962 On estimation of a probability density function and mode. *Annals of Mathematical Statistics* **33**(3), 1065–1076.

Peng H, Long F and Ding C 2005 Feature selection based on mutual information: criteria of max-dependency, max-relevance, and min-redundancy. *IEEE Transactions on Pattern Analysis and Machine Intelligence* **27**(8), 1226–1238.

Pernestal A 2007 *A Bayesian approach to fault isolation with application to diesel engines*. PhD thesis. KTH School of Electrical Engineering, Sweden.

Qi F and Huang B 2010 A Bayesian approach for control loop diagnosis with missing data. *AIChE Journal* **56**(1), 179–195.

Terrel G and Scott D 1992 Variable kernel density estimation. *Annals of Statistics* **20**(3), 1236–1265.

Tourassi G, Frenderick E, Markey M and Floyd C 2001 Application of the mutual information criterion for feature selection in computer-aided diagnosis. *Medical Physics* **28**(12), 2394–2402.

Wand M and Jones M 1995 *Kernel Smoothing*. Chapman & Hall/CRC.

9

Accounting for Sparse Data Within a Mode

9.1 Introduction

The Bayesian diagnostic methods discussed so far are all data-driven. To estimate the distribution of evidence, which is required by any Bayesian diagnostic method, sufficient historical samples must be collected. Otherwise, poor diagnostic results are inevitable. As discussed in previous chapters, the dimension of the evidence space will grow exponentially as the number of monitors increases. Thus, a large amount of historical data is required for a reasonable diagnosis in a medium- or large-scale system. However, the fault data in industry can be very sparse. In extreme cases, a fault may only appear once. Estimating the evidence likelihood with a limited number of historical samples is a challenging problem.

Knowing the nominal value of the monitor output, which is obtained from the sparse evidence samples, is nonetheless still not sufficient for the Bayesian diagnosis. Recall that the core of the Bayesian diagnosis is the estimation of the evidence likelihood. Only knowing the nominal value of the evidence will result in a likelihood of:

- one for the discrete bin into which the nominal value falls
- zeros for all the other bins.

Such a likelihood completely removes the uncertainty, and will not fit into the Bayesian diagnostic framework. In order to have a reasonable diagnostic result, the uncertainty should be estimated and incorporated into the diagnostic framework.

The uncertainty of monitor output, either continuous or discrete, originates from disturbance and the finite window length of the monitor calculation: if there is no disturbance, or the segment window for monitor calculation contains infinite process data samples, the monitor output is solely determined by the underlying fault. Otherwise, the uncertainty is unavoidable.

Process Control System Fault Diagnosis: A Bayesian Approach, First Edition. Ruben Gonzalez, Fei Qi and Biao Huang.
© 2016 John Wiley & Sons, Ltd. Published 2016 by John Wiley & Sons, Ltd.

To estimate the uncertainty, the distribution of continuous monitor output needs to be reconstructed. In this chapter, we will show how the monitor output distribution can be estimated using an analytical approach and a data-driven approach with a limited number of historical evidence samples. It should be noted that there is a large number of monitoring algorithms; we will focus on some of the critical ones as examples to demonstrate the proposed distribution function estimation approach. The general idea, which is to establish the relationship between the nominal monitor output value and the monitor output distribution, can be extended to other monitoring algorithms.

9.2   Analytical Estimation of the Monitor Output Distribution Function

The idea to be discussed in this section is about how to derive the monitor output distribution as an analytical function of the nominal monitor output (mean value), such that the distribution function can be estimated with even only one historical evidence sample. Recall that each historical evidence sample is calculated from a section or a number of process data sampled from physical process variables. Therefore, it is possible to derive the distribution of the corresponding evidence sample. In this section, several monitoring algorithms are selected for the analytical derivation of the distribution function to illustrate the proposed approach. These algorithms include the minimum variance control performance monitor (Huang and Shah 1999), the process model monitor (Ahmed et al. 2009) and the sensor bias monitor (Qin and Li 2001).

9.2.1   Control Performance Monitor

The univariate control performance assessment method considered in this chapter is the minimum variance benchmark (Huang and Shah 1999). The filtering and correlation (FCOR) algorithm is employed to compute the control performance index of single controlled variable (CV).

A stable, closed-loop process can be modeled as an infinite-order moving-average process:

$$y_t = (f_0 + f_1 q^{-1} + \cdots + f_{d-1} q^{-(d-1)} + f_d q^{-d} + \cdots) a_t \tag{9.1}$$

where d is the process delay, and a_t is white noise. Multiplying Eqn (9.1) by $a_t, a_{t-1}, \cdots, a_{t-d+1}$, respectively, and then taking the expectation of both sides of the equation yields

$$r_{ya}(0) = E[y_t a_t] = f_0 \sigma_a^2$$

$$r_{ya}(1) = E[y_t a_{t-1}] = f_1 \sigma_a^2$$

$$\vdots$$

$$r_{ya}(d-1) = E[y_t a_{t-d+1}] = f_{d-1} \sigma_a^2$$

The minimum variance or the invariant portion of output variance is (Huang and Shah 1999)

$$\sigma_{mv}^2 = (f_0^2 + f_1^2 + \cdots + f_{d-1}^2) \sigma_a^2 \tag{9.2}$$

We define the control performance index as

$$\eta(d) = \frac{\sigma_{mv}^2}{\sigma_y^2} \tag{9.3}$$

where σ_y^2 can be estimated from the output data.

In Desborough and Harris (1992), the values of the mean and variance of the performance index are estimated as

$$mean[\hat{\eta}(d)] = \eta(d) \tag{9.4}$$

$$var[\hat{\eta}(d)] = \frac{4}{n}(1 - \eta(d))^2 \left[\sum_{k=1}^{d-1}(\rho_k - \rho_{e,k})^2 + \sum_{k=d}^{\infty}\rho_k^2\right] \tag{9.5}$$

where d is the process time delay; ρ_k and $\rho_{e,k}$ represent the output and residual autocorrelations, respectively, and can be calculated from the available process data. Readers are referred to Desborough and Harris (1992) for the derivation details. It is also shown in Desborough and Harris (1992) that the control performance distribution can be approximated by a normal distribution. Therefore, with the nominal value from a single piece of evidence available, the variance of the monitor output can be calculated to give an estimation of the monitor output distribution.

9.2.2   Process Model Monitor

A local output error method has been employed to validate the nominal process model (Ahmed et al. 2009).

For a multi-input single-output (MISO) subsystem, we have

$$Y(s) = G_1(s)e^{-\delta_1 s}U_1(s) + \cdots + G_k(s)e^{-\delta_k s}U_k(s) + V(s) \tag{9.6}$$

where $U_i(s)$ is the the Laplace transform of the ith input, $G_i(s) = \frac{B_i(s)}{A_i(s)}$, $A_i(s) = a_{i,n}s^{n_i} + \cdots + a_{i,1}s + a_{i,0}$, $B_i(s) = b_{i,n}s^{n_i} + \cdots + b_{i,1}s + b_{i,0}$, $Y(s)$ is the Laplace transform of output and the model parameters are given by

$$\theta = [\theta_1^T, \theta_2^T, \cdots, \theta_k^T]^T$$

$$\theta_i = [a_{i,n_i}, \cdots, a_{i,0}, b_{i,n_i}, \cdots, b_{i,0}, \delta_i]^T$$

Defining the overall model output $\hat{y}(t|\theta) = \mathcal{L}^{-1}[\hat{Y}(s)]$ and the output corresponding to the ith input channel $\hat{y}_i(t|\theta) = \mathcal{L}^{-1}[\hat{G}_i(s)e^{-\delta_i s}U_i(s)]$, we have

$$\hat{y}(t|\theta) = \hat{y}_1(t|\theta_1) + \cdots + \hat{y}_k(t|\theta_k) \tag{9.7}$$

The primary residuals and improved residuals are defined as

$$\rho(\theta, x_t) = \varphi(t)(y(t) - \hat{y}_k(t|\theta)) = \varphi(t)e(t, \theta) \tag{9.8}$$

$$\xi_N(\theta) = \frac{1}{\sqrt{N}}\sum_{t-1}^{N}\varphi(t)e(t, \theta) \tag{9.9}$$

where

$$\varphi(t|\theta) = -\frac{\partial \hat{y}(t|\theta)}{\partial \theta} = \begin{pmatrix} -\frac{\partial \hat{y}_1(t|\theta_1)}{\partial \theta_1} \\ \vdots \\ -\frac{\partial \hat{y}_k(t|\theta_k)}{\partial \theta_k} \end{pmatrix} = \begin{pmatrix} \varphi_1(t|\theta_1) \\ \vdots \\ \varphi_k(t|\theta_k) \end{pmatrix}$$

$$\varphi_i(t|\theta_i) = [\underline{\hat{y}}_i^{(n_i)}(t), \cdots \underline{\hat{y}}_i(t), -\underline{\hat{u}}_i^{(n_i)}(t^{*i}), \cdots, -\underline{\hat{u}}_i^{(n_i)}(t^{*i}), \underline{\hat{y}}_i^1(t)]^T$$

$$t^{*i} = t - \delta_i$$

$$\underline{\hat{u}}_i^{(j)} = \mathscr{L}^{-1}\left[\frac{\varepsilon^j}{A_i(s)}U_i(s|\theta)\right]$$

$$\underline{\hat{y}}_i^{(j)} = \mathscr{L}^{-1}\left[\frac{s^j}{A_i(s)}Y_i(s|\theta)\right]$$

The generalized likelihood ratio test is defined as

$$I = \xi_N(\theta)^T \Sigma^{-1}(\theta)\xi_N(\theta)$$

where

$$\Sigma(\theta) = \sum_{t=-\infty}^{\infty} cov(\rho(\theta, x_t), \rho(\theta, x_t))$$

The test value I can be used as model validation monitor for a MISO system. The model monitoring index follows a χ^2 distribution when there is no model plant mismatch and follows a noncentral χ^2 distribution when a mismatch exists. Note that a multi-input, multi-output (MIMO) system can be divided into MISO systems, one for each output, allowing this technique to be easily applied to MIMO systems as well.

When there is a model plant mismatch, the improved residual, of dimension p, will follow a nonzero multivariate normal distribution

$$\xi(\theta) \sim N(\mu, \Sigma) \tag{9.10}$$

where $\mu \neq 0$. As a result, the model monitor index, which is calculated as

$$I = \xi^T S^{-1} \xi \tag{9.11}$$

will no longer follow χ^2 distribution. The distribution of I has to be remodeled.

Let us define d and M, according to Mardia et al. (1979), as

$$d = \Sigma^{-1/2}\xi$$

$$M = \Sigma^{-1/2} S \Sigma^{-1/2}$$

and then define α as

$$\alpha = (n-1)(d'M^{-1}d/d'd)d'd = (n-1)I \tag{9.12}$$

Knowing that M follows the Wishart distribution (Mardia et al. 1979),

$$M \sim W_p(\Sigma, n-1) \tag{9.13}$$

and that its distribution is independent with respect to the distribution of d, the distribution of $\beta = d'd/d'M^{-1}d$ is χ^2_{n-p}. Furthermore, the mean value of d is nonzero, and therefore the distribution of $d'd$ should be a noncentral χ^2 distribution with degree of freedom p, having a noncentral parameter

$$\lambda = n\mu^T\Sigma^{-1}\mu \tag{9.14}$$

Thus, α can be represented as the ratio of a noncentral χ^2 variable and a central χ^2 variable. This results in a noncentral F distribution, with a noncentral parameter λ.

$$\alpha = (n-1)\chi^2_{p,\lambda}/\chi^2_{n-p} = (n-1)p/(n-p)F_{p,n-p,\lambda} \tag{9.15}$$

As such, I can be defined as

$$I = \frac{1}{(n-1)}F_{p,n-p,\lambda} \tag{9.16}$$

According to Evans et al. (2000), the mean value of a noncentral $F_{p,n-p,\lambda}$ distribution is

$$\mu_F = \frac{(n-p)(p+\lambda)}{p(n-p-2)} \tag{9.17}$$

Consequently, we can estimate the mean value of I

$$E[I] = \frac{1}{n-p}\frac{(n-p)(p+\lambda)}{p(n-p-2)} \tag{9.18}$$

In Eqn (9.18), the only unavailable variables are λ and $E[I]$. We can substitute for $E[I]$ with the nominal monitor output. Once $E[I]$, n and p are all known, λ can be calculated using Eqn (9.18), and thereafter the distribution of I can reconstructed as per Eqn (9.16).

9.2.3 Sensor Bias Monitor

The sensor bias monitor discussed here is the algorithm proposed in Qin and Li (2001). A system with a sensor fault can be described by the following state-space model:

$$\begin{cases} x(t+1) &= Ax(t) + Bu(t) + d \\ y(t) &= Cx(t) + y^f + o \end{cases} \tag{9.19}$$

Let us define Y_s as

$$Y_s = [y(t-s), y(t-s-1), \cdots, y(t)]^T$$

where s is the observability index of the system. It should be noted that $y(t-i), i = 0, \cdots, s$ are actual outputs with the normal operating value y_o subtracted. Therefore, knowing the nominal operating point y_o is a necessary condition for sensor fault diagnosis.

For simplicity, we can assume that the state-space model is a minimal realization, say, $s = n$ for a single output system. Eqn (9.19) can be transformed as (Qin and Li 2001)

$$Y_s = \Gamma_x(t-s) + Y_s^f + H_sU_s + G_sD_s + Os \tag{9.20}$$

where

$$\Gamma_x = \begin{pmatrix} C \\ CA \\ \vdots \\ CA^s \end{pmatrix}$$

$$H_s = \begin{pmatrix} 0 & \cdots & \cdots & 0 \\ CB & \ddots & & \vdots \\ \vdots & \ddots & \ddots & \vdots \\ CA^{s-1}B & \cdots & \cdots & 0 \end{pmatrix}$$

$$G_s = \begin{pmatrix} 0 & \cdots & \cdots & 0 \\ C & \ddots & & \vdots \\ \vdots & \ddots & \ddots & \vdots \\ CA^{s-1} & \cdots & \cdots & 0 \end{pmatrix}$$

and where U_s, D_s are all defined similarly to Y_s. Denote

$$Z_s = \begin{bmatrix} Y_s \\ U_s \end{bmatrix} \tag{9.21}$$

Eqn (9.20) can be written as

$$\begin{bmatrix} I & -H_s \end{bmatrix} Z_s = \Gamma_x(t-s) + Y_s^f + G_s D_s + O_s \tag{9.22}$$

Let the characteristic polynomial of A be

$$|\lambda I - A| = \sum_{k=1}^{n} a_k \lambda^k \tag{9.23}$$

According to the Cayley–Hamilton theorem, we know that

$$\sum_{k=1}^{n} a_k A^k = a_0 I + a_1 A + \cdots + a_n A^n = 0 \tag{9.24}$$

Defining Φ as

$$\Phi = \begin{bmatrix} a_0 & a_1 & \cdots & a_n \end{bmatrix}$$

and applying this definition to Eqn (9.24) gives the following result:

$$\Phi \Gamma_s = C(a_0 + a_1 A + \cdots + a_n A^n) = 0$$

Multiplying both sides of Eqn (9.22) by Φ yields

$$\Phi \begin{bmatrix} I - H_s \end{bmatrix} Z_s = \Phi \Gamma_x(t-s) + Y_s^f + G_s D_s + O_s$$
$$= \Phi[Y_s^f + G_s D_s + O_s] \tag{9.25}$$

so that the unknown state $x(t - s)$ is completely removed. Now let us define $e^* = \Phi G_s D_s + \Phi O_s$, which is a scalar value. Then Eqn (9.25) is equal to

$$e = \Phi \begin{bmatrix} I & -H_s \end{bmatrix} Z_s = \Phi Y_s^f + e^* \tag{9.26}$$

where D_s and O_s are process noise and output noise, respectively, both of which follow a normal distribution. From this, we can see that e^* is a linear combination of D_s and O_s, so as a consequence, e^* also follows a normal distribution

$$e^* \sim N(0, R_{e^*})$$

where R_{e^*} is the variance of $e^*(t)$. If there is a sensor bias fault

$$E(Y_s^f) \neq 0$$

$$e = \Phi Y_s^f + e^*$$

will not have zero mean. Therefore, we can use the index $d = e^T R_e^{-1} e$ as an observation of the sensor bias monitor, where R_e is the variance of e.

Similar to the index used on the process model, the sensor bias index follows a χ^2 distribution when there is no sensor bias and a noncentral χ^2 distribution when sensor bias exists. The only difference lies in that the sensor bias monitor is calculated with a scalar residual, while the process model monitor is based on a vector residual. Thus we can consider the distribution of the sensor bias monitor as a special case of the process model monitor when $p = 1$. Following the same procedure as in Section 9.2.2, the distribution of the sensor bias monitor output can be estimated.

9.3   Bootstrap Approach to Estimating Monitor Output Distribution Function

In this section, a bootstrap method is introduced for the purpose of estimating monitor output distributions. In contrast to the analytical approaches discussed in Section 9.2, the bootstrap method provides an empirical estimate of the monitor distribution function. This approach is more useful as many monitors do not have analytical distributions. Furthermore, many monitors that do have analytical distributions make assumptions about the process behaviour that may not be accurate, particularly the nature of disturbances.

One monitor of interest, the valve stiction monitor, can be widely applied but does not have a distribution that can be derived analytically. Because of this, the valve stiction monitor will serve as an example of applying the bootstrap approach.

9.3.1   Valve Stiction Identification

In Lee et al. (2010), a method based on Hammerstein model identification is proposed for the estimation of valve stiction parameters. The process, excluding the sticky valve, is approximated by a linear transfer function model. The valve stiction model introduced by He et al. (2007) is chosen to describe the nonlinearities invoked by the sticky valve.

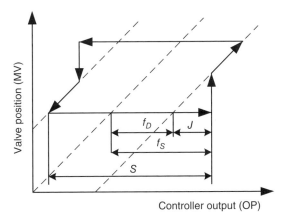

Figure 9.1   Operation diagram of sticky valve

Figure 9.1 shows the operation diagram of a sticky valve, where f_D is the kinetic friction band, f_S is the static friction band, S is the stick plus deadband,

$$S = f_S + f_D \tag{9.27}$$

and J is the slip jump,

$$J = f_S - f_D \tag{9.28}$$

If there is no stiction, the valve movement will follow the dashed line crossing the origin. Any change in the controller output – i.e. input to the valve – is matched exactly by the valve output – i.e. the valve movement. If there is stiction, the valve movement, will follow the solid line in Figure 9.1.

He et al. (2007) proposed a valve stiction model as illustrated by the flow chart in Figure 9.2. u_r is the residual force applied to the valve that has not moved yet; cum_u is an intermediate variable describing the current force acting on the valve based on the cumulative input of $u(t)$.

$$cum\_u(t) = u_r(t-1) + u(t) - u(t-1)$$

If cum_u is larger than the static friction f_S, the valve position will jump to $u(t) - f_D$; if not, the valve will remain in the same position and the residual force u_r will be set to cum_u and this value will be carried over to the next time instant.

Stiction estimation can be considered a Hammerstein model identification problem. The identification of the overall Hammerstein model is performed by a global optimization search for the stiction parameters, in conjunction with the identification of the linear transfer function model. In order to have an effective search for the optimal estimate, the search space needs to be specified. The bounded search space can be defined by analyzing the collected operation data and the relationship of the stiction parameters.

From Figure 9.1, it is noted that

$$f_D + f_S \le S_{max} \tag{9.29}$$

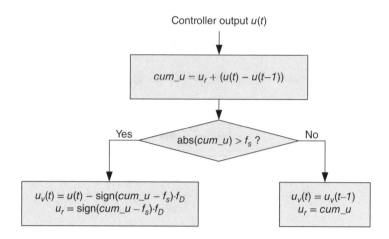

Figure 9.2   Stiction model flow diagram

where S_{max} is the span of OP. Also consider the relation

$$f_S = f_D + J \tag{9.30}$$

so

$$2f_D + J \leq S_{max} \tag{9.31}$$

Figure 9.3 shows the constrained search space for the stiction parameter set (f_S, f_D).

For each point in the stiction parameter space, a series of intermediate valve output data (denoted MV', which is input to the linear dynamic model) can be calculated from the collected OP data as per the stiction model. With MV' and the collected PV data, the linear dynamic model, which is chosen as a first- or second-order plus dead time transfer function,

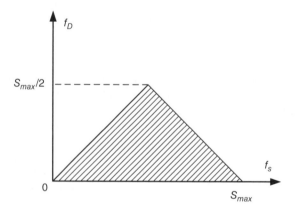

Figure 9.3   Bounded stiction parameter search space

is identified using least squares. A grid optimization method is applied to search within the bounded stiction parameter space for the minimal mean square error so as to obtain the corresponding linear dynamic model and the stiction model parameters.

9.3.2   The Bootstrap Method

The purpose of the bootstrap method is to estimate the distribution of parameter estimators (Chernick 2008). Suppose that we have a set of data $x = \{x_1, x_2, \cdots, x_N\}$ collected from the realizations of a random variable X, which follows the distribution F_X. Let θ be a parameter or statistic of the distribution F_X. $\hat{\theta}$ is a estimator of θ, which can be, for instance, the estimator for the mean value, $\hat{\theta} = \sum_{i=1}^{N} x_i/N$. If the distribution F_X is known and yet relatively simple, it will not be difficult to evaluate the distribution of the parameter estimator. A common example is the variance of the sample mean value of a normal distribution. However, if the distribution is unknown, or is too complicated to evaluate analytically, the bootstrap method provides a good alternative for estimating the distribution of the estimator by resampling the collected data.

Assume that the collected data samples $x = \{x_1, x_2, \cdots, x_N\}$ are independently and identically distributed (i.i.d.). The collected samples $x = \{x_1, x_2, \cdots, x_N\}$ are resampled to construct the bootstrap samples from the same distribution. Often the distribution is unknown and has to be estimated from the original samples.

To construct the bootstrap samples, one must estimate the distribution of the collected sample set and then draw from this distribution. One method to do this is to randomly sample values from the collected data with even probability and with replacement. An optional additional step is to add noise to each sample, which results in *smoothed bootstrapping*. The optimal variance of the added noise is nothing but the optimal bandwidth for the kernel density estimation described in Chapter 8. The bootstrap sample sets have the same size as the original sample set, and can be denoted

$$x_i^b = \{x_{i,1}^b, x_{i,2}^b, \cdots, x_{i,N}^b\} \tag{9.32}$$

where x_i^b is the ith bootstrap sample set, and $x_{i,j}^b$ is the jth sample in the ith bootstrap sample set. Each bootstrap set is considered a new set of data. Suppose that in total M sets of bootstrap samples are collected. For the ith bootstrap set, the bootstrap estimator of θ, $\hat{\theta}_i^b$ can be evaluated. Based on the M sets of bootstrap samples, we can have a group of bootstrap estimators,

$$\{\hat{\theta}_1^b, \hat{\theta}_2^b, \cdots, \hat{\theta}_M^b\} \tag{9.33}$$

With a sufficiently large M, the distribution of parameter estimator $\hat{\theta}$, $F_{\hat{\theta}}$, can be approximated by the distribution of $\hat{\theta}^b$, $F_{\hat{\theta}^b}$, which is determined from the bootstrap estimators $\hat{\theta}_i^b$, $i = 1, 2, \cdots, M$.

An important assumption for the bootstrap is that the data samples that are to be bootstrapped must be i.i.d. This assumption, however, has a clear limitation in control-related applications. If the collected data are dependent, the aforementioned bootstrap method will lead to an incorrect estimation result of $F_{\hat{\theta}}$. In many cases, the data exhibits autocorrelation, and for a closed-loop system with a sticky valve autocorrelation almost always exists. A solution for this is to whiten the data through a time series model and then bootstrap the whitened data (Zoubir and Iskander 2007), as elaborated below.

Suppose that a set of autodependent data $x = \{x_1, x_2, \cdots, x_N\}$ is collected. To estimate the distribution of the estimator $\hat{\theta}$, the first step is to fit the data into a time series model. With the fitted model, simulated model output can be obtained, $\hat{x} = \{\hat{x}_1, \hat{x}_2, \cdots, \hat{x}_N\}$. By subtracting the model output from the collected data, a set of i.i.d. residuals is obtained, $\hat{e} = \{e_1, e_2, \cdots, e_N\}$. Bootstrapping can be performed on these residuals, allowing us to generate M sets of bootstrap residual samples

$$e_i^b = \{e_{i,1}^b, e_{i,2}^b, \cdots, e_{i,N}^b\}, \qquad i = 1, \cdots, M \tag{9.34}$$

where e_i^b is the ith bootstrapped residual sample set and $e_{i,j}^b$ is the jth sample in the ith bootstrapped residual sample set. The new bootstrapped sample sets of x_i can be obtained by adding the bootstrapped residual noise to the simulated model output,

$$x_i^b = \hat{x} + e_i^b \tag{9.35}$$

It has been shown that with the above procedure, the bootstrap distribution is an asymptotically valid estimator of the parameter estimate distribution (Allen and Datta 2002; Kreiss and Franke 1992).

Controller Identification

The process model, which includes the linear dynamic model and the sticky valve model, can be identified by the method outlined in Section 9.3.1. In order to reconstruct the closed-loop model, the controller model is needed. If the controller is known, the closed-loop model can be readily constructed. However, in some scenarios when the controller is unknown, the controller has to be identified.

In this exercise, all controllers attached to the valves are assumed to be proportional integral (PI) or proportional integral derivative (PID) controllers. Thus, the controller model can be written as

$$u(t) = K_p \epsilon(t) + K_i \sum_{i=1}^{t} \epsilon(i) + K_d(\epsilon(t) - \epsilon(t-1)) \tag{9.36}$$

where $u(t)$ is the controller output and $\epsilon(t)$ is the controller error, $\epsilon(t) = r(t) - y(t)$. $r(t)$ is the control loop setpoint and $y(t)$ is the process output.

The PID controller parameters

$$\theta = (K_p, K_i, K_d)^T$$

can be identified with the collected process data. Let

$$U = (u(2)u(3) \cdots u(N))^T \tag{9.37}$$

and

$$X = \begin{pmatrix} \epsilon(2) & \sum_{i=1}^{2} \epsilon(i) & \epsilon(2) - \epsilon(1) \\ \epsilon(3) & \sum_{i=1}^{3} \epsilon(i) & \epsilon(3) - \epsilon(2) \\ \vdots & \vdots & \vdots \\ \epsilon(N) & \sum_{i=1}^{N} \epsilon(i) & \epsilon(N) - \epsilon(N-1) \end{pmatrix}. \tag{9.38}$$

Accordingly we have

$$U = X \cdot \theta \tag{9.39}$$

Following the least squares method, the PID parameters are calculated as

$$\theta = (X^T X)^{-1} X^T U \tag{9.40}$$

In a closed control loop, two equations exist

$$y(t) = H(S, J, u(t))G_p(q^{-1}) + G_l(q^{-1})e(t) \tag{9.41}$$

$$u(t) = G_c(q^{-1})(r(t) - y(t)) \tag{9.42}$$

where $H(S, J, u(t))$ is the sticky valve model, $G_p(q^{-1})$ is the linear dynamic model, $G_l(q^{-1})$ is the disturbance model, $e(t)$ is white noise, $G_c(q^{-1})$ is the controller model and $r(t)$ is the setpoint signal; in most valve stiction scenarios $r(t)$ is 0. These two equations can be written as

$$u(t) = F(S, J, (y(t) - G_l(q^{-1})e(t))G_p^{-1}(q^{-1})) \tag{9.43}$$

$$u(t) = -G_c(q^{-1})y(t) \tag{9.44}$$

where $F(S, J, \cdot)$ is the inverse function of $H(S, J, \cdot)$. The purpose of controller model identification is to find a model that can fit the process data $u(t)$ and $y(t)$,

$$u(t) = -\hat{G}_c(q^{-1})y(t) \tag{9.45}$$

In Eqn (9.43), $F(S, J, \cdot)$ is a nonlinear function due to valve stiction, and the output $u(t)$ is corrupted by noise; in Eqn (9.45) the controller model is linear and noise free, and the model structure is a perfect match of the real controller used, if correctly selected. Identifiability of $G_c(q^{-1})$ can be easily proved.

Although only the PID controller is considered, the discussion of this work can be extended to controllers of other structures by selecting the corresponding controller model structure appropriately. For example, some PID controllers have filters. In this case we can fit a second- or higher-order model, with the constraint that the denominator has a pole that equals one, to account for the integrator. With the controller model, the sticky valve parameters and the linear dynamic model, the closed loop model can be built and we are ready to generate bootstrap data, as discussed below.

Bootstrapping Sticky Valve Parameters

When identifying the valve stiction parameters and linear dynamic model parameters, a set of residuals is generated. However, we cannot bootstrap the identification residuals directly. The reason lies in the fact that the valve stiction identification algorithm outlined in Section 9.3.1 only estimates the linear dynamic model and the valve stiction model; the disturbance model has not been estimated. If the disturbance model is considered, the identification of the linear dynamic model becomes a nonlinear optimization problem (Ljung 1999). Considering the global grid search for stiction parameters and the need to identify the linear dynamic model from each grid search, the overall computation will be formidable. The generated identification residuals are not temporally independent and thus are not i.i.d., so whitening of the residuals is needed.

Let the residual be $e(t)$. Fit the residual by a time series model:

$$e(t) = \frac{C(q^{-1})}{B(q^{-1})} a(t) \qquad (9.46)$$

where $a(t)$ is i.i.d. white noise. Since $a(t)$ is i.i.d., the bootstrap can be applied. Let

$$a(t) \sim \mathcal{N}(\hat{\mu}, \hat{\sigma}^2) \qquad (9.47)$$

where $\hat{\mu}$ and $\hat{\sigma}^2$ can be calculated from the filtered residuals $a(t)$. New sets of bootstrapped $a^b(t)$ are generated from the following normal distribution:

$$a^b(t) \sim \mathcal{N}(\hat{\mu}, \hat{\sigma}^2) \qquad (9.48)$$

The bootstrapped $a^b(t)$ are then passed to the model identified in Eqn (9.46) to get bootstrap residuals $e^b(t)$.

The bootstrapped residuals $e^b(t)$ are added into the closed-loop model, to simulate new process data, including the process output (PV) and the controller output (OP). It should be noted that the newly simulated bootstrap dataset must have the same length as the original one. By using the valve stiction parameter identification method presented in Section 9.3.1, a new set of stiction parameters can be estimated from the resimulated closed-loop response data.

The above procedure must be repeated for a sufficient number of iterations; say, M times. The result is that M different sets of stiction parameters are estimated. With the newly bootstrapped stiction parameters, the distribution of the valve stiction parameters can be determined. The procedure is summarized in Figure 9.4.

9.3.3 Illustrative Example

To verify the proposed procedure, a single-input, single-output (SISO) closed control loop with sticky valve is simulated. The linear part of the process model, together with the disturbance model of Garatti and Bitmead (2009), is

$$y(t) = \frac{q^{-1}}{1 + q^{-1}} u_v(t) + (1 + q^{-1}) a(t) \qquad (9.49)$$

where $u_v(t)$ is the valve output (position) and $a(t)$ is normally distributed white noise, $a(t) \sim \mathcal{N}(0, 0.01)$. A sticky valve model with the structure described in Section 9.3.1 is used to convert the controller output $u(t)$ into a corresponding valve position $u_v(t)$. The valve stiction parameters are $S = 2$ and $J = 1$. The loop is controlled by a PI controller, with parameters $K_p = 0.5$ and $K_i = 0.1$.

In total, 2000 Monte-Carlo simulation runs are performed. Each simulation generates 1000 samples of process data. Based on each single simulation run, the parameters of the linear process model, disturbance model, valve stiction model and the controller are identified. The histograms of the 2000 sets of identified valve stiction parameters are shown in Figures 9.5 and 9.6.

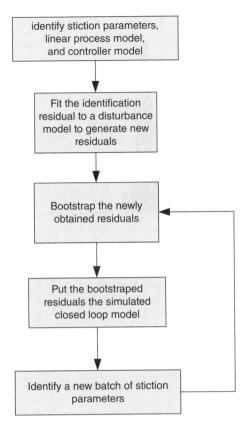

Figure 9.4 Bootstrap method flow diagram

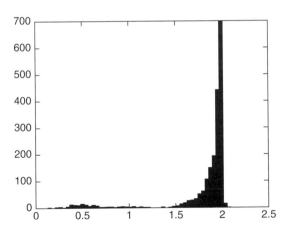

Figure 9.5 Histogram of simulated $\hat{S}$

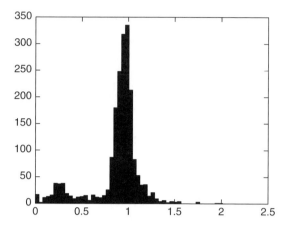

Figure 9.6   Histogram of simulated $\hat{J}$

From each of the simulations, a set of stiction and linear dynamic model parameters is estimated. With any set of:

- identified linear dynamic models
- stiction parameters
- disturbance model
- controller parameters
- whitened residuals

we can generate bootstrap samples to determine the empirical distribution of the identified stiction parameters. Before that, we also need to ensure that the whitened residuals are i.i.d. The autocorrelation coefficients of the filtered 1000 residuals are shown in Figure 9.7 and the histogram of the residuals is shown in Figure 9.8.

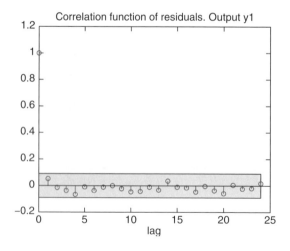

Figure 9.7   Auto-correlation coefficient of residuals

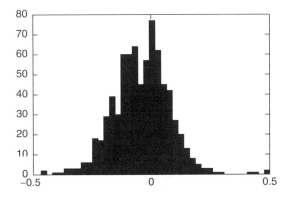

Figure 9.8   Histogram of residual distribution

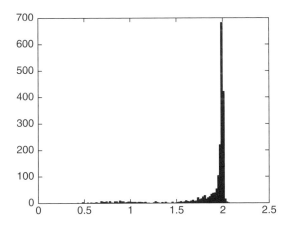

Figure 9.9   Histogram of $\hat{S}^b$

It can be observed that the whitened residuals are uncorrelated and have a distribution close to normal. Thus the filtered residual can be bootstrapped to generate new samples.

The distributions of parameters based on the bootstrap from one set of the identification results are compared with the results from 1000 Monte-Carlo simulations based on true stiction parameters, as presented in Figures 9.9 and 9.10. $\hat{S}^b$ and $\hat{J}^b$ are the stiction parameters estimated from the bootstrap samples based on one of the 1000 Monte-Carlo simulations. Comparing to Figures 9.5 and 9.6, the distributions of bootstrapped parameters $\hat{S}^b$ and $\hat{J}^b$ are close to the Monte-Carlo simulated results. This is also verified by comparing the sample standard deviations, as shown in Table 9.1.

To further quantify the accuracy of the bootstrap estimation, Kullback–Leibler (KL) divergence is employed to measure the distance between the the bootstrap distribution and the simulated distributions. In information theory, KL divergence is a measure of the difference between two probability distributions P and Q (Kullback 1959). It is defined as

$$D_{KL}(P\|Q) = \int_{-\infty}^{\infty} p(x) \log \frac{p(x)}{q(x)} dx \tag{9.50}$$

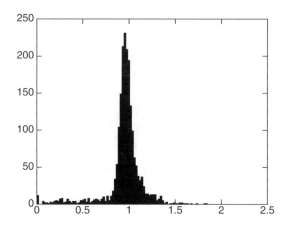

Figure 9.10   Histogram of $\hat{J}^b$

Table 9.1   Comparison of sample standard deviations

	$\sigma_{\hat{S}}$	$\sigma_{\hat{J}}$
Simulated	0.3614	0.2645
Bootstrapped	0.3013	0.2021

where p and q denote the densities of P and Q. While for probability distributions P and Q of a discrete random variable, the KL divergence of Q from P is defined as

$$D_{KL}(P\|Q) = \sum_i P(i) \log \frac{P(i)}{Q(i)} \tag{9.51}$$

Since the analytical density function is unavailable for both the simulated distribution and the bootstrapped distribution, the only way to calculate the KL divergence in the valve stiction case is to consider the two distributions as discrete distributions over a vector X. The elements of X are selected as a finite number of small consecutive intervals within the range of the stiction parameters. $P(i)$ or $Q(i)$ is the frequency over the ith interval. Choosing an interval size of 0.05, the KL divergences of the two sets of distribution pairs are

$$D_{KL}(\hat{S}\|\hat{S}^b) = 1.0285 \tag{9.52}$$

$$D_{KL}(\hat{J}\|\hat{J}^b) = 0.8695 \tag{9.53}$$

which further confirms the good estimation performance of the bootstrap method.

9.3.4   Applications

In this section, several industrial datasets are selected to further investigate the performance of the proposed method. These datasets have been used in Jelali and Huang (2009) and Thornhill et al. (2002) to test stiction detection performance.

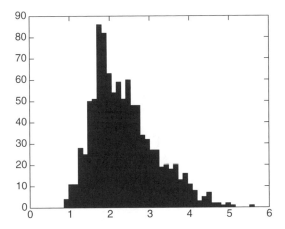

Figure 9.11   Histogram of bootstrapped $\hat{S}^b$ for Chemical 55

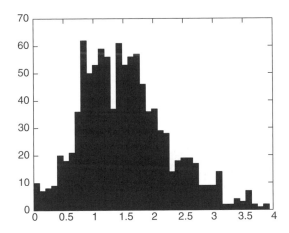

Figure 9.12   Histogram of bootstrapped $\hat{J}^b$ for Chemical 55

One thousand bootstrap datasets are simulated for each loop. The histograms of the boot-strapped parameters are shown in Figures 9.11–9.18. The estimated stiction parameters and 95% confidence intervals (CIs) of the estimated parameters are summarized in Table 9.2. The distribution of stiction parameter estimates provides valuable information to determine the stiction.

According to the data source (Thornhill et al. 2002), valve stiction occurs in the 'Chemical 55' dataset, and the estimated stiction parameters are nonzero for both S and J. The stiction identification methods in earlier sections only provide a point estimation, but not a confidence interval. According to the distribution provided by the bootstrap method, we can observe that the 95% CIs do not include zero. Thus, we can conclude that the valve does have a stiction problem with 95% confidence.

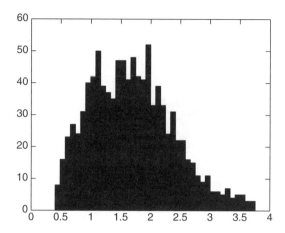

Figure 9.13   Histogram of bootstrapped $\hat{S}^b$ for Chemical 60

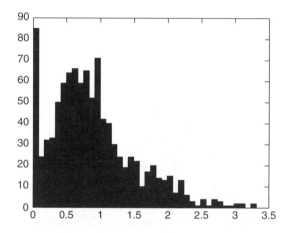

Figure 9.14   Histogram of bootstrapped $\hat{J}^b$ for Chemical 60

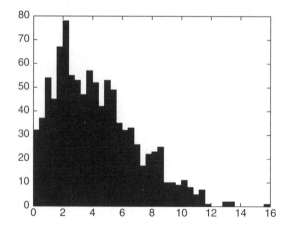

Figure 9.15   Histogram of bootstrapped $\hat{S}^b$ for Paper 1

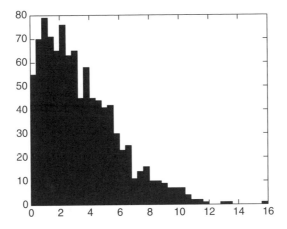

Figure 9.16 Histogram of bootstrapped $\hat{J}^b$ for Paper 1

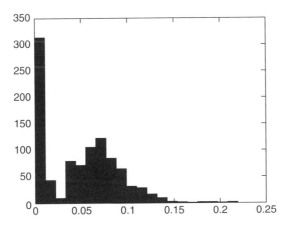

Figure 9.17 Histogram of bootstrapped $\hat{S}^b$ for Paper 9

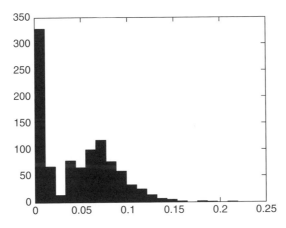

Figure 9.18 Histogram of bootstrapped $\hat{J}^b$ for Paper 9

Table 9.2 Confidence intervals of the identified stiction parameters

	Chemical 55	Chemical 60	Paper 1	Paper 9
Stiction?	Yes	Yes	Yes	No
$\hat{S}$	1.6672	1.8809	4.4006	0.015
$\hat{J}$	0.6669	0.8996	3.5204	0.010
95% CI of $\hat{S}^b$	[1.1795, 4.1668]	[0.5860, 3.2911]	[0.2090, 10.4878]	[0, 0.1306]
95% CI of $\hat{J}^b$	[0.2852, 3.1174]	[0, 2.3051]	[0.0713, 9.6138]	[0, 0.1277]

For the 'Chemical 60' dataset, the bootstrap results show that the 95% CI of $\hat{J}^b$ includes zero, although the point estimation suggests that $\hat{J}$ is nonzero. Therefore we can only conclude with 95% confidence that the valve has a deadband problem, i.e. $S > 0$, but J might be zero.

Both the two 95% CIs of 'Paper 5' do not include zero, indicating that the valve does have a stiction problem with 95% confidence. However, both the CIs of $\hat{S}^b$ and $\hat{J}^b$ are wide. The quantification of the valve stiction is of great uncertainty; in other words the extent of stiction estimated from this set of data may be unreliable.

For the 'Paper 9' dataset, the point estimation algorithm yields nonzero estimates for both $\hat{S}$ and $\hat{J}$, which does not agree with the fact that there is no stiction in the loop (Jelali and Huang 2009). However, in the 1000 bootstrap simulations, over 30% yield zero stiction parameters. The 95% CIs of both $\hat{S}$ and $\hat{J}$ cover zero. Thus zero is still within the possible range of the stiction parameters. When diagnosing the valve problem, even though the point estimate may indicate stiction, it is important to check its confidence interval to avoid misleading conclusions.

It should be noted that the bootstrap method is not only applicable to the valve stiction monitor. During the bootstrap procedure, the controller, valve stiction, process and disturbance models all have parameters to be identified and distributions for these parameters can all be estimated. While more complex and possibly less accurate than analytical methods, the bootstrap method has much greater flexibility; in particular in estimating the distribution of multiple monitors in the same control loop to take into account cross-dependency between monitors.

9.4   Experimental Example

In order to demonstrate the practicality of the monitor distribution estimation technique we have developed, we investigate its use on a real distillation column problem.

9.4.1   Process Description

A distillation column with a diameter of 0.3 m is used to separate a methanol and isopropanol mixture. A schematic diagram of the experimental setup is shown in Figure 9.19. The column contains five identical sieve trays spaced 0.457 m apart. Each tray is made of stainless steel and equipped with thermocouples at liquid sampling points on the tray outlets. The column is made

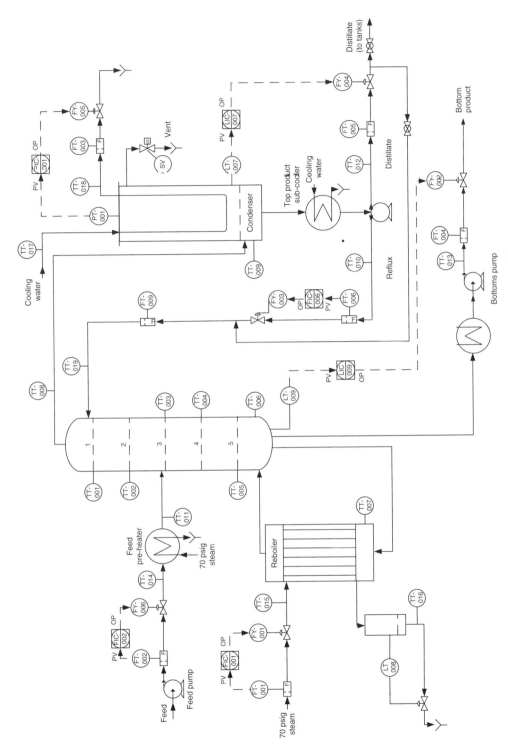

Figure 9.19 Schematic diagram of the distillation column

of Pyrex glass to enable observation of the vapor/liquid phenomena. Detailed dimensions of the column and trays are shown in Table 9.3. The total pressure drop for two trays is measured using a Rosemount differential pressure cell. A total condenser and a thermosiphon partial reboiler completes the distillation system. The column is designed for continuous unattended operation. An Opto-22 process I/O subsystem interface with a personal computer running LabView software is used for process control and data acquisition.

The column is started with total reflux operation and is then switched to continuous mode by introducing feed to the column and withdrawal of two products from the top and bottom of the column. In this study, a total of five different steady-state operating modes are carried out under ambient pressure using the methanol/isopropanol mixture (Olanrewaju et al. 2010). For each operating mode, the column-bottom level and the top reflux drum level are kept constant while the other variables, including feed rate, reflux rate, top pressure and steam rate are varied. Table 9.4 shows the operating variables for the five steady-state operating modes. In Table 9.4, all the process data are normalized for easier computation and comparison. When the flow rate and temperature profiles shown by the software remain constant for a period of 30 min, steady-state conditions are assumed for that particular mode. Liquid samples from each tray outlet and condenser bottom as well as one from the reboiler are taken and analysed to minimize the measurement uncertainty. During the steady-state operation, the sampling period is set to 3 s.

Table 9.3   Dimensions of the distillation column

Column diameter	0.3 m
Tray active area	0.0537 m^2
Hole diameter	4.76 mm
Open hole area	0.00537 m^2
Tray thickness	3.0 mm
Outlet weir height	0.063 m
Inlet weir height	0.051 m
Weir length	0.213 m
Liquid path length	0.202 m
Tray spacing	0.457 m

Table 9.4   Operating modes for the column

Variables	NF Benchmark	m_1 Feed	m_2 Reflux	m_3 Pressure	m_4 Steam
Feed	0.8	0.95	0.8	0.8	0.8
Reflux	0.15	0.15	0.3	0.15	0.15
Pressure	0.3	0.3	0.3	0.4	0.3
Steam	0.2	0.2	0.2	0.2	0.35
Bottom level	0.45	0.45	0.45	0.45	0.45
Top level	0.45	0.45	0.45	0.45	0.45

9.4.2 Diagnostic Settings and Results

Six monitors are commissioned to detect any changes in the process. For the model and sensor bias monitors, we use the variable that has the most significant direct impact on the CV as the model input. For example, the reflux flow rate is the variable used as input for tray 1 (top) temperature. The monitors are as presented in Table 9.5.

For each mode, 5 h of steady-state process data is collected. In total there are 6000 process data samples available for each mode. Every 50 process data samples are segmented for a calculation of one evidence/monitor data sample, resulting in 120 evidence samples for each mode. Out of the 120 evidence samples, 80 are designated as historical samples and the other 40 are used as cross-validation samples. Detailed diagnostic settings are summarized in Table 9.6.

Figure 9.20 shows the diagnostic results in terms of average posterior probability when all historical data samples are available for the five modes. It is observed that all the modes are assigned with the largest posterior probabilities, which are highlighted with gray.

Now assuming that only one historical evidence sample is available for each faulty mode, the proposed estimation techniques for monitor distributions are applied to the six monitors to generate diagnostic results. Monitors that consider the same process variables are considered to be cross-dependent (i.e. the pair of monitors π_1 and π_5, and the pair of monitors π_2 and π_6). The bootstrap approach is employed to estimate the joint distribution of the two pairs. For monitors π_3 and π_4, which do not have direct correlation with other monitors, analytical approaches are applied.

Table 9.5 Commissioned monitors for the column

Monitor	Description
π_1	Control performance monitor for tray 5 temperature
π_2	Control performance monitor for tray 1 temperature
π_3	Control performance monitor for tray 3 temperature
π_4	Control performance monitor for cooling water flow rate
π_5	Model monitor between steam flow rate and tray 5 temperature
π_6	Sensor bias monitor between reflux rate and tray 1 temperature

Table 9.6 Summary of Bayesian diagnostic parameters

Parameter	Summary
Discretization	$k_i = 2$, $K = 2^6 = 64$
Historical data	80 samples for each mode
Prior samples	Uniformly distributed with prior sample, $a_j = 1$, $A = 64$
Prior probabilities	Uniformly distributed for all modes
Evaluation data	40 samples for each mode

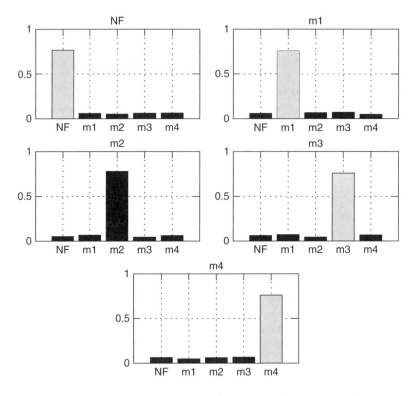

Figure 9.20   Distillation column diagnosis with all historical data

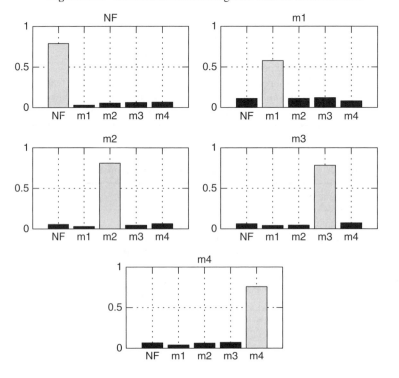

Figure 9.21   Distillation column diagnosis with only one sample from mode m_1

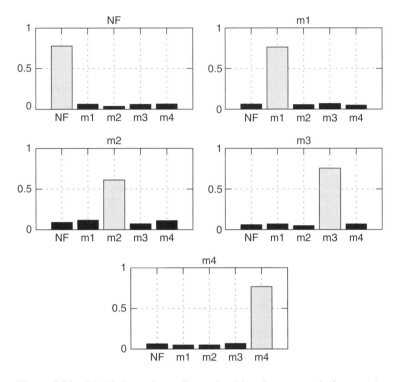

Figure 9.22   Distillation column diagnosis with only one sample from mode m_2

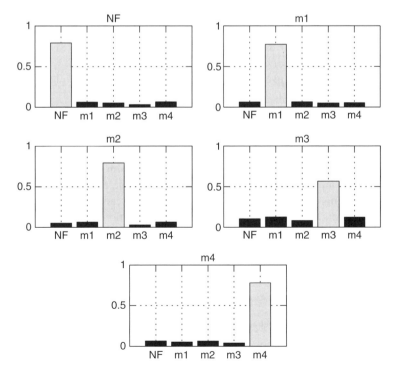

Figure 9.23   Distillation column diagnosis with only one sample from mode m_3

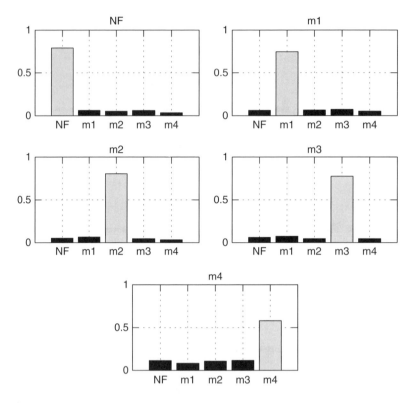

Figure 9.24   Distillation column diagnosis with only one sample from mode m_4

Figures 9.21 to 9.24 show the diagnostic results when only one historical sample is available for modes m_1 to m_4. The title of each subfigure indicates the mode from which the validation data come. It can be seen that even though the average posterior probabilities assigned to the true underlying mode are lower, the correct diagnosis is still made for all the five modes.

9.5   Notes and References

A large amount of literature has been devoted to valve stiction detection; Jelali and Huang (2009) provides a comprehensive summary of the work completed on valve stiction detection and diagnosis. For more details on the bootstrap approach, readers are referred to Chernick (2008) and Zoubir and Iskander (2007). This chapter's section on the bootstrap approach stemmed from Qi and Huang (2011).

References

Ahmed S, Huang B and Shah S 2009 Validation of continuous-time models with delay. *Chemical Engineering Science* **64**(3), 443–454.

Allen M and Datta S 2002 A note on bootstrapping m-estimators in ARMA models. *Journal of Time Series Analysis* **20**, 365–379.

Chernick M 2008 *Bootstrap methods: a guide for practitioners and researchers*, 2nd edn. Wiley-Interscience.

Desborough L and Harris T 1992 Performance assessment measure for univariate feedback control. *Canadian Journal of Chemical Engineering* **70**, 1186–1197.

Evans M, Hastings N and Peacock J 2000 *Statistical Distributions*. Wiley.

Garatti S and Bitmead R 2009 On resampling and uncertainty estimation in linear system identification *Proceeding of 15th IFAC symposium on system identification*, Saint-Malo, France.

He Q, Wang J, Pottmann M and Qin S 2007 A curve fitting method for detecting valve stiction in oscillation control loops. *Industrial Engineering Chemistry Research* **46**, 4549–4560.

Huang B and Shah S 1999 *Performance Assessment of Control Loops*. Springer-Verlag.

Jelali M and Huang B (eds) 2009 *Detection and Diagnosis of Stiction in Control Loops: State of the Art and Advanced Methods*. Springer.

Kreiss JP and Franke J 1992 Bootstrapping stationary autoregressive moving-average models. *Journal of Time Series Analysis* **13**, 297–316.

Kullback S 1959 *Information Theory and Statistics*. John Wiley.

Lee K, Ren Z and Huang B 2010 Stiction estimation using constrained optimisation and contour map. In: *Detection and Diagnosis of Stiction in Control Loops: State of the Art and Advanced Methods*. Springer London.

Ljung L 1999 *System Identification: Theory for the User*, 2nd edn. Prentice Hall.

Mardia K, Kent J and Bibby J 1979 *Multivariate Analysis*. Academic Press.

Olanrewaju M, Huang B and Afacan A 2010 Online composition estimation and experiment validation of distillation processes with switching dynamics. *Chemical Engineering Science* **66**, 1597–1608.

Qi F and Huang B 2011 Estimation of statistical distribution for control valve stiction quantification. *Journal of Process Control* **21**, 1208–1216.

Qin S and Li W 2001 Detection and identification of faulty sensors in dynamic process. *AIChE Journal* **47**(7), 1581–1593.

Thornhill N, Shah S, Huang B and Vishnubhotla A 2002 Spectral principal component analysis of dynamic process data. *Control Engineering Practice* **10**, 833–846.

Zoubir A and Iskander D 2007 Bootstrap methods and applications. *IEEE Signal Processing Magazine* **24**(4), 10–19.

10

Accounting for Sparse Modes Within the Data

10.1 Introduction

In this chapter, we consider how to address the problem of diagnosing a system when modes are missing entirely. Recall that operating modes must include information of all system components. As the number of components increases, the number of possible operating modes will tend to grow exponentially. Because of this, modes that are missing entirely is a very pertinent issue to systems having more than a small number of components, for example an eight component system will have at least $2^8 = 256$ modes.

This chapter discusses two approaches one can take to deal with the problem of missing operating modes.

1. The first approach is to focus on diagnosing the state of each component; this approach is fairly easy to perform in practice, as it is a simple restructuring of the original diagnosis problem.
2. The second approach is to use process modeling and bootstrapping to simulate the missing faulty scenarios; this approach is more difficult to implement in practice as it requires sufficient process knowledge to simulate faulty behaviour. Furthermore, if the system is large, a very large number of simulations will be required in order to cover every possible mode.

The two techniques are independent and do not interfere with each other. Consequently, applying both techniques simultaneously requires no modification of either technique.

10.2 Approaches and Algorithms

This chapter discusses two separate approaches: the first focuses on diagnosis in component space; the second focuses on generating data for unencountered modes.

Process Control System Fault Diagnosis: A Bayesian Approach, First Edition. Ruben Gonzalez, Fei Qi and Biao Huang.
© 2016 John Wiley & Sons, Ltd. Published 2016 by John Wiley & Sons, Ltd.

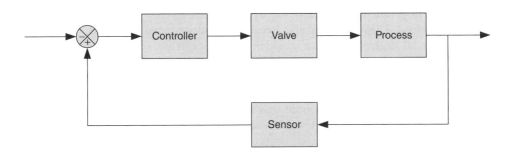

Figure 10.1   Overall algorithm

10.2.1   Approach for Component Diagnosis

Overview

Fault diagnosis focuses on collecting data to estimate the distributions $p(E|M)$. However, if there is a large number of possible modes in M, it will require a large number of data to cover all modes. Some modes will have abundant data, but many of them may have few data or no data at all. By contrast, if we evaluate $p(E|C)$, where C represents the state of the component of interest, there should exist data from each component.

Consider the basic control loop in Figure 10.1. Here, the components of interest are the control valve, the actuator, the sensor and the process (each could have multiple components itself).

The overall aim of component diagnosis is to diagnose the state of each component, and then diagnose the mode. The probability of the state for a component i can be calculated according to Bayes' rule.

$$p(C^i|E) = \frac{p(E|C^i)p(C^i)}{\sum_{C^i} p(E|C^i)p(C^i)}$$

If this system had three states for each component, the total number of modes (and required distributions) would be $3^4 = 81$. By contrast, considering each component one at a time would require the estimation of $4 \times 3 = 12$ distributions, which is a dramatic reduction in terms of the diagnosis space. While it may be a challenge to find data that would correspond to 81 modes (and hence 81 different conditions), it is much easier to find data for all states of a single component. When applied to our example, this would result in only 12 different conditions being needed. This analysis leads to the development of the component diagnosis technique:

1. **Reduction in problem complexity for modes:** The primary reason we consider the component diagnostic technique is that it reduces the number of conditions we have to diagnose. Using the mode-based diagnosis approach, the complexity of the problem grows exponentially with each new component. By contrast, using the component-based diagnosis approach, the complexity of the problem grows linearly with each new component. An eight-component system with two states each will have $2^8 = 256$ modes to diagnose under the mode-based approach; the same system will have $2 \times 8 = 16$ states to diagnose under the component-based approach. Since the component-based approach has one sixteenth as many conditions as the mode-based approach, it needs much fewer data for the diagnosis.

2. **Reduction in problem complexity for evidence:** Each piece of evidence tends to be sensitive toward a few components. Because the mode-based approach considers all components, all available evidence should be used; this can lead to fairly high-dimensional distributions. However, if we adopt the component-based approach, we only need to consider the evidence that is sensitive toward that component; the rest can be discarded. This allows us to effectively reduce the dimensionality of the evidence. Recall that in Chapter 8, evidence dimension space was a problem for kernel density estimation, and even more for discrete methods. Using the component-based approach allows us to reduce evidence dimensionality in a manner that is easier than testing and assuming independence (which was the solution suggested in Chapter 8).

While the component-based approach has its merits, it also has one key drawback. It assumes that the component states are independent of each other, which may not be true if, for example, a problem in one component tends to cause problems in another component. In such cases the mode-based approach will outperform the component-based approach if sufficient data is available for each mode.

Selecting Monitors for Components

When diagnosing between modes, any available sensor or monitor can be helpful as long as it can help distinguish at least one mode from another. Similarly, when diagnosing between components, one chooses a set of sensors or monitors that are sensitive to changes between any states in that component. Because modes include the states of all components, the number of sensors needed to distinguish a single component will tend to be fewer than the monitors needed to distinguish between modes.

In component diagnosis, it is important to select only monitors and sensors that are sensitive to changes in that particular component. Other sensors will be sensitive to other components and including them may result in misleading information, especially if not every mode is realized in the data. The use of fewer sensors and monitors for each component has the added advantage of dimension reduction. In most cases, using the MIC criterion to reduce dimensionality is unnecessary when component-based diagnosis is used. One can use the *false negative criterion* to determine how sensitive a particular sensor or monitor is to a mode. This can be calculated by following a series of steps.

1. Select a component of interest. For example, let us consider a four-component system $[C^1, C^2, C^3, C^4]$, where the first component C^1 is of interest.
2. Search for historical data for situations where only the component of interest changes. One would select modes that have different values in C^1 but the values of C^2, C^3, C^4 remain constant. It is best to select a set of constant values C^2, C^3, C^4 where abundant data is present for all values of C^1. Note that because we are only testing one evidence source at a time the dimension is small, making data requirements easier to meet; 100 data points is a good sample size, but 40 will suffice. This means we can be quite flexible with data collection for this purpose.

3. Group the data according to the different states in C^1; data from each state value in C^1 will be used to evaluate the likelihood $p(E^i|C^1, D)$, where D is the historical data.
4. Select a monitor/sensor of interest E^i and obtain the corresponding data D^i. Evaluate the set of historical data likelihoods $p(D^i|c_k^1, D)$ based on the monitor/sensor of interest.
5. Use the likelihoods from one state $p(D^i|c_k^1)$ to diagnose the state based on the maximum likelihood; $\hat{C}^1$ is the diagnosed state with the maximum likelihood.
6. Determine how frequently the data from state c_k^1 is diagnosed as some other state $n[\hat{C}^1 \neq c_k^1|c_k^1]_{E^i}$ when E^i is used to make the decision. This is referred to as a false negative frequency.
7. Obtain the false negative probability for each component state k by normalizing the false negative frequency

$$P[\hat{C}^1 \neq c_k^1|c_k^1]_{E^i} = \frac{n[\hat{C}^1 \neq c_k^1|c_k^1]}{n[c_k^1]_{E^i}}$$

8. The false negative rate is obtained as

$$FN_{E^i} = \frac{1}{n-1}\sum_{k=1}^{n} P[\hat{C}^1 \neq c_k^1|c_k^1]_{E^i} \tag{10.1}$$

where n is the number of states component C^i can take. $FN_{E^i} = 1$ indicates that E^i is perfectly uninformative, and $FN_{E^i} = 0$ indicates that E^i is a perfect classifier.
9. Based on FN_{E^i}, decide whether or not E^i should be included to estimate the state of the component of interest C^1. Generally, it is best to select E^i when FN_{E^i} is low, for example less than 0.5.
10. Repeat steps 4–9 for other monitors/sensors.
11. Repeat steps 1–10 for other components of interest.

Constructing and Evaluating Probabilities

After the included evidence has been selected for each component, we can construct their respective likelihood functions, either by discrete means or kernel density estimation. The simplest manner to construct the likelihood function $p(E|C)$ is to estimate the function based on data from all modes where C has the desired state. However, modes that occur frequently will be given a lot of weight in this function. It is often better to select the most likely mode where the component state $C(k)$ is true (denoted $M \supset C$), so that

$$p(E|C) = \max_{M \supset C} p(E|M)$$

Note that if all of the evidence E is selected for every component, then the component diagnosis results will be the same as the mode diagnosis results if the component states are independent. The improvement of the component diagnosis result stems from the fact that only evidence sensitive to the component is used to diagnose the component's state. When this is done, it is possible to evaluate any of the possible modes so long as all states for each component are present in the data.

The posterior probability of the component state is obtained using Bayes' theorem

$$p(C|E) = \frac{p(E|C)p(C)}{\sum_k p(E|c_k)p(c_k)} \tag{10.2}$$

where

$$p(C) = \sum_{M \supset C} p(M)$$

One can diagnose the most probable state as the state of component. Once all the components $C^1, C^2, \ldots, C^p$ are diagnosed (as $\hat{C}^1, \hat{C}^2, \ldots, \hat{C}^p$), we can diagnose the mode that contains the appropriate component states.

$$\hat{m} = [\hat{C}^1, \hat{C}^2, \ldots, \hat{C}^p]$$

By assuming all component states are independent, it is also possible to evaluate the posterior probability of the modes

$$p(M|E) = \prod_k p(C^k \subset M|E) \tag{10.3}$$

where $C^k \subset m$ indicates that C^k is contained in the mode M or, equivalently, C^k takes the value specified by that component of M. Obviously, the most probable mode is the one that contains all diagnosed component states.

10.2.2 Approach for Bootstrapping New Modes

Bootstrapping for new modes is done in a manner similar to the technique presented in Qi and Huang (2011). However, in this chapter the underlying fault parameters are varied in order to simulate new scenarios. Considering a control loop with various components, the algorithm can be performed by taking the following steps:

1. Create a model structure for each system component; this should include all relevant model and fault parameters.
2. If there are unknown model parameters, obtain data for model identification – for the most reliable results, open-loop testing – and use gray-box modeling to identify unknown parameters.
3. Obtain residual error information by subtracting predicted output from observed output.
4. Whiten residual errors by identifying an autoregressive (AR) model and applying its inverse to the residual errors.
5. Estimate a kernel density function from the whitened residual errors.
6. Simulate new process data using the identified model by manipulating the fault parameters. Disturbances can be generated by sampling from the kernel density function (smoothed bootstrapping) and applying the AR model to the sampled estimates.
7. Apply monitoring algorithms to both simulated and real process data. Monitor results are used as evidence data for training the likelihood.

Step 1: Create Model Structures

This step is highly process-dependent. If there are multiple components in the system, such as in the case of a control loop, they must all be modeled. The model structure should be such that fault-related parameters can be easily seen and manipulated. An example of this is given in Section 10.3 with respect to the hybrid tank system. Because the model in question must make use of parameters that have physical meaning, the model structure must be derived using a *white-box* or *gray-box* approach, where white-box models are based on first principles and have no unknown parameters, and where gray-box models are constructed based on first principles, but can have simplifications and unknown parameters that are to be estimated using data.

Step 2: Use Data to Identify Gray-box Models

If one cannot construct a white-box model – which is usually the case – one can use data to estimate unknown parameters in a gray-box model. Consider a dynamic model $f(x, u, \Theta)$ with states x, inputs u and unknown parameters Θ.

$$\frac{dx}{dt} = f(x, u, \Theta) + \varepsilon_x$$

This model has an observation function $h(x)$ to yield the observed values y

$$y = h(x) + \varepsilon_y$$

where the observation function $h(x)$ is assumed to be known. The predicted values of y can be obtained by using an ordinary differential equation solver, such as the RK45 method, to solve for x given u, and then predict y using the observation function $h(x)$. The estimated value of Θ can then be obtained by minimizing an error expression

$$\hat{x}(t) = \text{RK45}[f(x(t-1), u(t-1), \Theta)]$$

$$\hat{\Theta} = \arg\min_{\Theta} \sum_t [h(\hat{x}(t)) - y(t)]^T R^{-1} [h(\hat{x}(t)) - y(t)]$$

where R is a positive-definite (often diagonal) weighting matrix that represents the noise variance of each evidence-related variable. The purpose of R is to give an appropriate weight to each element of the observation vector $y(t)$. If $y(t)$ has only one element, $R = 1$ is sufficient, and if all elements of $y(t)$ take the same units, $R = I$ will also be sufficient. Otherwise, one might want to take the variance of each element in y in order to obtain a suitable metric for scaling.

Step 3: Obtain Residual Errors

Residual error calculations require estimates of the hidden state, which can be obtained using Kalman filtering. In general, the state models are nonlinear and thus an ordinary Kalman filter will not suffice. It is recommended that more advanced techniques such as the extended Kalman filter, the unscented Kalman filter (UKF), the ensemble Kalman filter, or the particle filter are used. Due to a combination of easy computation and effectiveness, the UKF is a popular choice for nonlinear state estimation problems.

The UKF estimates the state given a system that has the following model:

$$\frac{dx}{dt} = f(x, u, \Theta) + \varepsilon_x$$

$$y = h(x) + \varepsilon_y$$

$$\varepsilon_x \sim N(0, Q)$$

$$\varepsilon_y \sim N(0, R)$$

where $\varepsilon_x \sim N(0, Q)$ indicates that ε_x is Gaussian white noise with mean zero and covariance Q; similar conditions are assumed for $\varepsilon_y \sim N(0, Q)$. If one already has suitable values of Q and R, the residual errors of x and y can be obtained via the UKF. The residual errors on x (denoted $\hat{\varepsilon}_x$) can be obtained using

$$x(t) = \text{UKF}[f(x(t-1), u(t-1)), p(t-1), h(x), y(t), Q, R]$$

$$\hat{x} = \text{RK45}[f(x(t-1), u(t-1), \Theta)]$$

$$\hat{\varepsilon}_x = x - \hat{x}$$

where x is the updated estimate, as it uses observations y, and $\hat{x}$ is the predicted value. Meanwhile the residual errors on y (denoted $\hat{\varepsilon}_y$) can be obtained using

$$\hat{y} = h(x(t))$$

$$\hat{\varepsilon}_y = y - \hat{y}$$

The values for Q and R are often used as tuning parameters to express the reliability of the model and observations, respectively. Large values of Q assume the model is less reliable, while large values of R assume that the observations are unreliable. It is also possible to estimate Q and R from data using a technique that is similar to the EM algorithm. The technique makes use of the following steps.

1. Start with initial values for Q and R. If the system changes slowly and the model is fairly accurate, a good initial estimate of R can be obtained using

$$R_0 = \frac{1}{2(n-1)} \text{diag} \left[\sum_{i=2}^{n} (y(i) - y(i-1))^2 \right]$$

where $\text{diag}(x)$ is the same as the MATLAB command $\text{diag}(x)$, which takes a vector x and constructs a diagonal matrix out of it. An initial value of Q can be obtained by analysing R_0 and the observation function h. Generally, it is best to choose large values for Q initially, so that the observations carry more weight than the model.

2. Use Kalman filtering to estimate states given the values chosen for Q and R

$$f(x) = f(x, u(t-1), \Theta)$$

$$x(t) = \text{UKF}(f(x), x(t-1), p(t-1), h(x), y(t), Q_0, R_0)$$

where $u(t-1)$ is assumed to be a constant over the given sampling interval.

3. Estimate a new value for Q using

$$\hat{x} = \text{RK45}[f(x(t-1), u(t-1), \Theta)]$$

$$Q_1 = \frac{1}{n} \sum_{i=1}^{n} [x(i) - \hat{x}(i)] \times [x(i) - \hat{x}(i)]^T$$

4. Estimate a new value for R using

$$\hat{y} = h(\hat{x})$$

$$R_1 = \frac{1}{n} \text{diag} \left[\sum_{i=1}^{n} (y(i) - \hat{y}(i))^2 \right]$$

5. Repeat steps 2–4 until the log-likelihood of the data L converges

$$L = L_x + L_y$$

$$L_x = -\frac{n}{2} \log \left[(2\pi)^{d_x} |Q| \right] - \frac{n}{2} \sum_{i} [x(i) - \hat{x}(i)]^T Q^{-1} [x(i) - \hat{x}(i)]$$

$$L_y = -\frac{n}{2} \log \left[(2\pi)^{d_y} |R| \right] - \frac{n}{2} \sum_{i} [y(i) - \hat{y}(i)]^T R^{-1} [y(i) - \hat{y}(i)]$$

where d_x is the dimension of x and d_y is the dimension of y.

After obtaining Q and R, one can further tune these values to suit their needs. Scaling for reducing values of Q will result in smoother state estimates but will be less responsive to observations. Conversely, scaling for increasing values of Q result in rougher state estimates but will be more responsive to observations.

Step 4: Whitening Residual Errors

Bootstrapping (and its smoothed counterpart) requires that the residual errors be i.i.d. Often times, the residual errors are autocorrelated but can be whitened by using an AR model. This is done by first identifying an AR model for both $\hat{\varepsilon}_x$ and $\hat{\varepsilon}_y$ and then whitening the residuals by inverting the model so that

$$\hat{\varepsilon}_x(t) + A_1 \hat{\varepsilon}_x(t-1) + \ldots + A_n \hat{\varepsilon}_x(t-n) = \hat{\varepsilon}_x^w(t)$$

$$\hat{\varepsilon}_x = \frac{1}{A(z)} \hat{\varepsilon}_x^w$$

The whitened residuals are obtained by inverting the model so that

$$\hat{\varepsilon}_x^w = A(z) \hat{\varepsilon}_x$$

Note that the same whitening procedure should be performed on $\hat{\varepsilon}_y$.

Step 5: Kernel Density Estimation

Recall that the kernel density estimate can be calculated as

$$f(x) \approx \frac{1}{n} \sum_{i=1}^{n} \frac{1}{\sqrt{(2\pi)^d |H_i|}} \exp([x - D_i]^T H_i^{-1} [x - D_i])$$

where D represents the data points (in this case $D = \hat{\varepsilon}_x^w$), d is the dimension of x and H_i is the bandwidth at x_i. Also recall that if one uses a uniform bandwidth, $H_i = H$ can be calculated using the normal bandwidth reference rule

$$H_N = \left(\frac{4}{n(d+2)} \right)^{\frac{2}{d+4}} S \tag{10.4}$$

where S is the sample covariance matrix of the data $D = \hat{\varepsilon}_x^w$ or $D = \hat{\varepsilon}_y^w$. The kernel density estimate consists of data and bandwidth matrices corresponding to the data points. Note that the adaptive bandwidth technique mentioned in Chapter 8 can also be used.

For this step, since the data $D = \hat{\varepsilon}_x^w$ and $D = \hat{\varepsilon}_y^w$ are already obtained, one simply needs the bandwidth matrix to complete the kernel density estimation.

Step 6: Simulate New Data via Smoothed Bootstrapping

The kernel density estimation forms the basis of the smoothed bootstrap. One can sample from the kernel density estimate by means of a two-step process:

1. Randomly select a data point $\hat{\varepsilon}_x^w(i)$ or $\hat{\varepsilon}_y^w(i)$ from the history, where i is a random integer between 1 and n with a uniform distribution.
2. Add Gaussian noise to $\hat{\varepsilon}_x^w(i)$ or $\hat{\varepsilon}_y^w(i)$, with mean zero and covariance H_i. Note that adding Gaussian noise makes samples from a Gaussian distribution centred around the selected data point instead of sampling from the existing data points themselves; by sampling the kernel function around the data point, ordinary bootstrapping is converted to *smoothed bootstrapping*.

The sampling and noise addition process is repeated for the number of times as desired. After smoothed bootstrapping, the AR filter is used to generate disturbances that act similarly to the original one.

$$\hat{\varepsilon}_x = \frac{1}{A(z)} \hat{\varepsilon}_x^w$$

$$\hat{\varepsilon}_y = \frac{1}{A(z)} \hat{\varepsilon}_y^w$$

The disturbance sequences can be used in the process and observation models ($f(x, u, \Theta)$ and $h(x)$, respectively) to generate new samples for y. New modes can be created by varying the parameters Θ in the gray-box model $f(x, u, \Theta)$ simulating different faults, but the simulating of generating noise from bootstrapping remains the same for each simulated mode.

Step 7: Apply Monitoring Algorithms

The monitoring algorithms are process specific and can be applied to the additional simulated data in exactly the same manner as the original data.

10.3   Illustration

In this illustration, we consider a hybrid tank system, which will also be presented in a practical application in Section 10.4 as well as in Part II of this book in more detail; a schematic is available in Figure 10.2. The hybrid tank system has four components, each having two states, resulting in 16 modes in total. The four components are the flow sensor into Tank 1 (FM_1), the flow sensor into Tank 2 (FM_2), the valve between Tanks 1 and 2 (V_1), and the valve between Tanks 2 and 3 (V_2). Modes will be described in terms of bias in FM_1, FM_2 (B_1, B_2, respectively) and leaks caused by opening the valves V_1, V_2 (L_1, L_2, respectively); the mode vector in this example is $[B_1, B_2, L_1, L_2]$ where each component can take the state 0 (where the problem does not exist) or 1 (where the problem does exist).

The monitor selected for this system consists of an augmented Kalman filter and calculated pump model prediction errors. For the original filter, the state model is given as follows:

$$
\frac{d\,X_1}{d\,t} = A_c^{-1}\left[\frac{U_1}{B_1} - C_1\left[X_1^{1/2} + h_r\right] + L_1\frac{X_2 - X_1}{|X_2 - X_1|^{1/2}}\right] + \varepsilon_{X_1}
$$

$$
\frac{d\,X_2}{d\,t} = A_c^{-1}\left[-C_2\left[X_2^{1/2} + h_r\right] + L_1\frac{X_2 - X_1}{|X_2 - X_1|^{1/2}} + L_2\frac{X_2 - X_3}{|X_2 - X_3|^{1/2}}\right] + \varepsilon_{X_2}
$$

$$
\frac{d\,X_3}{d\,t} = A_c^{-1}\left[\frac{U_2}{B_2} - C_3\left[X_3^{1/2} + h_r\right] + L_1\frac{X_2 - X_3}{|X_2 - X_3|^{1/2}}\right] + \varepsilon_{X_3}
$$

$$
Y = IX + \varepsilon_Y
$$

where the state vector X consists of three level indicators and the input vector U consists of flow-rate measurements for the flows into Tanks 1 and 3. The state is augmented to include

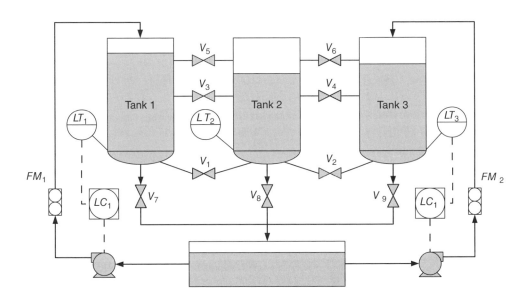

Figure 10.2   Hybrid tank system

four additional states, B_1, B_2, L_1, L_2, representing bias and leak parameters, which are ideally constant:

$$\frac{d\,B_1}{d\,t} = 0$$

$$\frac{d\,B_2}{d\,t} = 0$$

$$\frac{d\,L_1}{d\,t} = 0$$

$$\frac{d\,L_2}{d\,t} = 0$$

When augmented, the additional state covariance entries in Q are set to small values, as the monitored values are believed to be relatively constant. Small values in Q also mean that the monitored results are smooth.

The four monitored values are obtained by applying the UKF to the augmented model. A MATLAB implementation of the UKF is available on the MATLAB file exchange, courtesy of Yi Cao of Cranfield University, and is presented below.[1]

```
1   function [x,P]=ukf(fstate,x,P,hmeas,z,Q,R)
2   % UKF   Unscented Kalman Filter for nonlinear dynamic systems
3   % [x, P] = ukf(f,x,P,h,z,Q,R) returns state estimate, x and state covariance, P
4   % for nonlinear dynamic system (for simplicity, noises are assumed as additive):
5   %           x_k+1 = f(x_k) + w_k
6   %           z_k   = h(x_k) + v_k
7   % where w ¬ N(0,Q) meaning w is gaussian noise with covariance Q
8   %       v ¬ N(0,R) meaning v is gaussian noise with covariance R
9   % Inputs:   f: function handle for f(x)
10  %           x: "a priori" state estimate
11  %           P: "a priori" estimated state covariance
12  %           h: function handle for h(x)
13  %           z: current measurement
14  %           Q: process noise covariance
15  %           R: measurement noise covariance
16  % Output:   x: "a posteriori" state estimate
17  %           P: "a posteriori" state covariance
18  %
19  %
20  % By Yi Cao at Cranfield University, 04/01/2008
21  %
22  L=numel(x);                            %numer of states
23  m=numel(z);                            %numer of measurements
24  alpha=1e-3;                            %default, tunable
25  ki=0;                                  %default, tunable
26  beta=2;                                %default, tunable
27  lambda=alpha^2*(L+ki)-L;               %scaling factor
28  c=L+lambda;                            %scaling factor
```

---

[1] For the original version, one can visit http://www.mathworks.com/matlabcentral/fileexchange/18217-learning-the-unscented-kalman-filter.

```
29   Wm=[lambda/c 0.5/c+zeros(1,2*L)];                %weights for means
30   Wc=Wm;
31   Wc(1)=Wc(1)+(1-alpha^2+beta);                    %weights for covariance
32   c=sqrt(c);
33   X=sigmas(x,P,c);                                 %sigma points around x
34   [x1,X1,P1,X2]=ut(fstate,X,Wm,Wc,L,Q);              %unscented transformation of process
35   % X1=sigmas(x1,P1,c);                              %sigma points around x1
36   % X2=X1-x1(:,ones(1,size(X1,2)));                  %deviation of X1
37   [z1,Z1,P2,Z2]=ut(hmeas,X1,Wm,Wc,m,R);            %unscented transformation of ...
            measurments
38   P12=X2*diag(Wc)*Z2';                             %transformed cross-covariance
39   K=P12/P2;
40   x=x1+K*(z-z1);                                   %state update
41   P=P1-K*P12';                                     %covariance update
42
43   function [y,Y,P,Y1]=ut(f,X,Wm,Wc,n,R)
44   %Unscented Transformation
45   %Input:
46   %         f: nonlinear map
47   %         X: sigma points
48   %        Wm: weights for mean
49   %        Wc: weights for covraiance
50   %         n: numer of outputs of f
51   %         R: additive covariance
52   %Output:
53   %         y: transformed mean
54   %         Y: transformed smapling points
55   %         P: transformed covariance
56   %        Y1: transformed deviations
57
58   L=size(X,2);
59   y=zeros(n,1);
60   Y=zeros(n,L);
61   for k=1:L
62       Y(:,k)=f(X(:,k));
63       y=y+Wm(k)*Y(:,k);
64   end
65   Y1=Y-y(:,ones(1,L));
66   P=Y1*diag(Wc)*Y1'+R;
67
68   function X=sigmas(x,P,c)
69   %Sigma points around reference point
70   %Inputs:
71   %         x: reference point
72   %         P: covariance
73   %         c: coefficient
74   %Output:
75   %         X: Sigma points
76
77   A = c*chol(P)';
78   Y = x(:,ones(1,numel(x)));
79   X = [x Y+A Y-A];
```

In addition, two more monitors are included to estimate flow meters using pump-speed signals. Under normal conditions, the output error (OE) model is estimated in order to predict flow rates

$$U(t) = \frac{B(z)}{C(z)} S(t)$$

where $S(t)$ is the pump speed. The two additional monitors estimate bias through

$$B_1(t) = \frac{U_1(t)}{\frac{B(z)}{C(z)} S(t)}$$

$$B_2(t) = \frac{U_2(t)}{\frac{B(z)}{C(z)} S(t)}$$

10.3.1 Component-based Diagnosis

Offline Step 1: Select Components and Choose Reference Modes

In the case of our data, the most frequent mode is the no-fault mode $[0, 0, 0, 0]$. Thus for the first component, we will evaluate the false negative criterion (FN) between $m_1 = [0, 0, 0, 0]$ and $m_9 = [1, 0, 0, 0]$. Likewise, for the second component, we will evaluate FN between $m_1 = [0, 0, 0, 0]$ and $m_5 = [0, 1, 0, 0]$. This procedure is then repeated for the third and fourth components, which compare $m_3 = [0, 0, 1, 0]$ and $m_2 = [0, 0, 0, 1]$, respectively, with $m_1 = [0, 0, 0, 0]$. The list of modes for this system is as follows:

$$
\begin{bmatrix}
m_1 \\
m_2 \\
m_3 \\
m_4 \\
m_5 \\
m_6 \\
m_7 \\
m_8 \\
m_9 \\
m_{10} \\
m_{11} \\
m_{12} \\
m_{13} \\
m_{14} \\
m_{15} \\
m_{16}
\end{bmatrix}
=
\begin{bmatrix}
0 & 0 & 0 & 0 \\
0 & 0 & 0 & 1 \\
0 & 0 & 1 & 0 \\
0 & 0 & 1 & 1 \\
0 & 1 & 0 & 0 \\
0 & 1 & 0 & 1 \\
0 & 1 & 1 & 0 \\
0 & 1 & 1 & 1 \\
1 & 0 & 0 & 0 \\
1 & 0 & 0 & 1 \\
1 & 0 & 1 & 0 \\
1 & 0 & 1 & 1 \\
1 & 1 & 0 & 0 \\
1 & 1 & 0 & 1 \\
1 & 1 & 1 & 0 \\
1 & 1 & 1 & 1
\end{bmatrix}
= \texttt{ModeBV} \tag{10.5}
$$

Let us consider the first component, where the reference modes are $m_1 = [0, 0, 0, 0]$ and $m_9 = [1, 0, 0, 0]$. We use MATLAB as the computation engine, and assume that data has already been sectioned into different modes in a cell array labeled `Data` so that `Data{1}` contains data for $m_1 = [0, 0, 0, 0]$ and `Data{9}` contains data for $m_9 = [1, 0, 0, 0]$.

Offline Step 2: Calculate FN Criterion for each Monitor

The FN criterion for each monitor can be calculated according to the MATLAB code below, which follows steps 1–10 under 'Selecting monitors for components' in Section 10.2.1 (see p. 173).

```
1   function FN = FalseNegativeCriterion(Data)

2

3   ne = length(Data{1}(1,:)); %find the dimension of the data
4   ns = length(Data); %find the number of states for this component

5

6   %For each piece of evidence
7   for e = 1:ne
8       %For each state
9       for i = 1:ns
10          %Define the kernel density estimate
11          KDE(i,e) = fKernelEstimateNorm(Data{i}(:,e));
12      end
13  end

14

15

16  %For each component state
17  Pn = zeros(ns,ne);
18  for s = 1:ns
19      %For each piece of evidence
20      D = Data{s};
21      n = length(D);
22      Lik = zeros(n,ns);
23      for e = 1:ne
24          %Estimate the likelihood from the KDE for each component state
25          for k = 1:ns
26              Lik(:,k) = fKernelDensity(D(:,e),KDE(k,e));
27          end
28          %Diagnose most likely mode (here, max defines largest element in rows)
29          [¬,Diagnosis] = max(Lik,[],2);

30

31          %Define the proportion of false diagnosis results
32          Pn(s,e) = sum(Diagnosis ≠ s)/n;
33      end
34  end

35

36  %Define false negative criterion FN, a row vector with respect to evidence
37  FN = 1/(ns-1)*sum(Pn);
```

Given the selected modes for component 1 (m_1, m_9), one can obtain the FN criteria for each monitor using the following code.

```
1   FN = FalseNegativeCriterion(Data(1,9))
```

After calculating the FN criteria for all pieces of evidence, one can make a decision as to which pieces of evidence to include for that component.

Offline Step 3: Obtain Kernel Density Estimates Using Selected Monitors

For each mode that occurs in the data, we evaluate the kernel density based on the selected evidence. Consider the variable EvidenceSelection, which is a cell array that contains a vector of selected evidence in each cell. The kernel density estimate can be obtained through:

```
1   nc = 4; %number of components
2   nm = length(Data); %number of modes in the data
3
4   %for each component
5   for c = 1:nc
6       %obtain selected evidence
7       ind = EvidenceSelection{c};
8
9       %for each mode
10      for m = 1:nm
11          %estimante KDE using selected evidence
12          if ¬isempty(Data{m})
13              KDE(m,c) = fKernelEstimateNorm(Data{m}(:,ind));
14          end
15      end
16  end
```

Online Step 1: Calculate Likelihoods for a New Data Point

The Bayesian diagnosis algorithm starts off with evaluating likelihoods of modes. Here, however, we need to take into account different pieces of evidence selected for components. Thus the likelihood matrix Lik for evidence e will have columns pertaining to modes and rows pertaining to components.

```
1   [nm,nc] = size(KDE);
2
3   %for each mode
4   for m = 1:nm
5       %for each component
6       for c = 1:nc
7           %selected evidence index
8           ind = EvidenceSelection{c};
9           %Evaluate likelihood of mode if it appears in the data
10          if ¬isempty(KDE(m,c).bwm)
11              Lik(m,c) = fKernelDensity(e(ind),KDE(m,c));
12          end
13      end
14  end
```

Online Step 2: Calculate Component Likelihoods

The component state likelihoods are calculated by going through all the modes having that state and using the largest likelihood. Determining whether or not a mode has the right component state requires the use of the ModeBV variable, which is defined in Eqn (10.5). As an example, if we wish to figure out the modes in which component 1 is equal to 0, we search through all elements of the first column in ModesBV and note the rows where the element is 1. The code below performs this search for all applicable modes and all component states.

```
1   [nm,nc] = size(ModesBV);
2   %Find the number of component states for each column
3   mc = max(ModesBV);
4
5   ApplicableModes = cell(1,nc);
6   %For each component, create a new ApplicableModes matrix
7   for c = 1:nc
8       ApplicableModes{c} = zeros(nm,mc(c));
9       %For each component state, search ModesBV
10      for ci = 0:mc(c)
11          %A binary matrix having 1 where the mode matches component state
12          ApplicableModes{c}(:,(ci+1)) = ModesBV(:,c)==ci;
13      end
14  end
```

After the search for applicable modes has been performed, we can calculate the likelihoods for each component state by selecting the most probable likelihood out of all applicable modes.

```
1   LikC = cell(1,nc);
2   %For each component
3   for c = 1:nc
4       %For each component state
5       for ci = 1:(mc(c)+1)
6           %Use the maximum likelihood of applicable modes
7           aModes = find( ApplicableModes{c}(:,ci) );
8           LikC{c}(ci,1) = max(Lik(aModes,c));
9       end
10  end
```

The result `LikC` contains likelihoods for each component state under each component.

Online Step 3: Bayesian Inference

For each component, there are prior probabilities for each state. The posterior can be calculated using typical Bayesian methods.

```
1   %For each component
2   for c = 1:nc
3       Post{c} = LikC{c}.*Prior{c};
4       Post{c} = Post{c}/sum(Post{c});
5   end
```

Note that for each component c, `LikCc` contains column vectors of likelihoods, and `Priorc` contains column vectors of prior probabilities of the same size. The mode can be diagnosed by selecting the state values of highest probability for each component.

Online Step 4: Obtain Posterior for Modes (Optional)

If one wishes to display the posterior probability of the modes – assuming that the states are independent – one can use Eqn (10.3), which in MATLAB takes the following form:

```
 1   PostM = ones(nm,1);
 2   %For each mode
 3   for m = 1:nm
 4       %Find the component state indices
 5       %because zero is first index, (index = value +1)
 6       CompInd = ModeBV(m,:)+1;
 7       %For each component
 8       for c = 1:nc
 9           %Product of applicable posterior probabilities
10           PostM(m) = PostM(m)*Post{c}(CompInd(c));
11       end
12   end
```

10.3.2   Bootstrapping for Additional Modes

Here we discuss the technique of bootstrapping for more data. Because the end-goal is to produce more learning data, this entire technique is performed offline. If combining the two techniques, the bootstrapping procedure is completed before the first step is taken in the component-space approach.

Step 1: Identify the Model

The original model is defined in MATLAB according to the function `Tanks`, which yields the differential dX given:

- the input flow rate U
- current level X
- and parameters:
  - tank cross-sectional area A_c
  - coefficients for the drain C_1, C_2, C_3
  - original bias or scaling parameters for the flow rates B_1, B_2.

```
 1   function dX = Tanks(U,X,Ac,C1,C2,C3,B1,B2,Ho)
 2       dX(1,1) = ( B1*U(1) - C1*(X(1)+Ho)^(0.5) )/Ac;
 3       dX(2,1) = (-C2*(X(2)+Ho)^(0.5))/Ac;
 4       dX(3,1) = ( B2*U(2) - C3*(X(3)+Ho)^(0.5) )/Ac;
 5   end
```

The level prediction is obtained using the RK45 method

```
1   function Y = HybridTanks(U,X0,Ac,C1,C2,C3,B1,B2,Ho,Ts)
2   %Restrict the values from being negative
3   options = odeset('NonNegative',[1,2,3]);
4   %u has rows for t and columns for components (in this case 2)
5   [nt,nu] = size(U);
6   X = X0;
7
8   for t = 1:nt
9       u = OP(k,:);
10      dXu = @(t,X) Tanks(u,X,AC,C1,C2,C3,B1,B2,Ho);
11
12      [~,vX] = ode45(dXu,[0,Ts],X,options);
13      X = vX(end,:);
14      X(X>100) = 100; %Upper limit on X
15      Y(k,:) = X;
16  end
```

The parameter of the model `Theta = [Ac,C1,C2,C3,B1,B2,Ho]` can be identified by the optimization technique of your choice. For example, in MATLAB one can use `fminunc`.

```
1   %U, Y X0 and Ts are already defined earlier
2   Theta0 = [Ac,C1,C2,C3,B1,B2]; %Initial Parameters
3
4   %Set up objective function
5   Yhat = @(Th) HybridTanks(U,X0,Th(1),Th(2),Th(3),Th(4),Th(5),Th(6),Th(7),Ts);
6   Objective = @(Th) sum( sum( (Yhat (Th) - Y ).^2 ) );
7
8   Theta = fminunc(Objective,Theta0);
```

The function `fminunc` is used for this purpose, but it may be desirable to include constraints on parameter values. Furthermore, C_1, C_2, C_3 could be forced to be equal, as the tank outlets are identical. This helps reduce the dimensionality of the search.

Step 2: Obtain Residual Errors

When identifying model error covariances, it is required that the Q and R matrices are identified first. The first step is to set up the UKF so that state errors $Xe = \varepsilon_x$ and observation errors $Ye = \varepsilon_y$ are reported.

```
1   function [X,P,Xe,Ye] = TankUKF(X0,Y1,U0,P0,Theta,Q,R,Ts)
2   %Set up differential equation model
3   dTank = @(t,X) Tanks(X,u,Th(1),Th(2),Th(3),Th(4),Th(5),Th(6),Th(7),Ts);
4
5   %Set up predictor for differential equation model
```

```
6    options = odeset('NonNegative',[1,2,3],'InitialStep',(Ts/10));
7    function Xf = Ftank(X,Ts)
8        X = X.*(X>0);
9        [¬,vX] = ode45(dTank,[0,Ts],X,options);
10       Xf = (vX(end,:))';
11   end
12
13   %state transition and observation functions
14   ftank = @(X) Ftank(X,Ts);
15   htank = @(X) X;
16
17   %UKF result
18   [X,P] = ukf(ftank,X0,P0,htank,Y1,Q,R);
19
20   %Residual error in X
21   Xe = X-ftank(X0);
22
23   %Residual Error in Y
24   Ye = Y1-htank(X);
```

Now that we have a function to define prediction errors in X and Y, we can iteratively estimate Q and R :

```
1    %Y, U, Ts and Theta are already defined
2    %Initial values for $Q$ and $R$ are already chosen
3    [nt,¬] = size(U);
4    X = Y(:,1);
5    P = Q;
6
7    %Obtain covariance constant for x and y log likelihoods
8    K = -0.5*(  log( (2*pi)^(length(Q))*det(Q) ) + log( (2*pi)^(length(R))*det(R) ) );
9
10   %Cholesky decomposition for easy inversion
11   Qr = chol(Q);
12   Rr = chol(R);
13
14   LogLik0 = -1/0;
15   dLog = 100;
16
17   %Iterate the estimation of Q and R until log likelihood converges
18   while dLog > 1e-5
19       LogLik1 = 0;
20       for t = 2:nt
21           [X,P,xe,ye] = TankUKF(X,Y(t,:),U(t-1,:),P,Theta,Q,R,Ts);
22           Xe(t,:) = xe;
23           Ye(t,:) = ye;
24
25           %Use a more accuate expression for inversion
26           %   sum((Xe\Qr).^2) = Xe*inv(Qr)*Xe'
27           LogLikX = -0.5*sum((Xe(t,:)\Qr).^2);
28           LogLikY = -0.5*sum((Ye(t,:)\Rr).^2);
29           LogLik1 = LogLik1 + K + LogLikX + LogLikY;
```

```
30          end
31          Q = cov(Xe);
32          R = cov(Ye);
33          dLog = LogLik1 - LogLik0;
34      end
```

This procedure estimates covariances Q and R, but the residual errors $\varepsilon_x = $ Xe and $\varepsilon_y = $ Ye can also be obtained from the final iteration.

Step 3: Whitening Residual Errors

Residual error whitening can be performed by estimating an AR model and applying it to the data. There are many techniques that can be used to estimate an AR model, and MATLAB has a command Model = ar(y,n) strictly for this purpose. It can be applied to all elements of the noise in X and Y as follows

```
1   %n is your selected model order
2   for i = 1:length(Q)
3           ARx{i} = ar(Xe(:,i),n);
4   end
5   for i = 1:length(R)
6           ARy{i} = ar(Ye(:,i),n);
7   end
```

After the model has been estimated, it is possible to whiten the residual errors by applying the model. First, we want to create a data object so that

$$\varepsilon = [\varepsilon_t, \varepsilon_{t-1}, \varepsilon_{t-2}, \ldots, \varepsilon_{t-n}]$$

This can be done using the following code:

```
1   for k = 1:length(Q)
2           %Construct the desired noise sequence for each residual sequence k
3           XEk = zeros(length(Xe(:,k))-(n-1),n);
4           %For each coefficient of A
5           for i = 1:n
6               xe = Xe(:,k);
7               %remove the first n - i data points
8               xe(1:(n-i)) = [];
9               %remove the last i-1 data points
10              xe((end-i+2):end) = [];
11              %Place the result in the kth XE matrix
12              XEk(:,i) = xe;
13          end
14          XE{k} = XEk;
15  end
```

The output of the AR modeling step in MATLAB is such that

$$\varepsilon_t^w = A(1)\varepsilon_t + A(2)\varepsilon_{t-1} + \ldots + A(n)\varepsilon_{t-(n-1)}$$

$$= \varepsilon_t \times A$$

In MATLAB, given the XE variable, the whitened output is obtained as:

```
1   %For each residual error sequence
2   for k = 1:length(Q)
3       A = (ARx{k}.a)';
4       Xew(:,k) = XE{k}*A;
5   end
```

The final product Xew contains the whitened residuals of Xe. This procedure is repeated for residual error sequences in Y as well.

Step 4: Bootstrapping Residual Errors

When running a new simulation, new residual error sequences must be generated. This can be done by first obtaining the kernel density estimate:

```
1   %For each X residual error component
2   for k = 1:length(Q)
3       KDE_xew(k) = fKernelEstimateNorm(Xew(:,k));
4   end
5
6   %For each Y residual error component
7   for k = 1:length(R)
8       KDE_yew(k) = fKernelEstimateNorm(Yew(:,k));
9   end
```

Then, data points in the history can be randomly selected:

```
1    %nt is the number of time samples desired in the simulation
2    XewBS = zeros(nt,length(Q));
3    %For each X residual error component
4    for k = 1:length(Q)
5            %Select data from KDE object
6        Data = KDE_xew(k).data;
7        nd = length(Data);
8        %For as many time instances desired
9        for t = 1:nt
10           %Select random index from data
11           ind = round(nd*rand(1));
12           XewBS(t,k) = Data(ind,:);
13       end
14   end
```

If these data points were directly used to simulate the noise, the process would be regular bootstrapping, which samples directly from the historical data. Smoothed bootstrapping, on the other hand, samples from the kernel density estimate. This can be done by first selecting the random piece of historical data, as done in bootstrapping, and then adding noise that is sampled from the kernel function. For the Gaussian kernel, one simply samples from the standard normal distribution and multiplies the result by the square-root of the bandwidth. The result is a smoothed bootstrap sample, which is obtained by implementing the following MATLAB code:

```
1   for k = 1:length(Q)
2          %Select bandwidth matrix from KDE object
3       H = KDE_xew(k).bwm;
4       %Smooth the bootstrap with Gaussian noise
5       XewBS(:,k) = XewBS(:,k) + randn(nt,1)*H^0.5;
6   end
```

The smoothed bootstrapping procedure is performed on residual errors in X above, but the process should also be repeated in residual errors on Y.

Finally, one has to invert the AR model to predict a new sequence. The reverse model takes the form

$$A(1)\varepsilon_t = -A(2)\varepsilon_{t-1} - \ldots - A(n)\varepsilon_{t-(n-1)} + \varepsilon_t^w$$

where $A(1) = 1$ by convention. This inverse AR model can be applied using the following MATLAB code:

```
1   XeBS = zeros(nt,length(Q));
2   %For each component
3   for k = 1:length(Q)
4       %Obtain AR parameters and set up noise sequence
5       A = ARx{k}.a;
6       na = length(A);
7       y = zeros(nt+na,1);
8
9       %Inverse model: y = -A(2) y-1 - ... - A(n) y-(n-1) + XewBS
10      %Time starts at 1+length(A), as A pertains to past values
11      for t = ((1:nt)+na)
12          %Select past inputs for reverse model
13          %wrev inverts vector (we can only construct ascending)
14          ind = wrev( (t-(na-1)):(t-1) );
15          %Predict new output from inverse model
16          y(t) = -A(2:end)*y(ind) + XewBS(t-na);
17      end
18      %remove the first na inputs as they are not used
19      y(1:na) = [];
20      XeBs(k,:) = y;
21  end
```

This inverse modeling procedure should likewise be performed on residual errors in Y in addition to X, as was performed above.

Step 5: Simulating New Data from the Bootstrap

Once the bootstrapped noise sequence has been generated, this sequence has to be inserted into a simulation. The simulation itself must contain all aspects of the control loop. For the hybrid tank system, the control loop is shown in Figure 10.3. In this model, the appropriate places to add the noise are indicated, as well as the appropriate places to sample outputs. Simulation can be done using MATLAB, Simulink or any other desired simulation software.

Step 6: Generate New Data from Monitoring

From the simulated data, monitoring algorithms can be applied. In this case, the monitors include an augmented UKF for the four components (two sensors and two valves) and a pump prediction model (for the two sensors).

10.4 Application

In this chapter, only the hybrid tank system is considered, for the following reasons:

1. The model is simple enough for most audiences to grasp the essences of the bootstrapping technique.
2. The monitors for this system have already been explained, so that the monitor selection results for this system will be most meaningful.
3. Unlike the industrial system considered in Chapter 11, we have control over what modes appear in the data so that we can easily validate performance for modes that are missed in the historical data.

Experimental data was obtained for the 16 possible modes; roughly one hour's worth of data for each mode. The nonlinear model was estimated using the prediction error method, based on the lab data without faults. The closed-loop behaviour was then simulated for each of the 16 possible modes.

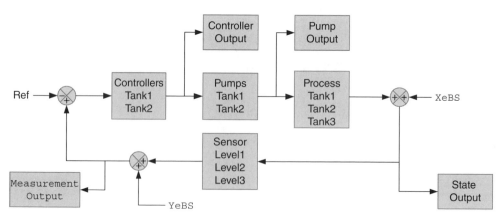

Figure 10.3 Hybrid tank control system

The first purpose of this experiment is to assess how the bootstrap simulation approach performs. The second purpose is to assess how well the mode-space method and component-space method are improved by replacing simulated data with real process data. For the mode-space method, replacing simulated data with process data for a single mode will improve the result for that mode, resulting in a significant but localized improvement. Conversely, for the component-space method, the same data replacement for the same mode should improve results in diagnosing all components, resulting in a slight but more widespread improvement. For realism, real data is only applied to the most common modes, namely normal operation and modes with a single fault occurrence.

Because the system has already been modeled, the mode-space and component-space methods are compared to another intuitive method: the model-based approach. The model-based method functions similarly to the component-space method, except that only the corresponding UKF fault-parameter estimate is used to diagnose the fault. If the UKF reading for fault parameters exceeds a certain boundary, the corresponding fault is considered to exist.

10.4.1 Monitor Selection

When setting up the component-space diagnosis approach, the monitors had to be selected based on their sensitivity toward that component's faulty state. It was found that leaks did not affect the pump-based bias monitors and thus these monitors were discarded when diagnosing bias. Additionally, because of the complex interactions between bias and leaks when implementing the UKF, both bias and leak parameters were used for Tank 1 when diagnosing leaks and bias in that tank. Likewise, both UKF parameters are included when diagnosing bias and leaks in Tank 2. A summary of monitor inclusion is given in Table 10.1

10.4.2 Component Diagnosis

The three diagnosis methods were tested using monitor results obtained from the laboratory setup. For the first trial, diagnosis was performed using only simulated data for training. For the second trial, diagnosis was performed using experimental data from normal operations. Finally, for the third trial, experimental data was included from all modes with a single fault. In all runs, each method attempted to diagnose the mode and the individual faults. Mode diagnosis results are shown in Table 10.2 and component diagnosis results are shown in Table 10.2.

Table 10.1 Included monitors for component space approach

	Bias Tank 1	Bias Tank 2	Leak Tank 1	Leak Tank 2
UKF B_1	yes	no	yes	no
UKF B_2	no	yes	no	yes
UKF L_1	yes	no	yes	no
UKF L_2	no	yes	no	yes
Pump B_1	yes	no	no	no
Pump B_2	no	yes	no	no

Table 10.2 Misdiagnosis rates for modes

	All simulated	Real data: $n_f \leq 0$	Real data: $n_f \leq 0$
Model-driven	42.7%	38.5%	30.3%
Mode-space	50.3%	40.3%	22.1%
Component-space	42.6%	37.9%	23.1%

Table 10.3 Misdiagnosis rates for component faults

	All simulated	Real data: $n_f \leq 0$	Real data: $n_f \leq 1$
Model-driven	13.4%	12.6%	9.12%
Mode-space	20.1%	15.8%	8.9%
Component-space	12.3%	11.3%	6.6%

One of the first observations is that it was more difficult to diagnose the mode than it was to diagnose the individual component faults. This was expected, as the mode will be misdiagnosed if all components are correctly diagnosed except one. For example, if the three out of four faults were correctly diagnosed, the misdiagnosis of a single fault would mean that the entire mode is misdiagnosed. Misdiagnosis rates are therefore expected to be higher for modes than they are for components.

It was also found that the performance of the mode-space method was inferior to that of the component-space method. This is best explained by the fact that diagnosing based on component space is simpler than diagnosing based on modes. There are fewer distributions to estimate when diagnosing the presence of each fault. Furthermore, monitors not sensitive to the fault can be discarded, further reducing the dimensionality of the distributions to be estimated.

While the component-based method has its merits in this application, we should note that this method assumes that component states are independent of each other. That assumption holds true for this system as the leaks and bias were introduced independently. The mode-space approach does not make such assumptions, so if sufficient data is available for all modes, it becomes more practical, as it can better take into account interactions between states of different components.

Adding experimental data to the learning data set (first the normal mode, and then single-fault modes) was found to improve performance, especially in the mode-space approach. Improvements from adding experimental data indicated that the model was not completely able to replicate the process, but simulated data did provide valuable information for diagnosis, as performance was still acceptable when only simulated data was used as a reference. Adding experimental data had a more balanced effect when it was introduced in the component-space approach. Every time experimental data was included for a mode, half of the distributions in the component-space approach were affected, but only one distribution in the mode-space was affected. Thus, the component-space approach had a more evenly spread benefit.

These trends can be observed in Figure 10.4, where including the normal operating mode (m_1) decreased the misdiagnosis rate but left other modes fairly untouched. This is also true

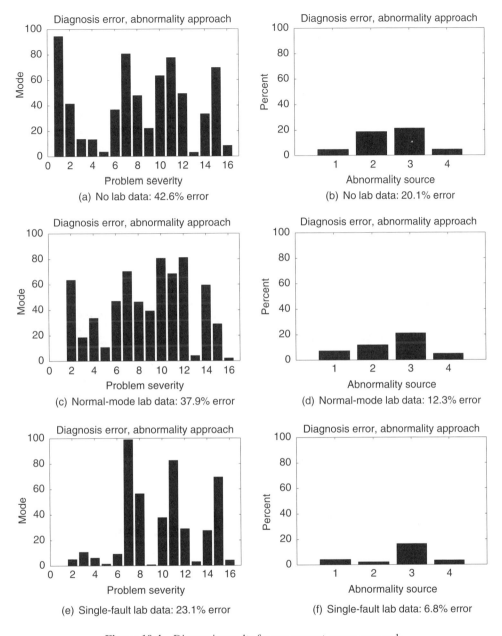

Figure 10.4 Diagnosis results for component-space approach

when experimental data from single-fault modes (m_2, m_3, m_5, m_9) was added: misdiagnosis rates for these modes decreased but the other modes were left untouched. Conversely, in Figure 10.5, including experimental data from these modes affected the misdiagnosis rate of all modes, usually resulting in a decreased misdiagnosis rate for modes on average.

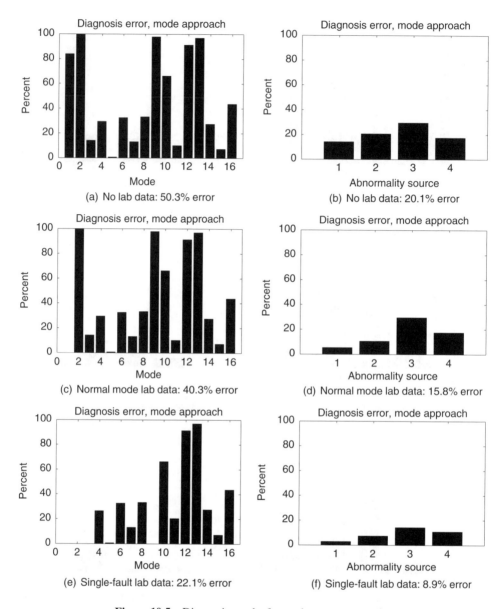

Figure 10.5 Diagnosis results for mode-space approach

10.5   Notes and References

Much of the material on bootstrapping on closed-loop systems originated from Qi and Huang (2011), while information on the smoothed bootstrapping procedure stemmed from Silverman (1987). Gray-box modeling, including residual error covariance estimation, has been extensively covered in Murphy (2002).

References

Murphy K 2002 *Dynamic Bayesian networks, representation, inference and learning*. PhD thesis, University of California, Berkeley.

Qi F and Huang B 2011 Estimation of distribution function for control valve stiction estimation. *Journal of Process Control* **28**(8), 1208–1216.

Silverman W 1987 The bootstrap: to smooth or not to smooth? *Biometrika* **74**(3), 469–479.

Part Two

Applications

11

Introduction to Testbed Systems

11.1 Simulated System

The system used for simulation is the well-known Tennessee Eastman process (Downs and Vogel 1993), which has been the testbed for a host of process control and fault diagnosis techniques (Chiang et al. 2000; Ku et al. 1995; McAvoy and Ye 1994; Qi and Huang 2011). A schematic is presented in Figure 11.1. In this system, the gaseous reactants A, D and E are fed directly into the reactor along with inert gas B; C comes in through the recycle stream. In the reactor, G and H, which are liquids under standard conditions, are formed according to the following reactions:

$$A(g) + C(g) + D(g) \rightarrow G(l)$$

$$A(g) + C(g) + E(g) \rightarrow H(l)$$

These reactions are irreversible, exothermic and approximately first-order with respect to reactant concentrations. Product from the reactor is condensed and sent to a separator to remove the reactants, which are much more volatile than the products. Liquid separator product is then stripped using reactant C as the stripping agent. The liquid product from the stripper is the final product. Meanwhile, the gaseous products from the stripper and separator are recycled to the reactor, with some of the separator product being purged in order to prevent a build-up of inert gas B.

The code used for simulation allows for 15 known pre-programmed process faults and uses the decentralized control strategy outlined by Ricker (1996). The code is available in the Tennessee Eastman Challenge Archive (2011). For our application, we consider a normal operating mode and seven other modes where each fault happens one at a time; modes with multiple faults are not considered. A list of these modes is presented in Table 11.1.

11.1.1 Monitor Design

In order to demonstrate the superiority of the Bayesian approach, monitors are chosen arbitrarily and some therefore have high false-alarm/misdetection rates. By selecting monitors with mediocre performance, the merits of our proposed methods can be more clearly seen.

Process Control System Fault Diagnosis: A Bayesian Approach, First Edition. Ruben Gonzalez, Fei Qi and Biao Huang.
© 2016 John Wiley & Sons, Ltd. Published 2016 by John Wiley & Sons, Ltd.

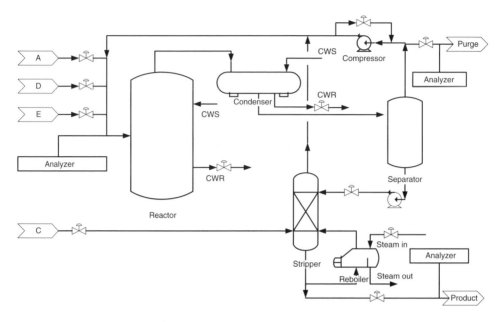

Figure 11.1   Tennessee Eastman process

Table 11.1   List of simulated modes

Variable number	Process variable	Type
NF	N/A	N/A
IDV 1	A/C feed ratio B composition constant (stream 4)	Step
IDV 2	B composition, A/C ratio constant (stream 4)	Step
IDV 7	C header pressure loss, reduced availability	Step
IDV 8	A, B, C feed composition (stream 4)	Variation
IDV 9	D feed temperature (stream 2)	Variation
IDV 12	Reactor cooling water inlet temperature	Variation
IDV 14	Reactor cooling water valve	Sticking

Control Performance Monitor

Six univariate control performance monitors are commissioned to monitor the control performance of the six key PVs. The FCOR algorithm (Huang and Shah 1999) is employed to compute control-performance indices based on univariate CVs.

Valve stiction monitor

According to Downs and Vogel (1993), the reactor cooling-water valve and the condenser cooling-water valve both have the potential to develop stiction. Two valve stiction monitors are commissioned to monitor these problems.

For illustrative purposes, we consider the following simplified scenario: if a control loop has oscillation, then the oscillation is caused either by valve stiction or by an external oscillatory disturbance. The latter has sinusoid form while the former does not.

If the CV and the manipulated variable (MV) of a control loop oscillate sinusoidally, an ellipse will be obtained when plotting CV versus MV. It has been observed that an ellipse will be distorted if the oscillation is caused by valve stiction. The method adopted here is based on the evaluation of how well the shape of the CV vs. MV plot can be fitted by an ellipse. An empirical threshold of distance between each data point and the ellipse is used to determine the goodness-of-fit, and thereafter the existence of valve stiction.

Process Model Validation Monitor

In addition to the control performance monitors, three additional model validation monitors are commissioned to monitor the change in the model of reactor level, separator level and stripper level.

The local approach based on the OE method (Ahmed et al. 2009) is employed to validate the nominal process model. This method applies to MISO systems; note that any MIMO system can be separated into several MISO subsystems. Models of each MISO part can be monitored with the local approach.

11.2   Bench-scale System

The hybrid tank system is a bench-scale process; a schematic is shown in Figure 11.2. The system consists of three tanks that can be interconnected through a system of valves. The two outer tanks have water inlets that can be used to control the water level, while all three tanks have outlets leading to the main water basin. In this book, only the lower valves (Valves 1 and 2) are used to interconnect the tanks. Furthermore, the outlet valve for the middle tank (Valve 8) is closed, while the other outlets remain open. Valve 1 and Valve 2 are opened and closed in order to create different operating modes. In addition, bias is added to Flow Meters 1 and 2 in order to create more operating modes (biased and unbiased). When coupled with the open and closed modes for the two valves, the system has 16 possible operating modes.

In addition, the system is equipped with different measuring devices. Flow Meters 1 and 2 have already been mentioned; these instruments measure water flow rates into Tanks 1 and 2 respectively. Control signals to the pumps are also measured (for Pumps 1 and 2), as well as the levels in the three tanks (Level Transmitters 1, 2 and 3).

Detection of different operating modes is performed using model validation algorithm. A gray box model is obtained for each tank, given as

$$\frac{d L_1}{d t} = A_c^{-1} \left(F_1 - C_1 L_1^{1/2} \right) \tag{11.1}$$

$$\frac{d L_2}{d t} = -A_c^{-1} \left(C_2 L_2^{1/2} \right) \tag{11.2}$$

$$\frac{d L_3}{d t} = A_c^{-1} \left(F_2 - C_3 L_3^{1/2} \right) \tag{11.3}$$

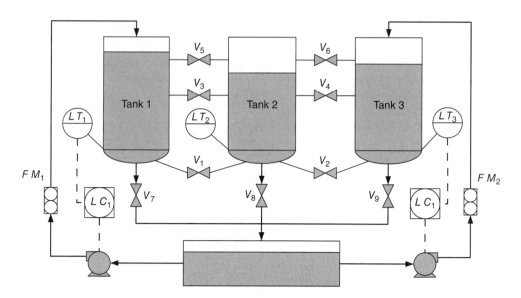

Figure 11.2   Hybrid tank system

where A_c is the cross-sectional area of the tanks, F_1, F_2 are flow rates into Tanks 1 and 3, respectively, C_1, C_2, C_3 are flow coefficients for each tank outlet, and L_2, L_2, L_3 are the tank levels for each of the tanks.

Once this model has been estimated, it is perturbed in order to include bias terms B_1, B_2 for the flow-rate measurements of water going into Tanks 1 and 3, and leak terms C_{L_1} and C_{L_2}, which are flow constants for Valves 1 and 2 when they are open.

$$\frac{d\,L_1}{d\,t} = A_c^{-1}\left[B_1F_1 - C_1L_1^{1/2} + C_{L_1}F_L(L_1, L_2)\right]$$

$$\frac{d\,L_2}{d\,t} = A_c^{-1}\left[-C_2L_2^{1/2} + C_{L_1}F_L(L_2, L_1) + C_{L_2}F_L(L_2, L_3)\right]$$

$$\frac{d\,L_3}{d\,t} = A_c^{-1}\left[B_2F_2 - C_3L_3^{1/2} + C_{L_2}F_L(L_3, L_2)\right]$$

where $F_L(L_1, L_2)$ is a function given as

$$F_L(L_1, L_2) = (L_2 - L_1)|L_2 - L_1|^{-1/2}$$

Using the UKF, the monitor estimates parameters $B_1, B_2, C_{L_1}, C_{L_2}$ as additional hidden states.

In addition, modeling flow rate as a function of pump signal gives additional estimates for bias terms B_1, B_2. The estimate is given as

$$B_1 = \frac{\hat{F}_1(U_1)}{F_1}$$

$$B_2 = \frac{\hat{F}_2(U_2)}{F_2}$$

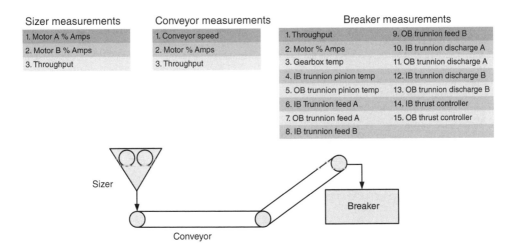

Figure 11.3   Solids handling system

where U_1 and U_2 are pump-control signals to Pumps 1 and 2, respectively, and $\hat{F}_1(U_1)$, $\hat{F}_2(U_2)$ are the respective flow predictions based on the pump signals. When these are combined with the level measurements there is a total of nine monitoring inputs to this diagnostic system.

11.3   Industrial Scale System

The industrial system considered as an example in this book is part of a solids handling facility in Canada's oil sands industry, similar to the one found in Gonzalez and Huang (2014). As part of the preparation for oil production, mined oil sand is crushed, sized, and made into slurry though a breaker system. Data obtained from this facility does not have faults, but each of the subsystems (the sizing subsystem, the conveyor and the breaker) operates under modes 'on' and 'off'. This leads to eight possible modes overall, but certain modes occur very infrequently due to the safety systems installed. For example, the conveyor is turned off if the breaker is turned off, and the sizer is turned off when the conveyor is turned off. Nevertheless, as rare as these modes can be, they do occur occasionally during transitional periods.

The system makes use of 20 unique measurements; the measurements along with a simple system schematic are given in Figure 11.3.

References

Ahmed S, Huang B and Shah S 2009 Validation of continuous-time models with delay. *Chemical Engineering Science* **64**(3), 443–454.

Chiang L, Russell E and Braatz R 2000 Fault diagnosis in chemical processes using Fisher discriminant analysis, discriminant partial least squares, and principal component analysis. *Chemometrics and Intelligent Laboratory* **50**, 243–252.

Downs J and Vogel E 1993 A plant-wide industrial process control problem. *Computers and Chemical Engineering* **17**, 245–255.

Gonzalez R and Huang B 2014 Control-loop diagnosis using continuous evidence through kernel density estimation. *Journal of Process Control* **24**(5), 640–651.

Huang B and Shah S 1999 *Performance Assessment of Control Loops*. Springer.

Ku W, Storer R and Georgakis C 1995 Disturbance detection and isolation by dynamic principal component analysis. *Chemometrics and Intelligent Laboratory Systems* **30**, 179–196.

McAvoy T and Ye N 1994 Base control for the Tennessee Eastman problem. *Computers and Chemical Engineering* **18**, 383–413.

Qi F and Huang B 2011 Estimation of distribution function for control valve stiction estimation. *Journal of Process Control* **28**(8), 1208–1216.

Ricker N 1996 Decentralized control of the Tennessee Eastman Challenge process. *Journal of Process Control* **6**, 205–221.

Tennessee Eastman Challenge Archive 2011. Available at http://depts.washington.edu/control/LARRY/TE/download .html.

12

Bayesian Diagnosis with Discrete Data

12.1 Introduction

In this chapter, all necessary process variables are assumed to be available for mode estimation. We further assume that all or some monitors are available for the components of interest in the process. Very often these monitors are designed specifically for certain problems. They are all subject to disturbances and thus false alarms, and each monitor can be sensitive to abnormalities of other problem sources. For instance, a controller performance monitor can also be sensitive to valve stiction. It is challenging to determine where the problem source is with several monitors issuing simultaneous alarms. Our goal is to determine the underlying mode of the process based on the outputs of all monitors.

There are several challenging issues in monitor synthesis according to Huang (2008). First, although the problem sources may be different, the symptoms can be similar. For instance, oscillations in a process can be introduced by a sticky valve or an improperly tuned controller. Second, all processes operate in an uncertain environment to some extent, and there are uncertainties in the links between the underlying process status and monitor readings due to disturbances. No monitor has a 100% successful detection rate or 0% false alarm rate, and thus a probabilistic framework should be built to describe the uncertainties. Lastly, it is also worth considering how to incorporate prior knowledge in the mode estimation system to improve diagnostic performance. Most of the existing monitoring methods are data-driven. However, incorporating a prior knowledge such as causal relations between variables is not only helpful, but is also necessary for an accurate diagnosis Huang (2008).

In the presence of disturbances, Bayesian inference provides a good framework to solve the mode-detection problem, and quantify the uncertainty in its conclusion. In the work of Pernestal (2007) and Pernestål and Nyberg (2007), a Bayesian approach is presented for diesel engine diagnosis based on a complete set of sensor readings. This chapter adopts this approach to the mode-estimation problem based on the readings of process monitors.

Process Control System Fault Diagnosis: A Bayesian Approach, First Edition. Ruben Gonzalez, Fei Qi and Biao Huang.
© 2016 John Wiley & Sons, Ltd. Published 2016 by John Wiley & Sons, Ltd.

12.2 Algorithm

Historical data samples may be dependent or independent, depending on how they are sampled and also how the disturbances affect the monitors. Monitor readings are often calculated or affected by a window of process data. If the windows are sampled far enough apart to avoid overlap and give sufficient space between windows, the monitor readings can be considered independent with respect to time. In this chapter, it is assumed that historical monitor readings are independent with respect to time:

$$p(\mathcal{D}) = p(d^1, d^2, \cdots, d^{\hat{N}}) = p(d^1)p(d^2)\cdots p(d^{\hat{N}}) \tag{12.1}$$

Given evidence E and the historical dataset $\mathcal{D}$, Bayes' rule can be formatted as follows:

$$p(M|E, \mathcal{D}) = \frac{p(E|M, \mathcal{D})p(M|\mathcal{D})}{p(E|\mathcal{D})} \tag{12.2}$$

where $p(M|E, \mathcal{D})$ is the posterior probability, being the conditional probability of mode M given evidence E, and the historical dataset $\mathcal{D}$. $p(E|M, \mathcal{D})$ is likelihood, or the probability of evidence E, conditioning on mode M with historical data $\mathcal{D}$. $p(M|\mathcal{D})$ is the prior probability of mode M, and $p(E|\mathcal{D})$ is a scaling factor, which can be calculated as

$$p(E|\mathcal{D}) = \sum_M p(E|M, \mathcal{D})p(M|\mathcal{D})$$

Note that historical data are selectively collected under a specific mode. Therefore they provide no information on the prior probability of mode, $p(M|\mathcal{D}) = p(M)$. As a result, Eqn (12.2) can be written as

$$p(M|E, \mathcal{D}) \propto p(E|M, \mathcal{D})p(M) \tag{12.3}$$

Since prior probability is determined by prior information, the main task of building a Bayesian mode-estimation system boils down to the estimation of the likelihood probability $p(E|M, \mathcal{D})$, and the objective is to make the estimated likelihood probability consistent with historical data $\mathcal{D}$.

As discussed in the Chapter 3 of this book, the likelihood estimation can be calculated using the following equation:

$$p(E = e_i|M = m_j, \mathcal{D}) = \frac{n_{i|m_j} + a_{i|m_j}}{N_{m_j} + A_{m_j}} \tag{12.4}$$

where $n_{i|m_j}$ is the number of historical samples with the evidence $E = e_i$, and mode $M = m_j$; $a_{i|m_j}$ is the number of prior samples that is assigned to evidence e_i under mode m_j; additionally,

$$N_{m_j} = \sum_i n_{i|m_j}$$

$$A_{m_j} = \sum_i a_{i|m_j}$$

Readers are referred to previous theoretical chapters of this book for the derivation of the above equation.

According to Eqn (12.4), the likelihood probability of evidence is determined by both prior samples and historical samples. As the number of historical data sample increases, the likelihood probability will shift to the relative frequency determined by the historical data samples, and the influence of the priors will diminish. The number of prior samples can be interpreted as the prior belief of the likelihood distribution. It is important to set nonzero prior sample numbers; otherwise the mode-estimation system may yield undesired results (Pernestal 2007). Consider an extreme case, for example, when there is only one sample in the historical dataset. Without any prior samples defined, the likelihood for the evidence corresponding to the single historical data sample will be one, while the likelihood of all the other evidence will be zero. We may consider that the numbers of the prior samples represent how strong the belief of the prior likelihood is. The larger the prior sample numbers are, the more confidence in the prior likelihood. If there is no prior information available, the numbers of prior samples of all evidences are set to the same value as a noninformative prior. In likelihood-estimation problems, $a_{i|m_j}$ are often assigned with a value of one.

To conclude, the Bayesian algorithm for process-mode estimation can be summarized as follows.

1. Build monitors to detect the status of the specific process components of interest.
2. Collect necessary process data samples from historical data records, where the underlying process mode is available.
3. Categorize the collected training data into different groups according to the underlying modes.
4. Choose a window size and segregate the data into sections without overlap.
5. Generate evidence by sending the segmented process data into the designed monitors.
6. Discretize the generated monitor readings according to predefined thresholds.
7. An evidence library can be built to include all discretized evidence results that appeared in the historical dataset.
8. Assign prior samples number $a_{i|m_j}$ for the combination of the ith evidence in the evidence library and mode m_j. A_{m_j} can be calculated as $A_{m_j} = \sum_i a_{i|m_j}$.
9. Count the numbers of each evidence for each mode in the historical dataset, namely, $n_{i|m_j}$ for the ith evidence in the library under mode m_j.
10. After obtaining $n_{i|m_j}$ for all evidences and modes, N_{m_j} can be readily calculated, $N_{m_j} = \sum_i n_{i|m_j}$.
11. Collect the process variable data that are to be diagnosed, and segregate the data into windows without overlap according to the window size determined in step 4.
12. Generate discrete diagnosis evidence with the monitors and discretization thresholds used in previous steps.
13. Find the diagnosis evidence in the evidence library. If the new evidence does not exist in the evidence library, then the corresponding historical evidence number is zero. Calculate the likelihood of the evidence for all modes with Eqn (12.4).
14. Choose appropriate prior probabilities for the modes.
15. Use Eqn (12.3) to calculate the posterior probabilities given the evidence. The mode with the largest posterior probability is then diagnosed as the underlying mode.
16. If there is more than one window of diagnosis data, calculate the average posterior probabilities for all windows. Pick the mode with largest average as the diagnosed mode.

The flow is shown in Figure 12.1.

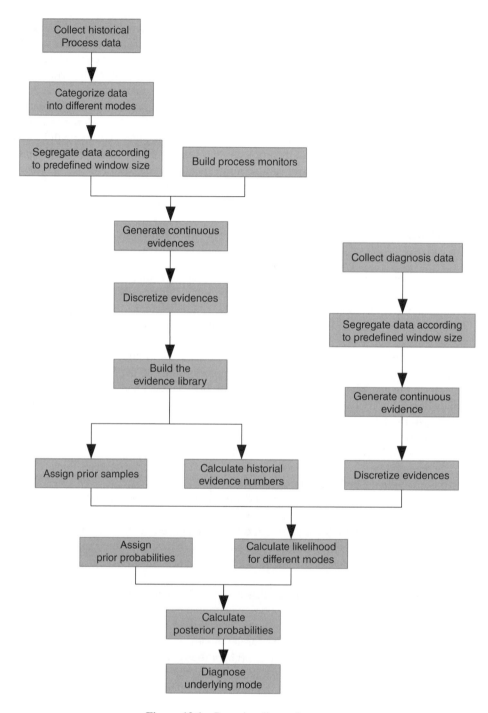

Figure 12.1   Bayesian diagnosis process

12.3   Tutorial

Here we use a process with two components – a valve subject to possible stiction, and a sensor subject to a possible bias problem – as an example for the Bayesian mode-estimation algorithm, as shown in Figure 12.2. Without loss of generality, we assume that the two faults will not occur simultaneously. Therefore, there are three operating modes for the process: the normal mode, the sticky valve mode, and the sensor bias mode.

Following the routines of the algorithm described in the previous section, the first step is to design monitors to detect the modes that are of interest to the users. In this example, the problems that we want to detect are the valve stiction and sensor bias. To pinpoint the status of the two components, a stiction-detection algorithm and a sensor-bias detection algorithm are used. The two monitors require controller output and process variable as inputs to the monitoring algorithm to perform diagnosis. Thus these two variables will be collected.

1000 s of historical process data are collected for each mode with sample rate of 1 s. Here we choose a window size of 100 s. Out of the 1000 s of process data, 10 windows of process data can be generated without overlap. When segregating the process data into evidence windows, there is a trade off between the window size and evidence number. The smaller the window, the more evidence samples can be obtained, which usually means a more accurate likelihood estimation. However, a smaller window size will include fewer process data in one piece of evidence, and thus produce a less accurate evidence result; in other words a worse detection rate.

In the next step, we use the 10 windows of process data as input to the two monitoring algorithms, and 10 evidence samples can be obtained. Since the Bayesian mode-estimation algorithm in this chapter is derived from discrete evidence, the continuous evidence obtained must be discretized. Generally the discretization threshold is chosen based on the specific problem that the monitor is designed to detect. In this tutorial, each continuous monitor output is discretized into two bins: 0 or 1. Using more bins can result in higher resolutions in terms of the evidence severity. However, too many discrete bins may lead to insufficient numbers of historical samples being assigned to each evidence bin, thus resulting in inaccurate likelihood estimation. In general, two bins are sufficient for most diagnosis problems.

Readings from each monitor are discretized into two bins with predefined thresholds; therefore there are four different evidence values in the evidence library, namely (0,0), (0,1), (1,0) and (1,1). The overall evidence space consists of four bins, as shown in Figure 12.3.

We use noninformative prior samples in this example. Each evidence bin is assigned with one prior sample, $a_{j|m} = 1$, $A_m = 4$, as shown in Figure 12.4.

The number of discrete evidence samples for each mode are shown in Table 12.1.

As an example, the evidence space for the biased sensor mode is shown with both prior samples and historical evidences in Figure 12.5.

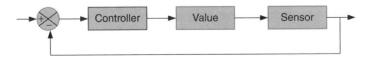

Figure 12.2   Illustrative process

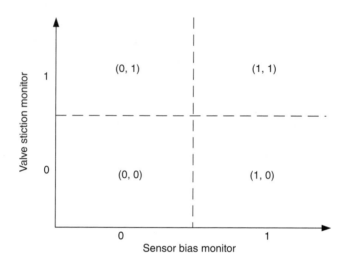

Figure 12.3   Evidence space with only prior samples

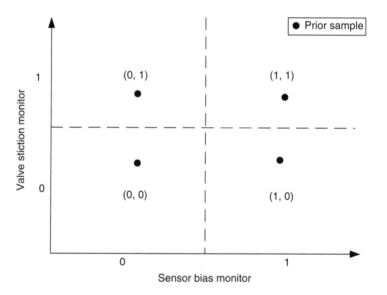

Figure 12.4   Evidence space with prior samples and historical samples

Table 12.1   Numbers of historical evidences

	Normal	Sticky valve	Biased sensor
(0,0)	10	0	0
(0,1)	0	1	7
(1,0)	0	8	1
(1,1)	0	1	2

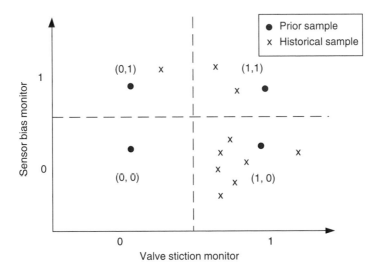

Figure 12.5   Evidence space with historical data

Table 12.2   Updated likelihood with historical data

	Normal	Sticky valve	Biased sensor
(0,0)	11/14	1/14	1/14
(0,1)	1/14	1/7	4/7
(1,0)	1/14	9/14	1/7
(1,1)	1/14	1/7	3/14

With all the variables in Eqn (12.4) available, we are ready to collect new process data that are to be diagnosed. The window size for the diagnosis data must be the same as the historical data. In this example, the diagnosis data will be segregated into windows of 100 samples without overlap. Using the same monitor algorithms and discretization thresholds as they are the historical data, continuous process data are converted into discrete evidence samples.

Before historical evidence data is collected, the likelihood of all four pieces of evidence for this mode is 1/4. With the historical data collected under the same underlying mode m, the likelihood probabilities can be updated according to Eqn (12.4), as presented in Table 12.2.

Consider the diagnosis of evidence (1,0). The likelihood of this evidence under different modes can be calculated according to Eqn (12.4). The likelihoods of this evidence under different modes are $p((1,0)|normal) = 1/14$, $p((1,0)|stickymode) = 9/14$, and $p((1,0)|biasedsensor) = 1/7$.

The next step is to assign a prior probability to each mode. The user can incorporate prior knowledge when deciding the priors. For instance, if we know that the valve has been in service for an extended period of time, then, based on our knowledge of how valves deteriorate, we can assign a higher prior probability to the "sticky valve" mode. In this case, the prior probabilities are assigned as $p(normal) = 1/4$, $p(stickyvalve) = 1/2$, and $p(biasedsensor) = 1/4$.

With the estimated likelihood probabilities $P(E|m_i, \mathcal{D})$ for current evidence E under different modes m_i and the user-defined prior probabilities $p(m_i)$, the posterior probabilities of each mode $m_i \in \mathcal{M}$ can be calculated according to Eqn (12.3). Among these modes, the one with the largest posterior probability is selected as the underlying mode based on the maximum a posteriori (MAP) principle. As an example, the posterior probability of different modes given evidence $(1, 0)$ can be calculated as

$$p(normal|(1,0)) \propto p(normal) \cdot p((1,0)|normal)$$

$$= 1/4 \cdot 1/14 = 1/56 \tag{12.5}$$

$$p(sticky\ valve|(1,0)) \propto p(sticky\ valve) \cdot p((1,0)|sticky\ valve)$$

$$= 1/2 \cdot 9/14 = 9/28 \tag{12.6}$$

$$p(biased\ sensor|(1,0)) \propto p(biased\ sensor) \cdot p((1,0)|biased\ sensor)$$

$$= 1/4 \cdot 1/7 = 1/28 \tag{12.7}$$

The sticky valve, has the largest posterior probability, and is then diagnosed as the underlying process mode.

12.4   Simulated Case

The Bayesian mode-estimation approach is applied to the Tennessee Eastman example that was described in the previous chapter. The diagnostic settings are summarized in Table 12.3.

The diagnostic results in Figure 12.6 are obtained from evaluation (cross-validation) data that are generated independently of historical samples. In Figure 12.6, the title of each plot denotes the true underlying mode, and the numbers on the horizontal axis stand for the diagnosed six possible modes numbered according to the sequence shown in the process description chapter. In each plot, the posterior probability corresponding to the true underlying mode is highlighted with gray bars, while the others are in black bars. The diagnostic conclusion is determined by selecting the mode with the largest posterior probability. If the largest probability happens to be the gray one, then the problem source is correctly identified. From Figure 12.6, we can see that all the true underlying modes are assigned with the largest posterior probabilities.

To further assess the performance of the Bayesian diagnosis algorithm the correct diagnosis rate of each point diagnosis is calculated for each mode. Table 12.4 shows the correct diagnosis rate for each mode.

Table 12.3   Summary of Bayesian diagnostic parameters for TE simulation problem

Discretization	$k_i = 2, K = 2^{13} = 8192$
Historical data	300 samples each mode
Prior samples	Uniformly distributed with prior sample $a_j = 1, A = 2^{13} = 8192$
Prior probabilities	Noninformative priors, $p(m) = 1/6$
Evaluation data	300 samples each mode

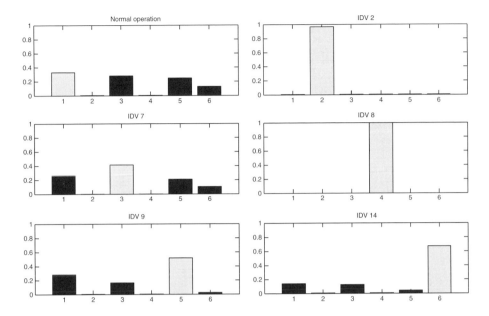

Figure 12.6   Posterior probability assigned to each mode for TE simulation problem

Table 12.4   Correct diagnosis rate

Mode	NF	IDV2	IDV7	IDV8	IDV9	IDV14
Diagnosis rate (%)	89.3	99.7	98.3	99.7	99	99.3

Even in the presence of low-performance monitors, the Bayesian approach can synthesize information from these monitors to provide good diagnostic results.

12.5   Bench-scale Case

The data-driven Bayesian diagnostic approach is also on the experimental three tank process. Table 12.5 summarizes parameters for the Bayesian diagnosis.

With the data-driven Bayesian approach, the diagnostic results shown in Figure 12.7 are obtained for the cross-validation data. In each plot, the posterior probability corresponding to the true underlying mode is shown with gray bars, while the others are in black. Thus if the

Table 12.5   Summary of Bayesian diagnostic parameters for pilot experimental problem

Historical data	1500 evidence samples for each mode
Prior samples	Uniformly distributed with prior sample $a_j = 1, A = 2^7 = 128$
Prior probabilities	Noninformative priors, $p(m) = 1/16$
Evaluation data	1500 evidence samples mode

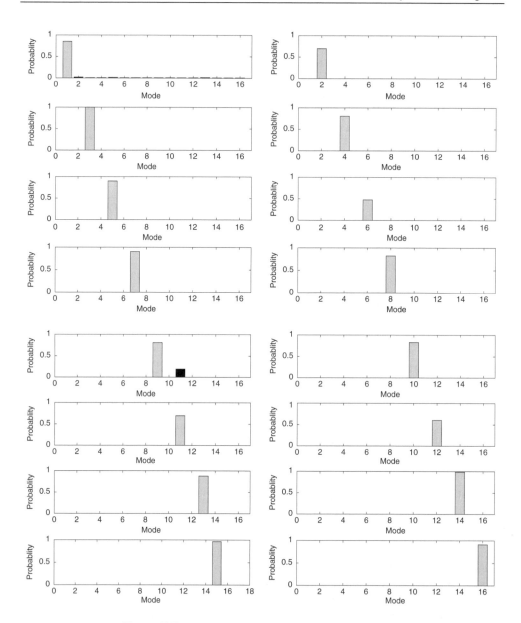

Figure 12.7 Posterior probability assigned to each mode

gray bar is highest then the correct diagnosis is obtained. From Figure 12.7, we can see that
all the true underlying modes are assigned with the largest probabilities. Thanks to abundant
historical data, the probabilities assigned to each underlying mode are all significantly larger
than the other probabilities. As a result, the correct point diagnosis is very high. Except for
Mode (6), which has a 98.1% correct diagnosis rate, the other modes all have a diagnosis rate
higher than 99%.

12.6   Industrial-scale Case

We also applied the Bayesian diagnosis approach to the industrial data. Four modes that appeared in both historical and validation datasets are selected for demonstrating the Bayesian diagnostic method. The diagnosis settings are summarized in Table 12.6.

The average posterior probabilities are shown in Figure 12.8. Due to the presence of significant noise and disturbances associated with Mode (3), the underlying mode is not assigned with the highest posterior probability. This demonstrates a drawback of the discrete Bayesian approach: the diagnosis performance is largely affected by the evidence discretization. If the distribution of an evidence is corrupted by external noise or disturbance, it will be difficult to find a proper discretization threshold. The problem, however, can be solved with the continuous Bayesian diagnosis approach, as discussed in other chapters of this book. The point correct diagnosis rate also agrees with the posterior probability results, as summarized in Table 12.7. The other three modes all have very high diagnosis rate, except for Mode (3).

Table 12.6   Summary of Bayesian diagnostic parameters for industrial problem

Prior samples	Uniformly distributed with prior sample $a_j = 1, A = 2^7 = 128$
Prior probabilities	Noninformative uniform distribution $p(m) = 1/4$

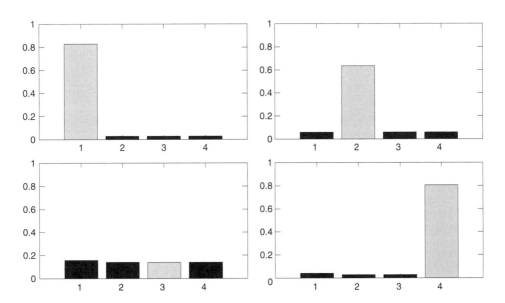

Figure 12.8   Posterior probability assigned to each mode for industrial process

Table 12.7   Correct diagnosis rate

Mode	Mode 1	Mode 2	Mode 3	Mode 4
Diagnosis rate (%)	100	93.4	0	100

12.7   Notes and References

For more detail on Bayesian diagnosis on control systems, the readers are referred to Huang (2008). The main references of this chapter are Qi and Huang (2008) and Qi et al. (2010).

References

Huang B 2008 Bayesian methods for control loop monitoring and diagnosis. *Journal of Process Control* **18**, 829–838.

Pernestal A 2007 *A Bayesian approach to fault isolation with application to diesel engine diagnose* PhD thesis KTH School of Electrical Engineering, Sweden.

Pernestål A and Nyberg M 2007 Using prior information in Bayesian inference - with application to diagnosis MaxEnt, Bayesian inference and maximum entropy methods in science and engineering. Albany, NY, USA.

Qi F and Huang B 2008 Data-driven Bayesian approach for control loop diagnosis *Proceeding of Amecican Control Conference*, Seattle, U.S.A.

Qi F, Huang B and Tamayo E 2010 A Bayesian approach for control loop diagnosis with missing data. *AIChE Journal* **56**, 179–195.

13

Accounting for Autodependent Modes and Evidence

13.1 Introduction

In statistics, it is a common assumption that the underlying mode (e.g. a fault) is independently identically distributed (i.i.d.). However, this is not always the case in engineering practice. For example, the occurrence of a fault can be dependent on process operation conditions. When a process operates near a high-risk region, it is more likely that faults will occur. On the other hand, if an instrument is already faulty, it is unlikely to revert to a normal condition without being repaired. Occurrence of faults can also follow a seasonal pattern. So a mode can be dependent on its past values. This dependence can actually provide useful information for diagnosis.

In some applications, and with some monitoring algorithms in particular, evidence is collected and summarized over windows of data. The independence among evidence samples depends on how the evidence data is sampled – for example, the sampling rate – and how the disturbance affects the monitor outputs. If the evidence samples are collected at sufficiently large time intervals, or if the disturbance has little to no correlation among the evidence samples, the evidence samples can be considered independent. Generally, the first requirement regarding the sampling interval can be easily satisfied by leaving a sufficient gap between consecutive monitor readings. If disturbances have long-term autocorrelation and the gap between consecutive monitor readings is not large enough, then the assumption of independence cannot apply to the evidence. The limitation of the conventional Bayesian method without considering dependence of evidence is that the temporal information has not been completely explored, leading to less efficient diagnostic performance. In summary, it is desirable to take temporal evidence into consideration when building the diagnostic model.

By considering both mode and evidence dependence, this chapter provides a comprehensive Bayesian diagnosis solution when the evidence exhibits temporal dependence.

Process Control System Fault Diagnosis: A Bayesian Approach, First Edition. Ruben Gonzalez, Fei Qi and Biao Huang.
© 2016 John Wiley & Sons, Ltd. Published 2016 by John Wiley & Sons, Ltd.

13.2 Algorithms

By considering both mode and evidence dependence, a dynamic Bayesian model can be formulated, as discussed in Ghahramani (2002). By adding dependency in both the modes and evidence, one will obtain a Bayesian network that resembles Figure 13.1, which is also known as the autoregressive hidden Markov model (Murphy 2002).

The inference of temporal-dependent models can be solved recursively to determine the posterior of the current mode M^t,

$$p(M^t|E^1, \cdots, E^t) \propto p(E^t|M^t, E^{t-1}) \sum_{M^{t-1}} p(M^{t-1}|E^1, \cdots, E^{t-1}) p(M^t|M^{t-1}) \quad (13.1)$$

To calculate the posterior given the previous mode posterior $p(M^{t-1}|E^1, \cdots, E^{t-1})$, two additional transition probabilities are required: the evidence transition probability $p(E^t|M^t, E^{t-1})$ and the mode transition probability $p(M^t|M^{t-1})$.

13.2.1 Evidence Transition Probability

The intent behind estimating the evidence transition probability is to obtain probabilities that are consistent with the historical data $\mathcal{D}$ when evidence is autodependent. The target result is the likelihood of evidence E^t given current underlying mode M^t and previous evidence E^{t-1}. Thus, every evidence transition sample should include these three pieces of information:

$$d_E^{t-1} = \{M^t, E^{t-1}, E^t\} \quad (13.2)$$

By gathering this information over the available history, we obtain an evidence transition dataset $\mathcal{D}_E$

$$\mathcal{D}_E = \{d_E^1, \cdots, d_E^{\hat{N}-1}\}$$
$$= \{(M^2, E^1, E^2), \cdots, (M^{\hat{N}}, E^{\hat{N}-1}, E^{\hat{N}})\} \quad (13.3)$$

which may also be called the transition dataset for simplicity. Figure 13.2 depicts how the collected historical evidence is divided to form transition samples. Two disjointed transition samples are indicated by hatched L-shaped regions and another overlapping sample by shaded nodes. These three regions are considered as three distinct samples under Eqn (13.2).

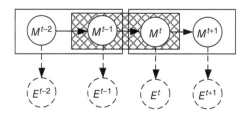

Figure 13.1 Dynamic Bayesian model that considers both mode and evidence dependence

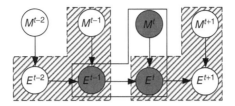

Figure 13.2   Illustration of evidence transition samples

Let us consider the transition probability from $E^{t-1} = e_s$ to $E^t = e_r$ under mode $M^t = m_c$ from a transition dataset

$$p(E^t|E^{t-1}, M^t, \mathcal{D}_E) = p(e_r|e_s, m_c, \mathcal{D}_E) \tag{13.4}$$

where

$$e_s, e_r \in \mathcal{E} = \{e_1, \cdots, e_K\} \tag{13.5}$$

and

$$m_c \in \mathcal{M} = \{m_1, \cdots, m_Q\} \tag{13.6}$$

It is clear that the evidence transition probability $p(e_r|e_s, m_c, \mathcal{D}_E)$ can only be estimated from the subset $\mathcal{D}_{E|m_c}$ where the mode $M^t = m_c$ occurs; that is,

$$p(e_r|e_s, m_c, \mathcal{D}_E) = p(e_r|e_s, m_c, \mathcal{D}_{E|m_c}, \mathcal{D}_{E|\neg m_c})$$
$$= p(e_r|e_s, m_c, \mathcal{D}_{E|m_c}) \tag{13.7}$$

where $\mathcal{D}_{E|\neg m_c}$ represents the set of historical transition data when the underlying mode M^t is not m_c. For simplicity of notation, the subscript m_c will be omitted when it is clear from the context.

The evidence transition probability can be estimated as

$$p(e_r|e_s, m_c, \mathcal{D}_E)$$
$$= \frac{\tilde{n}_{s,r} + b_{s,r}}{\tilde{N}_s + B_s} \tag{13.8}$$

Rewriting the evidence likelihood equation without considering evidence autodependence, we have

$$p(e_i|m_c, \mathcal{D}) = \frac{n_i + a_i}{N + A} \tag{13.9}$$

By comparing Eqns (13.8) and (13.9), we can see that the evidence transition probability is also determined by both prior samples and historical samples, similar to the evidence likelihood calculation with no autodependence. The difference lies in how the numbers of prior and historical samples are counted. Both equations use counts under a specific mode, but in Eqn (13.9) the prior and historical samples are single evidence counts, while in Eqn (13.8) the prior and historical samples are evidence transition counts.

As the number of historical transition samples increases, the transition probability will converge to the relative frequency determined by the historical samples, reducing the influence of prior samples. It is important to set nonzero prior sample numbers; otherwise unexpected results may occur in the case of limited historical samples (Pernestal 2007).

It should be noted that the evidence transition space is much larger than the single evidence space (the transition-space size is actually the square of the single-evidence space). Such large dimensions could make the data collection requirements quite problematic. However, by considering independence between evidence sources, the dimension can be reduced. Consider a diagnostic system with ten monitors,

$$E = \{\pi_1, \pi_2, \cdots, \pi_{10}\}$$

where each single monitor output is discretized into two bins. The total number of possible evidence values equals

$$K = 2^{10} = 1024 \tag{13.10}$$

and the total number of possible evidence transitions equals

$$K' = 2^{10} \cdot 2^{10} = 1048576 \tag{13.11}$$

When considering evidence from the previous time sample, the evidence space and the number of parameters (corresponding the probability for each possible evidence), has been squared. The significant increase in the estimated parameters will require a corresponding increase in the amount of data needed to reliably estimate such a likelihood distribution.

Obtaining such a large number of historical samples is challenging, especially for an industrial process where the available historical data corresponding to faulty modes may be sparse. To mitigate this problem, we propose an approximate solution based on correlation ratio analysis of the evidence. This serves to reduce the dimension of the evidence transition space and brings some relief to the large historical evidence data requirement. By analysing dependence of the evidence data, we can divide the monitor readings into two groups: the autodependent ones and the auto-independent ones.

The correlation ratio, which measures the functional dependence of two random variables, is used to quantify the dependence between the monitors. In contrast to the correlation coefficient, the correlation ratio is capable of detecting almost any functional dependence, not only a linear dependence (Duluc et al. 1996).

For illustration, consider a diagnostic system with one monitor, $E = \{\pi\}$. Suppose that N evidence samples are collected and are represented as

$$\{(\pi^1), (\pi^2), \cdots, (\pi^N)\}$$

where π^t is the ith collected sample of π. Assume that the monitor has two discrete-valued outcomes: 0 and 1. To estimate the correlation ratio of π^t on π^{t-1}, shift the data sequence of π by one sample and then create a set of paired data as follows:

$$\{(\pi^1, \pi^2), (\pi^2, \pi^3), \cdots, (\pi^{N-1}, \pi^N)\}$$

The collected samples of π^t are classified into two categories according to first value of each pair:

$$The\ 1st\ value\ of\ the\ pair = 0: \quad \pi^{0,1}\ \pi^{0,2} \cdots \pi^{0,n_0}$$

$$The\ 1st\ value\ of\ the\ pair = 1: \quad \pi^{1,1}\ \pi^{1,2} \cdots \pi^{1,n_1}$$

where $\pi^{i,j}$ is the second element of the jth pair whose first value is i.

Following Matthaus et al. (2008), the correlation ratio of π^t on the π^{t-1} can be calculated as

$$\eta(\pi^t|\pi^{t-1}) = \sqrt{\frac{\sum_{i=0}^{1} n_i(\overline{\pi}^i - \overline{\pi})^2}{\sum_{i,j}^{1,n_i}(\pi^{i,j} - \overline{\pi})^2}} \tag{13.12}$$

where $i = 0, 1$, where $\overline{\pi}^i$ is the average of the monitor readings in the ith category, and where $\overline{\pi}$ is the average of all the collected monitor readings. The calculated correlation ratio quantifies the dependence of π^t on π^{t-1}, namely the effect of the value of π^{t-1} on the value of π^t. This value is an indication of how the previous π value affects current π value. $\eta(\pi^t|\pi^{t-1}) = 0$ means that π is auto-independent, while $\eta(\pi^t|\pi^{t-1}) = 1$ indicates that current output of π is completely determined by its previous value. Generally if $\eta(\pi^t|\pi^{t-1})$ is sufficiently small, π can be treated as auto-independent. Similarly, we can calculate the cross-correlation ratio between different monitor outputs, say π_1 and π_2, by creating a new set of paired data

$$\{(\pi_1^1, \pi_2^1), (\pi_1^2, \pi_2^2), \cdots, (\pi_1^N, \pi_2^N)\}$$

By calculating the autocorrelation ratio of all the monitors, the temporal dependence of each may be determined. Now let us assume that the first q monitors generate auto-independent outputs, while the remaining monitors do not. The body of evidence can now be divided into two portions,

$$E = \{\pi_1, \pi_2, \cdots, \pi_q \vdots \pi_{q+1}, \cdots, \pi_L\} = \{E_c, E_{ic}\} \tag{13.13}$$

where $E_c = \{\pi_1, \pi_2, \cdots, \pi_q\}$ is the set of autodependent monitors and E_{ic} is the set of auto-independent monitors. Using the newly defined notation, the evidence transition probability can be written as

$$p(E^t|E^{t-1}, M^t) = p(E_c^t, E_{ic}^t|E_c^{t-1}, E_{ic}^{t-1}, M^t) \tag{13.14}$$

By assuming that E_c and E_{ic} are mutually independent – this may be tested approximately via the cross-correlation ratio – the probability expression reduces to

$$p(E_c^t, E_{ic}^t|E_c^{t-1}, E_{ic}^{t-1}, M^t) = p(E_c^t|E_c^{t-1}, E_{ic}^{t-1}, M^t)p(E_{ic}^t|E_c^{t-1}, E_{ic}^{t-1}, M^t) \tag{13.15}$$

Recall that E_{ic}^t only depends on the underlying mode M^t and E_c^t depends on both the underlying mode M^t and the previous evidence, E_c^{t-1}. Eqn (13.15) becomes

$$p(E_c^t, E_{ic}^t|E_c^{t-1}, E_{ic}^{t-1}, M^t)$$

$$= p(E_c^t|E_c^{t-1}, E_{ic}^{t-1}, M^t)p(E_{ic}^t|E_c^{t-1}, E_{ic}^{t-1}, M^t)$$

$$= p(E_c^t|E_c^{t-1}, M^t)p(E_{ic}^t|M^t) \tag{13.16}$$

which means that the probability of the evidence transition is the product between the likelihoods of the temporally dependent and temporally independent bodies of evidence. Thus the bodies of temporally dependent and independent evidence can be estimated separately. Let us consider a set of monitors in which each monitor π_i has k_i discrete outputs. The total number of possible complete evidence transitions is

$$K' = \prod_{i=1}^{L} k_i \cdot \prod_{i=1}^{L} k_i = \prod_{i=1}^{L} k_i^2 \tag{13.17}$$

In Eqn (13.16), the number of partial evidence transitions is $\prod_{i=1}^{q} k_i^2 < K'$. Consequently fewer historical data samples are required to calculate the evidence transition probability. Better diagnostic results are expected with the same quantity of historical data samples.

In summary, the following routine can be used to estimate the evidence transition probability.

1. Build monitors to detect the status of specific process components of interest (if monitor data is not already available).
2. Collect the necessary process data samples from the historical data record, where the underlying process mode is available.
3. Categorize the collected training data into different groups according to the underlying modes.
4. If monitoring data is not directly available, generate evidence by sending the process data into the designed monitors.
5. Discretize the generated monitor readings according to predefined thresholds.
6. Calculate the autocorrelation ratio of each monitor, and if any monitors exhibit autodependence, divide the evidence into the autodependent and auto-independent groups.
7. Calculate the correlation ratio between the monitor groups.
8. If the two monitor groups do not pass the correlation ratio test, the evidence space reduction cannot be achieved. Eqn (13.9) has to be strictly followed to calculate the complete evidence transition likelihood.
9. Count the numbers of evidence transitions $\tilde{n}_{i,j}$ from evidence $\tilde{e}_i$ to $\tilde{e}_j$ under mode m_k.
10. Assign a prior sample number $b_{i,j|m_k}$ for each transition. B_{m_k} can be calculated as $B_{m_k} = \sum_{i,j} b_{i,j|m_k}$.
11. After obtaining $\tilde{n}_{i,j|m_k}$ and $\eta_{i|m_k}$ for all evidence and modes, N_{m_k} can be calculated as $N_{m_k} = \sum_i \eta_{i|m_k} + \sum_j \eta_{i,j|m_k}$.
12. If the two groups are independent, the evidence space can be reduced. The transition likelihood of the full evidence can be calculated by multiplying the likelihoods of the independent groups.
13. The combined likelihood can then be used with the prior to obtain a posterior probability.

13.2.2 Mode Transition Probability

Accounting for autodependent modes, much like autodependent evidence, requires the collected data to have two elements:

$$d_M^{t-1} = \{M^{t-1}, M^t\} \tag{13.18}$$

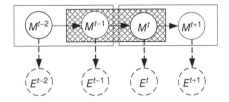

Figure 13.3   Historical composite mode dataset for mode transition probability estimation

It should be noted that since our focus is only on the mode transition probability and not the evidence transition, the composite mode sample d_M^t includes only the mode transitions. Accordingly, the new composite mode dataset $\mathcal{D}_M$ is assembled from the historical evidence set $\mathcal{D}$ to estimate the mode transition probability. This composite dataset is expressed as

$$\mathcal{D}_M = \{d_M^1, \cdots, d_M^{\tilde{N}-1}\}$$
$$= \{(M^1, M^2), \cdots, (M^{\tilde{N}-1}, M^{\tilde{N}})\} \tag{13.19}$$

Figure 13.3 depicts how the collected historical evidence data is organized to form composite mode samples. Here, three pairs of nodes are highlighted; two distinct pairs are boxed by white rectangles and an overlapping pair is boxed by a cross-hatched rectangle. Each these three pairs constitute a single composite mode sample, as described by Eqn (13.18).

The mode transition probability can be derived as

$$p(m_v | m_u, \mathcal{D}_M) = \frac{\hat{n}_{u,v} + c_{u,v}}{\hat{N}_u + C_u} \tag{13.20}$$

where $\hat{n}_{u,v}$ is the number of mode transitions from m_u to m_v in the historical composite mode dataset, $\hat{N}_u = \sum_j \hat{n}_{u,j}$ is the total number of mode transitions from mode m_u to any other mode, $c_{u,v}$ is the number of prior samples for the mode transition from m_u to m_v and $C_u = \sum_j c_{u,j}$ is the total number of prior mode transitions from m_u to any other mode.

The space-reduction solution that was used for estimating the evidence transition probability is usually unnecessary for mode transitions. There are generally two reasons for this:

1. We are often only interested in a small set of the possible modes, namely the modes that have already appeared in the data. Due to the fact that modes change much less frequently than evidence, the space of interest will often be smaller.
2. Modes represent a set of states given by process components – dealing with components is discussed in detail in Chapter 10 – and most of these process component states can be treated as independent of each other. Even if dependence exists, the effect is often weak, for example the failure of one engine component can cause stress to another, but it may take a considerable time for that stress to translate to a failure. In such a case, dependence between components would be weak enough not to warrant taking it into account. In cases where dependence is strong and the failure of one component often quickly leads to the failure of another, these two components can often be lumped together as a single component.

In summary, the following routine can be used to estimate the mode transition probability. Note that some of the common steps to collect historical data and build monitors are not included.

1. Build the composite mode sample data from the historical data following Figure 13.3.
2. Count the number of mode transitions $n_{i,j}$, from mode m_i to m_j.
3. Assign prior samples number $a_{i,j}$ for the transition from the mode m_i to mode m_j. A_{m_i} can be calculated as $A_{m_i} = \sum_j a_{i,j}$.
4. After obtaining $n_{i,j}$ and $\eta_{i|m_k}$ for all evidence and modes, N_{m_k} can be calculated as $N_{m_k} = \sum_i \eta_{i|m_k} + \sum_j \eta_{i,j|m_k}$.
5. The mode transition probability for mode transition mode m_i to m_j can be calculated using Eqn (13.20).

With both evidence and mode transition probability estimated, the posterior probability for the underlying mode can be readily calculated with Eqn (13.1).

13.3 Tutorial

As an example, consider a process with two monitors, π_1 and π_2. Both monitors have two discrete outputs, 0 and 1. The process has two modes, m_1 and m_2. There are four different possible evidence values in total, namely (0,0), (0,1), (1,0) and (1,1).

The evidence transitions between two evidence samples can have $4^2 = 16$ combinations. The evidence transition data $\{M^t, E^{t-1}, E^t\}$ collected when mode $M^t = m_1$ is summarized as follows:

$$
\begin{array}{ccccc}
 & (0,0) & (0,1) & (1,0) & (1,1) \\
(0,0) & 1 & 9 & 1 & 9 \\
(0,1) & 8 & 2 & 9 & 1 \\
(1,0) & 2 & 8 & 0 & 10 \\
(1,1) & 9 & 1 & 8 & 2
\end{array}
\tag{13.21}
$$

Following the steps in the previous section, the first step is to calculate the correlation ratio for each monitor, π_1 and π_2. Here we use π_2 to show how the correlation ratio is calculated. There are four value transition combinations. The number of each change can be calculated from the matrix in Eqn (13.21),

$$
\begin{array}{ccc}
 & 0 & 1 \\
0 & 4 & 37 \\
1 & 34 & 6
\end{array}
\tag{13.22}
$$

With Eqn (13.12), the correlation ratio of π_2 is calculated as

$$
\eta(\pi^t | \pi^{t-1}) = \sqrt{\frac{\sum_{i=0}^1 n_i (\overline{\pi}^i - \overline{\pi})^2}{\sum_{i,j}^{1,n_i} (\pi^{i,j} - \overline{\pi})^2}},
\tag{13.23}
$$

$$
= \sqrt{\frac{n_0 * (\overline{\pi}^0 - \overline{\pi})^2 + n_1 * (\overline{\pi}^1 - \overline{\pi})^2}{\sum_{i,j}^{1,n_i} (\pi^{i,j} - \overline{\pi})^2}}
\tag{13.24}
$$

$$= \sqrt{\frac{41 \cdot (0.9 - 0.53)^2 + 40 \cdot (0.15 - 0.53)^2}{(37 + 6)(1 - 0.53)^2}} \qquad (13.25)$$

$$= 0.753 \qquad (13.26)$$

The correlation ratio is close to 1, so it can be determined that π_2 is autodependent. A similar process can be applied to π_1. The autocorrelation ratio in this example is 0 indicating no autocorrelation; the cross correlation was also 0. Thus π_1 does not have to be considered in the evidence transition probability estimate. In other words, the evidence transition probability for m_1, $p((\pi_1^t, \pi_2^t)|(\pi_1^{t-1}, \pi_2^{t-1}), m_1)$ can be simplified as

$$p((\pi_1^t, \pi_2^t)|(\pi_1^{t-1}, \pi_2^{t-1}), m_1) = p(\pi_2^t | \pi_2^{t-1}, m_1) \cdot p(\pi_1^t | m_1) \qquad (13.27)$$

As an illustration, consider the evidence transition probability of $p((0,1)|(0,0), m_1)$. The probability can be written as

$$p((0,1)|(0,0), m_1) = p(\pi_2^t = 1 | \pi_2^{t-1} = 0, m_1) \cdot p(\pi_1^t = 0 | m_1) \qquad (13.28)$$

Following Eqn (13.8), the transition probability of $\pi_2^{t-1} = 0$ to $\pi_2^t = 1$ can be calculated as

$$p(\pi_2^t = 1 | \pi_2^{t-1} = 0, m_1) = \frac{\tilde{n}_{0,1} + b_{0,1}}{\tilde{N}_0 + B_0} \qquad (13.29)$$

In the collected historical data for π_2, there are 37 transition samples from 0 to 1 and 4 samples from 0 to 0. Thus $\tilde{n}_{0,1} = 37$, $\tilde{n}_{0,0} = 4$ and $\tilde{N}_0 = \tilde{n}_{0,1} + \tilde{n}_{0,0} = 41$. Next, we assign one prior sample to each of the possible transitions, i.e. $b_{0,1} = b_{0,0} = 1$ and $B_0 = b_{0,1} + b_{0,0} = 2$. As a result, we have

$$p(\pi_2^t = 1 | \pi_2^{t-1} = 0, m_1) = \frac{\tilde{n}_{0,1} + b_{0,1}}{\tilde{N}_0 + B_0} \qquad (13.30)$$

$$= \frac{37 + 1}{41 + 2} = 0.8837 \qquad (13.31)$$

The estimation procedure for $p(\pi_1^t = 0 | m_1)$ is described in Chapter 12, and will not be repeated in this chapter. With $p(\pi_1^t = 0 | m_1)$ and $p(\pi_2^t = 1 | \pi_2^{t-1} = 0, m_1)$ both calculated from the historical data, the evidence transition probability from (0,0) to (0,1) can be estimated using Eqn (13.28). Here we only calculate four-monitor transition probabilities for π_2 and two single-monitor likelihood probabilities for π_1. If the whole evidence transition including both π_1 and π_2 is to be considered, in total there will be 16 different evidence transition probabilities to estimate. It should be noted that the correlation ratio analysis should be performed for every mode. Due to possible changes in disturbance dynamics under different modes, bodies of evidence that are independent under one mode may not retain independence under others.

The estimation of the mode transition probability is more straightforward than that of the evidence transition. The collected mode transition data is summarized as follows:

$$
\begin{array}{c c c}
 & m_1 & m_2 \\
m_1 & 40 & 5 \\
m_2 & 2 & 30
\end{array}
\qquad (13.32)
$$

To estimate the transition from mode m_1 to m_2, for example,

$$p(m_2|m_1) = \frac{\hat{n}_{1,2} + c_{1,2}}{\hat{N}_1 + C_1} \tag{13.33}$$

where $\hat{n}_{1,2} = 5$, $\hat{N}_1 = \hat{n}_{1,2} + \hat{n}_{1,1} = 45$. Assign one prior sample to each mode transition:

$$c_{1,1} = c_{1,2} = c_{2,1} = c_{2,2} = 1$$
$$\hat{N}_1 = c_{1,1} + c_{1,2} = 2$$

Then the mode transition probability of $p(m_2|m_1)$ is calculated as

$$p(m_2|m_1) = \frac{\hat{n}_{1,2} + c_{1,2}}{\hat{N}_1 + C_1}$$

$$= \frac{5+1}{45+2} = 0.1277 \tag{13.34}$$

With both the evidence transition probability and mode transition probability obtained, the posterior probability of the current mode to be diagnosed can be calculated with Eqn (13.1).

$$p(M^t|E^1, \cdots, E^t) \propto p(E^t|M^t, E^{t-1}) \sum_{M^{t-1}} p(M^{t-1}|E^1, \cdots, E^{t-1})p(M^t|M^{t-1}) \tag{13.35}$$

Note that when the first sample at $t = 1$ is collected, the sample at 0 time stamp does not exist. Thus the posterior probability becomes

$$p(M^1|E^1) \propto p(E^1|M^1) \cdot p(M^1) \tag{13.36}$$

which is the same calculation that is performed when mode and evidence correlation are both ignored.

For evidence samples collected after the first one, the posterior of mode m_1, can be calculated as

$$\begin{aligned} p(m_1|e^1, \cdots, e^t) &\propto p(e^t|e^{t-1}, m_1) \cdot (p(m_1|e^1, \cdots, e^{t-1}) \cdot p(m_1|m_1) \\ &\quad + p(m_1|e^1, \cdots, e^{t-1}) \cdot p(m_1|m_2)) \\ &\propto p(\pi_1^t|m_1)p(\pi_2^t|\pi_2^{t-1}, m_1) \cdot (p(m_1|e^1, \cdots, e^{t-1}) \cdot p(m_1|m_1) \\ &\quad + p(m_1|e^1, \cdots, e^{t-1}) \cdot p(m_1|m_2)) \end{aligned} \tag{13.37}$$

Similarly the posterior of mode m_2 can also be calculated. The mode with higher probability is to be picked as the current underlying mode.

The algorithms presented in this chapter apply to smaller scale diagnostic problems better. For a large scale problem, unless the evidence space can be separated into smaller dimension groups through the cross dependence test, these algorithms may suffer from the curse of dimension problem and generate poorer diagnostic performance. An alternative solution is discussed in Chapter 18. This does not discretize the data, but instead uses continuous data directly, by means of kernel density estimation. Kernel density estimation methods do not suffer from evidence space issues to the degree that discrete methods do, and thus applying dynamic methods to kernel density estimation methods yields improved results. Readers are referred to Chapter 18 for application examples of the dynamic Bayesian methods for auto-dependent mode and evidence.

13.4   Notes and References

Qi and Huang (2011) details the solution for dependent evidence, and Qi and Huang (2010) provides a comprehensive result by considering both dependent evidence and mode. Readers can refer to Murphy (2002) for further details on the hidden Markov model and to Ghahramani (2002) for a dynamic Bayesian model.

References

Duluc J, Zimmer T, Milet N and Dom J 1996 A wafer level reliability method for short-loop processing. *Microelectronics Reliability* **36**(11/12), 1859–1862.

Ghahramani Z 2002 An introduction to hidden Markov models and Bayesian networks. *International Journal of Pattern Recognition and Artifical Intelligence* **15**(1), 9–42.

Matthaus L, Trillegberg P, Fadini T, Finke M and Schweikard A 2008 brain mapping with transcranial magnetic stimulation using a refined correlation ratio and Kendall's τ. *Statistics in Medicine* **27**, 5252–5270.

Murphy K 2002 *Dynamic Bayesian networks: representation, inference, and learning*. PhD thesis. University of California, Berkeley.

Pernestal A 2007 *A Bayesian approach to fault isolation with application to diesel engines*. PhD thesis. KTH School of Electrical Engineering, Sweden.

Qi F and Huang B 2010 Dynamic Bayesian approach for control loop diagnosis with underlying mode dependency. *Industrial and Engineering Chemistry Research* **46**, 8613–8623.

Qi F and Huang B 2011 Estimation of statistical distribution for control valve stiction quantification. *Journal of Process Control* **21**, 1208–1216.

14

Accounting for Incomplete Discrete Evidence

14.1 Introduction

Although a data-driven Bayesian procedure for process-mode estimation has been established, practical problems remain. An outstanding problem of the procedure in chapters 12 and 13 is its inability to handle incomplete historical evidence, which can easily occur in an industrial setting. Due to instrument reliability, heavy control-network traffic, or historical data storage problems, some key process values may not be available in the historical data. Depending on the monitor calculation algorithm, missing process variables will often lead to incomplete historical evidence where some of the monitor readings are missing. In the Bayesian diagnostic method discussed in chapters 12 and 13, only complete historical evidence samples can be used. Therefore, if there are any missing components in a piece of sampled evidence the entire sample must be removed. This can pose a problem with systems having a large number of variables as it would increase the probability of at least one measurement being missing, resulting in a significant amount of discarded data. It can also be very problematic in many practical applications where certain modes only appear infrequently. This chapter will provide a practical solution to the incomplete historical evidence problem.

14.2 Algorithm

14.2.1 Single Missing Pattern Problem

In the following discussions, the subscript for denoting the mode M will be omitted for simplicity without causing confusion.

In the presence of missing monitor readings, the historical dataset can be segregated into two parts, the complete and incomplete evidence samples:

$$\mathcal{D} = \{\mathcal{D}_c, \mathcal{D}_{ic}\}$$

Process Control System Fault Diagnosis: A Bayesian Approach, First Edition. Ruben Gonzalez, Fei Qi and Biao Huang.
© 2016 John Wiley & Sons, Ltd. Published 2016 by John Wiley & Sons, Ltd.

where $\mathcal{D}_c$ is the dataset with complete evidence, and $\mathcal{D}_{ic}$ is the remaining dataset with incomplete evidence. The two datasets are therefore termed as the complete and incomplete data sets respectively.

Let us consider a complete evidence sample consisting of L monitor readings,

$$E = (\pi_1, \pi_2, \cdots, \pi_L)$$

where each single monitor reading π_i has k_i different discrete values. In reality, the missing data problem may occur in any of these monitors. A concept named the 'missing pattern' needs to be introduced to classify the missing monitor readings problem.

A missing pattern is determined by the location of the missing monitor; in other words, how the missing monitor readings occur in an evidence sample. If two incomplete evidence samples have missing data from the same monitors, they belong to the same missing pattern, regardless of the value of the available monitor readings. For example, two evidence readings $(\times, \times, 0)$ and $(\times, \times, 1)$ belong to the same missing pattern, where $\times$ denotes the missing value.

If there is any difference in the missing evidence components, they are said to fall into different missing patterns. For instance, the two evidence readings $(\times, 1, 0)$ and $(1, \times, 1)$ belong to two different missing patterns; it should be noted that two evidence readings $(\times, \times, 0)$ and $(1, \times, 1)$ are also from two distinct missing patterns. By enumerating the numbers of missing patterns in the historical dataset, the incomplete evidence problem can be classified as a single missing pattern problem or a multiple missing pattern problem.

In a multiple missing pattern problem, the incomplete historical dataset is divided into groups of different single missing patterns. For each single missing pattern, all the missing data are from the same monitors and missing data occur across these problematic monitors simultaneously.

Without loss of generality, assume that the first q monitor readings have missing data for a given missing pattern. Thus an incomplete evidence in this missing pattern can be represented as

$$E = (\times, \cdots, \times, \pi_{q+1}, \cdots, \pi_L)$$

Each of the available monitor readings $\pi_{q+1}, \cdots, \pi_L$ can take one of its k_i discrete values so there are, in total,

$$S = \prod_{i=q+1}^{L} k_i$$

different combinations of incomplete evidence samples. Each missing reading of the first q monitors could have been any one of its possible output values, so each incomplete evidence may have come from one of the $R = \prod_{i=1}^{q} k_i$ underlying complete evidence values.

A unique underlying complete evidence matrix (UCEM) can be formed for each missing pattern. This matrix is constructed by enumerating all possible incomplete evidence values in a column in front of the matrix, and then listing all possible underlying complete evidence values corresponding to each incomplete evidence in the same row. Therefore the size of a UCEM is $S \times R$. An example of such a matrix is shown below for one of single missing patterns

$$
\begin{matrix}
(\times, \times, 0) \\
(\times, \times, 1)
\end{matrix}
\begin{bmatrix}
(0,0,0) & (0,1,0) & (1,0,0) & (1,1,0) \\
(0,0,1) & (0,1,1) & (1,0,1) & (1,1,1)
\end{bmatrix}
\tag{14.1}
$$

Each row of UCEM contains all possible underlying complete evidence values corresponding to the incomplete evidence value on the left. All elements of this matrix are unique, so there should be no coincidence between any two different rows in a UCEM. Therefore, all underlying complete evidence values can be located in the UCEM uniquely.

In the following discussion, consider a single missing pattern corresponding to a UCEM; the notation is defined as follows:

- The underlying complete evidence with location (i, j) in the kth UCEM is denoted $e_{i,j}$.
- The number of historical samples with this complete evidence is denoted $n_{i,j}$.
- The corresponding incomplete evidence is denoted e_i.
- The historical sample number of this incomplete evidence is denoted n_i.

For instance, in the UCEM in Eqn (14.1), the evidence (0,0,0) is denoted as $e_{1,1}$ and the number of historical samples with this evidence is denoted as $n_{1,1}$. The corresponding incomplete evidence, $(\times, \times, 0)$, is denoted as e_1 and the number of historical samples with the evidence being $(\times, \times, 0)$ is denoted n_1.

Recall the Bayesian method for calculating a posterior

$$p(M|E, \mathcal{D}) \propto p(E|M, \mathcal{D})p(M) \tag{14.2}$$

The prior probability $p(M)$ is determined by *prior* information, often obtained from process experts. The likelihood is obtained from process history. The objective is to derive the likelihood from historical data $p(\mathcal{D}|\Theta, M)$ in the presence of incomplete evidence.

For the single missing pattern problem, the likelihood of evidence $e_{s,r}$ can be determined as

$$p(e_{s,r}|M, \mathcal{D}) = \frac{n_{s,r} + a_{s,r}}{N + A} \cdot \frac{\sum_{j=1}^{R}(n_{s,j} + a_{s,j}) + n_s}{\sum_{j=1}^{R}(n_{s,j} + a_{s,j})}$$

$$= \frac{n_{s,r} + a_{s,r}}{N + A} \cdot \left(1 + \frac{n_s}{\sum_{j=1}^{R}(n_{s,j} + a_{s,j})}\right) \tag{14.3}$$

where $N = \sum_i n_i + \sum_i \sum_j n_{i,j}$ is the total number of historical data samples for mode M, including both complete and incomplete samples; $A = \sum_i \sum_j a_{i,j}$ is the total number of prior samples, which is, however, only applicable to the complete evidences.

The estimated likelihood in Eqn (14.3) has an intuitive explanation. It can be rewritten as

$$p(e_{s,r}|M, \mathcal{D}) = \frac{n_{s,r} + a_{s,r} + n'_{s,r}}{N + A} \tag{14.4}$$

where

$$n'_{s,r} = n_s \cdot \frac{n_{s,r} + a_{s,r}}{\sum_{j=1}^{R}(n_{s,j} + a_{s,j})} \tag{14.5}$$

can be interpreted as the expected number of samples with evidence $e_{s,r}$ in the incomplete dataset. See Figure 14.1 for an illustration, where the sensor-bias monitor reading is assumed to be missing in some historical samples. The incomplete data samples are located on the boundary between the 'biased' and 'unbiased' zones; so the underlying missing values could indicate that the sensor is biased or unbiased. With two samples in the bin labelled (*no stiction,*

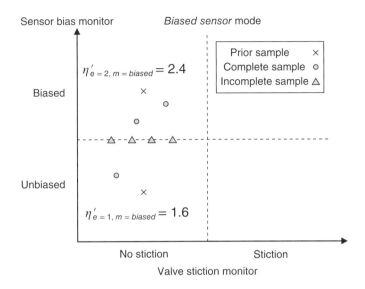

Figure 14.1   Estimation of expected complete evidence numbers out of the incomplete samples

unbiased), and three samples in the bin labelled (*no stiction, biased*), Eqn 14.5 can be used to determine how the incomplete evidence samples spanning these bins are to be distributed. In this case, the missing evidence values are expected to be distributed as a ratio of $2:3$, or 40% (*no stiction, unbiased*) and 60% (*no stiction, biased*) as shown in Figure 14.1. In this way, among the four incomplete evidence values, 1.6 are expected to be in the bin (*no stiction, unbiased*) and 2.4 are expected to be in the bin (*no stiction, biased*).

The newly allocated incomplete sample counts are used together with the complete sample counts to calculate the final likelihood according to Eqn (14.4).

Summary

In summary, the following routine can be developed to estimate the likelihood from historical data with incomplete evidence samples.

1. Build monitors to detect the status of the desired process components.
2. Collect necessary process data samples from historical data record, where the underlying process mode is available.
3. Categorize the collected training data into different groups according to the underlying modes.
4. Choose a window size, and segregate the data into sections without overlap.
5. Generate evidence by sending the segmented process data into the designed monitors.
6. Discretize the generated monitor readings according to predefined thresholds.
7. Segregate the historical evidence into complete and incomplete samples.
8. Observe the missing patterns in the incomplete evidence, and build the UCEM accordingly (note this method only works for a single missing pattern, and a multiple missing pattern technique is given in the next subsection).

9. Calculate the sample number pertaining to each incomplete evidence value, $n_{i|m_k}$, and the sample number for each complete evidence value, $n_{i,j|m_k}$ under mode m_k.
10. Assign prior sample numbers $a_{i,j|m_k}$ for each complete evidence value in the UCEM (under mode m_k). A_{m_k} can be calculated as $A_{m_k} = \sum_{i,j} a_{i,j|m_k}$.
11. After obtaining $n_{i,j|m_k}$ and $n_{i|m_k}$ for all evidence and mode values, N_{m_k} can be calculated, $N_{m_k} = \sum_i n_{i|m_k} + \sum_j n_{i,j|m_k}$.
12. Collect the process variable data that is to be diagnosed, and segregate the data into windows without overlap according to the window size determined in step 4.
13. Generate discrete diagnosis evidence using monitors and the corresponding discretization thresholds used in previous steps.
14. Find the diagnosis evidence in the UCEM. Calculate the likelihood of the evidence for all modes with Eqn (14.4).
15. Choose appropriate prior probabilities for the modes.
16. Calculate the posterior probabilities of the modes given the evidence according to Eqn (14.2). The mode with largest posterior probability is then diagnosed as the underlying mode.
17. If there is more than one window of diagnosis data, calculate the average posterior probabilities for all windows. Pick the mode with largest average as the diagnosed mode.

The flow diagram is shown in Figure 14.2.

14.2.2   Multiple Missing Pattern Problem

Solution Overview

The multiple missing pattern problem can be solved using the EM algorithm as described in Section 2.3, in which it was found that after defining the Q Function, the maximization step had an analytical solution. By applying this analytical EM solution to the missing evidence problem, the EM algorithm can be implemented as follows:

1. Estimate the likelihood using complete data

$$\hat{p}(E|M, \mathcal{D}) = p(E|M, \mathcal{D}_c)$$

2. Obtain the expected number of data samples for each missing data pattern (or UCEM) and sum them

$$n'_{s,r} = \sum_{e=\text{UCEM}_k} n_s \cdot \frac{n_{s,r} + a_{s,r}}{\sum_{j=1}^{R}(n_{s,j} + a_{s,j})}$$

$$= \sum_{e=\text{UCEM}_k} n_s \cdot \frac{\hat{p}(e_{s,r}|M, \mathcal{D})}{\sum_{j=1}^{R} \hat{p}(e_{s,r}|M, \mathcal{D})}$$

3. Re-estimate the likelihood using these new data samples

$$\hat{p}(e_{s,r}|M, \mathcal{D}) = \frac{n_{s,r} + a_{s,r} + n'_{s,r}}{N + A}$$

4. Iterate between steps 2 and 3 until $\hat{p}(E|M, \mathcal{D})$ converges.

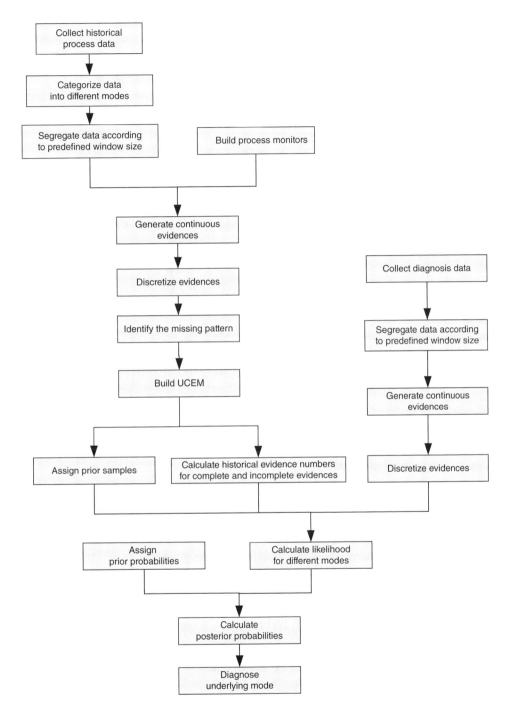

Figure 14.2 Bayesian diagnosis process with incomplete evidences

One might observe that the first iteration is identical to the single missing pattern solution. If the EM algorithm is applied to the single missing pattern problem, the algorithm converges in a single step and yields a result identical to the previous single missing pattern solution. In this sense, the EM algorithm is a generalization of the single missing pattern solution.

Practical Issues

When implementing this generalized solution, there are a few issues that the readers should be aware of.

Dimensionality

The multiple missing pattern solution suffers from *the curse of dimensionality*. The parameters estimated in the EM algorithm are the probabilities of the discrete evidence elements $p(E|M, \mathcal{D})$. As the number of monitors increases, the number of possible discrete evidence values increases exponentially. Thus the number of parameters estimated by the EM algorithm will also grow exponentially. As a result, in higher dimensions, very large datasets will be required to accommodate estimates with the exponentially growing number of parameters.

Independence

The multiple missing pattern solution works best when the monitors exhibit strong dependence. When evidence is strongly dependent, one can easily infer the missing piece of evidence from the other elements. If a monitor with a missing value is independent from the others, likelihoods affected by this missing piece of evidence can be unreliable; one should consider discarding such data.

Number of missing patterns

A large number of missing patterns will increase the complexity of the problem, significantly increasing the number of iterations required for convergence.

14.3 Tutorial

We apply the same example used in Chapter 12 to illustrate the single missing pattern algorithm. Following the same settings two monitors are designed specifically for each of the abnormal modes. The reading of each monitor is discretized into two bins with predefined thresholds; therefore there are four different evidence values in the evidence library, namely, (0,0), (0,1), (1,0), and (1,1). Noninformative prior samples are used. Each evidence bin is assigned with one prior sample, $a_{j|m} = 1$, $A_m = 4$.

Assume some of the first monitor readings are unavailable. The number of discrete evidence values (including complete and incomplete), for each mode, is shown in Table 14.1.

As an example, the evidence space for the biased sensor mode is shown in Figure 14.3 with samples given by prior information, complete evidence and incomplete evidence.

A UCEM can be built to cover all the incomplete and complete evidence samples.

$$\begin{array}{c} (\times, 0) \\ (\times, 1) \end{array} \begin{bmatrix} (0,0) & (1,0) \\ (0,1) & (1,1) \end{bmatrix} \tag{14.6}$$

Table 14.1   Number of historical evidence samples

	Normal	Sticky valve	Biased sensor
(0,0)	8	0	0
(0,1)	0	1	5
(1,0)	0	3	1
(1,1)	0	2	1
($\times$, 0)	2	4	0
($\times$, 1)	0	0	3

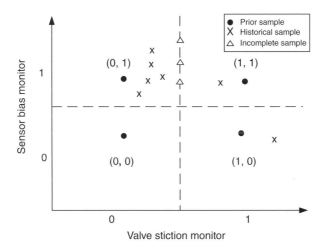

Figure 14.3   Evidence space with all samples

The incomplete evidence samples are then incorporated into the likelihood calculation. As discussed previously, the distribution of the incomplete evidence samples should follow the distribution of the complete evidence samples. As such, the expected number of each evidence in the incomplete samples can be calculated using Eqn (14.5). For instance, the expected number of evidence samples in (0,0) out of (0,0) and (1,0) is calculated as

$$n'_{s,r} = n_s \cdot \frac{n_{s,r} + a_{s,r}}{\sum_{j=1}^{R}(n_{s,j} + a_{s,j})}$$

$$= 2 \cdot \frac{8+1}{8+4} = 1.5 \tag{14.7}$$

The new sample numbers, including complete and incomplete samples are shown in Table 14.2.

By considering the incomplete samples, how will the likelihood be changed? Consider a dataset with incomplete samples. Table 14.3 contains the details of the dataset. The prior samples are uniformly distributed with one sample for each complete evidence.

The results of the reallocation are displayed in Table 14.4.

Table 14.2   Numbers of estimated sample numbers

	Normal	Sticky valve	Biased sensor
(0,0)	9.5	0.4	0.273
(0,1)	0.167	1.8	6.636
(1,0)	0.167	4.6	1.545
(1,1)	0.167	3.2	1.545

Table 14.3   Summary of historical and prior samples

Evidence	(0,0)	(0,1)	(1,0)	(1,1)	(0,×)	(1,×)
Historical sample	2	3	4	5	7	11
Prior sample	1	1	1	1	N/A	N/A

Table 14.4   Estimated likelihood with different strategy

Evidence	(0,0)	(0,1)	(1,0)	(1,1)
Likelihood(ignore)	0.2593	0.2222	0.1583	0.3333
Likelihood(consider)	0.3484	0.2989	0.1261	0.2269

With the difference being significant, it is demonstrated that the information from the incomplete evidence is useful for estimating the likelihood. Simply omitting all the incomplete evidence will result in losing a significant amount of information.

An interesting question at this point is when the proposed approach does not change the likelihood estimation. From Eqn (14.3), if the *missing ratio*

$$\rho_s^{miss} = \frac{n_s}{\sum_{j=1}^{R}(n_{s,j} + a_{s,j})} \tag{14.8}$$

is identical for all rows in UCEM, namely for different s, the likelihood of evidence $e_{s,r}$ is

$$p(e_{s,r}|M,\mathcal{D})$$

$$= \frac{1}{N+A} \cdot \left[(n_{s,r} + a_{s,r}) + n_s \cdot \frac{n_{s,r} + a_{s,r}}{\sum_{j=1}^{R}(n_{s,j} + a_{s,j})} \right]$$

$$= \frac{n_{s,r} + a_{s,r}}{N+A} \cdot [1 + \rho_s^{miss}]$$

$$= \frac{n_{s,r} + a_{s,r}}{\sum_{s,j}(n_{s,j} + a_{s,j}) + \rho_s^{miss}\sum_{s,j}(n_{s,j} + a_{s,j})} \cdot [1 + \rho_s^{miss}]$$

$$= \frac{n_{s,r} + a_{s,r}}{\sum_{s,j}(n_{s,j} + a_{s,j})} = \frac{n_{s,r} + a_{s,r}}{N_c + A} \tag{14.9}$$

Table 14.5 Summary of historical and prior samples

Evidence	(0,0)	(0,1)	(1,0)	(1,1)	(0,×)	(1,×)
Historical sample	2	3	4	5	7	11
Prior sample	1	1	1	1	N/A	N/A

which is the same as the result obtained with only consideration of the complete evidences. Thus, it can be concluded that if the missing ratios ρ_s^{miss} are the same in all the incomplete evidence samples, consideration of incomplete evidence will not introduce extra information, and hence the same likelihood will be obtained. For example, consider a set of historical evidence samples summarized in Table 14.5, with one prior sample being assigned to each complete evidence.

According to the definition of the missing ratio in Eqn (14.8),

$$\rho_{(0,\times)}^{miss} = \frac{7}{(2+1) + (3+1)} = 1$$

and

$$\rho_{(1,\times)}^{miss} = \frac{11}{(4+1) + (5+1)} = 1$$

In this example, introduction of the incomplete evidence samples will not change the estimation of the likelihood. Apart from this special case, the proposed method will generate different likelihood estimations from that obtained by merely using complete evidence.

14.4 Simulated Case

In the simulated case, we assume that duty valve stiction monitor reading π_7 and process model monitor reading π_{11} have missing data in some historical samples collected under the IDV8 mode (see Table 11.1). The incomplete evidence can be denoted

$$e = (\pi_1, \cdots, \pi_6, \times, \pi_8, \pi_9, \pi_{10}, \times, \pi_{12}, \cdots, \pi_{14})$$

The simulation is set up so that π_7 and π_{11} tend to be missing when the discrete output of the pressure sensor bias monitor reading is 'high'. Under such conditions, π_7 and π_{11} have 90% probability of being missing. When the discrete output of the pressure sensor bias monitor indicates 'low', the probability that monitors π_7 and π_{11} are missing is 10%. In the original historical dataset for IDV8 mode, π_{11} in most of the samples indicates 'high'. While most of such data samples are incomplete, the distribution of the complete evidence for IDV8 is distorted which will cause a deterioration in the diagnostic performance.

We use the approach adopted for the single missing pattern to tackle this problem. A UCEM is constructed, and the expected number of each possible underlying complete evidence is calculated according to Eqn (14.5). These estimated samples are added to the complete evidence samples as a new set of historical data. In this way, we will use the proposed approach to recover the missing data.

Consider a figure with the horizontal axis indicating different realizations of complete evidence (different combinations of monitor readings to form evidence that have appeared in data

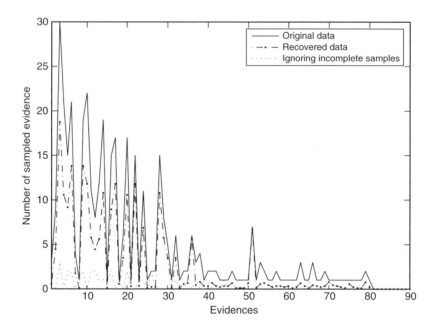

Figure 14.4   Comparison of complete evidence numbers

records) and the vertical axis indicating the number of corresponding occurrences. The original dataset (with no missing values), the recovered dataset and the dataset with missing values discarded are shown in Figure 14.4.

Clearly, the evidence distribution is seriously distorted if the incomplete evidence samples are discarded. Although the proposed approach cannot recover all underlying missing evidence, the trend of the recovered data can follow the original data trend fairly well, implying a satisfactory likelihood recovery.

Figure 14.5 shows the performance when attempting to diagnose IDV8 of the original data and the performance from the two incomplete data treatment strategies: ignoring incomplete samples and the proposed approach. By ignoring all the incomplete samples, the posterior probability assigned to mode IDV8, which is the true underlying mode, is only 0.1606. This probability is not the largest one being assigned, which means that the diagnostic system generates an erroneous result. By using the proposed approach, the posterior assigned to IDV8 mode is 0.3736, which is the largest probability assigned to all the potential modes. Thus the proposed approach generates the correct result.

To further assess the performance of the Bayesian diagnosis algorithm, the correct diagnosis rate is calculated for mode IDV 8. Figure 14.6 shows the correct diagnosis rate for different data handling strategies. From Figure 14.6, one can see that the diagnosis rate is much higher when incomplete samples are considered.

14.5   Bench-scale Case

The hybrid tank process is used as a second testbed for the proposed approach. Some of the evidence readings from monitor 3 and monitor 5 under mode 12 are masked to simulate

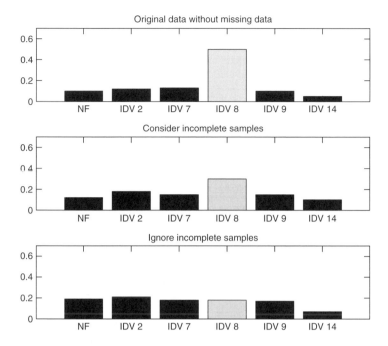

Figure 14.5   Diagnostic results with different dataset

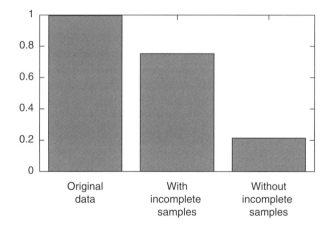

Figure 14.6   Diagnostic rate with different datasets

incomplete evidence readings. The masking process is simulated based on the rule that the two monitors have a 70% probability of being missing if one of the two original readings is normal.

With the proposed diagnostic algorithm, the diagnostic results shown in Figure 14.7 are obtained for mode 12. In each plot, the posterior probability corresponding to the true underlying mode is shown with gray bars, while others are in black bars. Thus, if the gray bar is

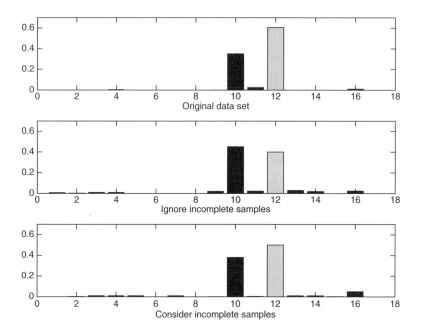

Figure 14.7   Posterior probability assigned to each mode

highest then the correct diagnosis is obtained. For a comparison, the diagnosis results from the original dataset and the diagnosis by ignoring incomplete evidence are also shown in the figure.

Without considering the incomplete samples, the average posterior probability assigned to the true underlying mode is not the highest, which clearly indicates a performance deterioration. However, with the incomplete samples incorporated into the diagnosis, the Bayesian framework is able to assign the highest posterior probability to the true underlying mode. The result can be further verified by comparing the correct diagnosis rate for the mode as shown in Figure 14.8.

14.6   Industrial-scale Case

We generate the missing data mechanism such that the 14th and 18th monitors have a 90% chance of being missing when either of the readings indicates 'abnormal' operation under mode 2. This again results in a distorted historical evidence distribution. Keeping the diagnosis settings the same as in chapter 12, the diagnosis results in terms of posterior probability are shown in Figure 14.9. The diagnosis results using the original complete dataset as well as simply ignoring all incomplete evidences are also shown.

Despite the incomplete samples being ignored, the underlying mode is still assigned the highest probability. However, the posterior value is smaller than in the case where incomplete samples are considered. The difference between the probabilities assigned to the true underlying mode and other modes is also smaller, which affects the individual diagnosis rate, as shown in Figure 14.10.

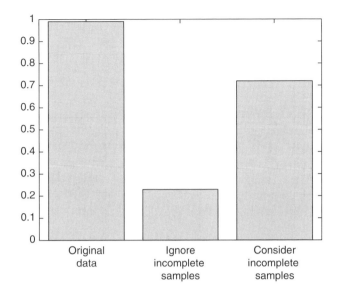

Figure 14.8 Diagnostic rate with different dataset

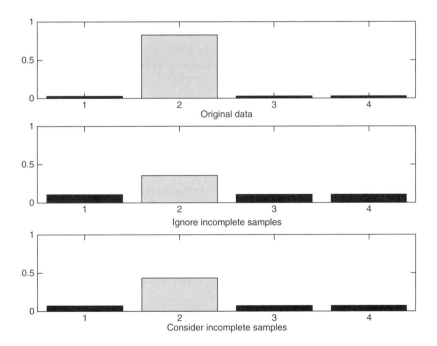

Figure 14.9 Posterior probability assigned to each mode for industrial process

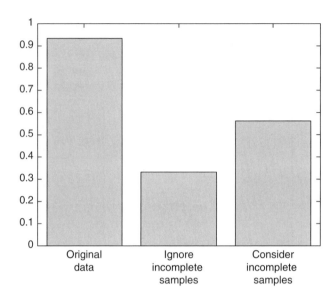

Figure 14.10   Diagnostic rate with different dataset

14.7   Notes and References

A similar approach for incomplete evidence handling is the Gibbs sampling approach. Readers are referred to Geman and Geman (1984) and Korb and Nicholson (2004) for more information. The main reference of this chapter is Qi et al. (2010), while the EM-algorithm solution stems from Zhang et al. (2014).

References

Geman S and Geman D 1984 Stochastic relaxation, gibbs distributions, and the Bayesian restoration of images. *IEEE Transactions on Pattern Analysis and Machine Intelligence* **6**, 721–741.

Korb K and Nicholson A 2004 *Bayesian Artificial Intelligence*. Chapman & Hall/CRC.

Qi F, Huang B and Tamayo E 2010 A Bayesian approach for control loop diagnosis with missing data. *AIChE Journal* **56**, 179–195.

Zhang K, Gonzalez R, Huang B and Ji G 2014 An expectation maximization approach to fault diagnosis with missing data. *Industrial Electronics, IEEE Transactions on* **62**(2), 1231–1240.

15

Accounting for Ambiguous Modes in Historical Data: A Bayesian Approach

15.1 Introduction

This chapter discusses the application of the second-order parametrization method, which is useful when the historical data contains ambiguous operating modes (i.e. uncertain cases where more than one mode is possible). Ambiguity occurs when information is missing from one or more problem sources. For example, let us say that there are two components: a sensor and a valve. The sensor can develop bias, while the valve can become sticky. If bias is known to exist, but it is unknown whether the valve is sticky or not, then there are two possible modes, one mode having bias without stiction, and the other having bias with stiction.

When ambiguous modes are in the data, the resulting probabilities can have ranges. The second-order method is used to combine prior probabilities along with one or more likelihoods in order to obtain a final result with probability boundaries. The approach consists of the following steps:

1. If there is a large number of monitors, group monitors into independent groups.
2. Set up a method for calculating likelihoods given θ: $p(E|M, \Theta)$.
3. Calculate the second-order approximation of the likelihood expression.
4. Combine likelihood expressions with the second-order combination rule to obtain the final diagnosis result.

Process Control System Fault Diagnosis: A Bayesian Approach, First Edition. Ruben Gonzalez, Fei Qi and Biao Huang.
© 2016 John Wiley & Sons, Ltd. Published 2016 by John Wiley & Sons, Ltd.

15.2 Algorithm

15.2.1 Formulating the Problem

When ambiguous modes M are present in the data, the likelihood expression takes the following form:

$$p(E|M,\Theta) = \frac{S(E|M)n(M) + \sum\limits_{m_k \supset M} \theta\{\frac{M}{m_k}\}S(E|m_k)n(m_k)}{n(M) + \sum\limits_{m_k \supset M} \theta\{\frac{M}{m_k}\}n(m_k)} \tag{15.1}$$

where $n(m_k)$ is the number of times m_k appears in the history and $S(E|m_k)$ is the support function for evidence E given m_k; here, the term $S(E|m_k)$ is calculated in the same manner as likelihood

$$p(E|M) = \frac{n(E|M)}{n(M)}$$

$$S(E|M) = \frac{n(E|M)}{n(M)}$$

The only difference from the likelihood is that m_k can be an ambiguous mode. The terms $\theta\{\frac{M}{m_k}\}$ are unknown and represent the proportions of data in an ambiguous mode m_k that belongs to one of its specific modes $M \subset m_k$.

$$\theta\{\frac{M}{m_k}\} = \frac{n(M|m_k)}{n(m_k)} = p(M|m_k)$$

The unknown parameters $\theta\{\frac{M}{m_k}\}$ can be used to define the probability boundaries, namely the plausibility (the maximum probability) and the belief (minimum probability):

$$Bel(E|M) = \min_{\Theta} p(E|M,\Theta)$$

$$Pl(E|M) = \max_{\Theta} p(E|M,\Theta)$$

The challenge with the expression in Eqn (15.1) is that it is difficult to use Bayesian methods to combine likelihoods with priors, as the resulting expressions are increasing in complexity with respect to Θ (where Θ represents the collection of all θ parameters). In order to manage this complexity, the second-order Taylor series approximation is used. When this approximation is applied, a simple updating rule can be derived and the resulting belief and plausibility are relatively easy to obtain.

15.2.2 Second-order Taylor Series Approximation of $p(E|M,\Theta)$

The Taylor series approximation makes use of differentiation in order to make an approximation around a reference point $\hat{\Theta}$. The univariate expression is given by

$$f(x) \approx f(\hat{x}) + \frac{d\,f(x)}{d\,x}\bigg|_{\hat{x}} (x - \hat{x}) + \frac{1}{2}\frac{d^2\,f(x)}{d\,x^2}\bigg|_{\hat{x}} (x - \hat{x})^2 + \dots \tag{15.2}$$

where $\hat{x}$ is a reference point around which the approximation is centred. However, the expression $p(E|M, \Theta)$ is a function of multiple variables Θ, thus a multivariate Taylor series approximation is needed. In this case, a second-order multivariate Taylor series approximation is desired:

$$\Delta\Theta = \hat{\Theta} - \Theta$$

$$p(E|M, \Theta) = p(E|M, \hat{\Theta}) + J\Delta\Theta + \frac{1}{2}\Delta\Theta^T H\Delta\Theta \qquad (15.3)$$

where J and H are Jacobian and Hessian matrices

$$J = \left[\frac{\partial \, p(E|M,\Theta)}{\partial \, \theta\{\frac{M}{m_1}\}} \quad \frac{\partial \, p(E|M,\Theta)}{\partial \, \theta\{\frac{M}{m_2}\}} \quad \cdots \quad \frac{\partial \, p(E|M,\Theta)}{\partial \, \theta\{\frac{M}{m_N}\}} \right]$$

$$H = \begin{bmatrix} \frac{\partial^2 \, p(E|M,\Theta)}{\partial \, \theta\{\frac{M}{m_1}\}^2} & \frac{\partial^2 \, p(E|M,\Theta)}{\partial \, \theta\{\frac{M}{m_1}\}\theta\{\frac{M}{m_2}\}} & \cdots & \frac{\partial^2 \, p(E|M,\Theta)}{\partial \, \theta\{\frac{M}{m_1}\}\theta\{\frac{M}{m_N}\}} \\[2mm] \frac{\partial^2 \, p(E|M,\Theta)}{\partial \, \theta\{\frac{M}{m_2}\}\partial \, \theta\{\frac{M}{m_1}\}} & \frac{\partial^2 \, p(E|M,\Theta)}{\partial \, \theta\{\frac{M}{m_2}\}^2} & \cdots & \frac{\partial^2 \, p(E|M,\Theta)}{\partial \, \theta\{\frac{M}{m_2}\}\theta\{\frac{M}{m_N}\}} \\[2mm] \vdots & \vdots & \ddots & \vdots \\[2mm] \frac{\partial^2 \, p(E|M,\Theta)}{\partial \, \theta\{\frac{M}{m_N}\}\partial \, \theta\{\frac{M}{m_1}\}} & \frac{\partial^2 \, p(E|M,\Theta)}{\partial \, \theta\{\frac{M}{m_N}\}\theta\{\frac{M}{m_2}\}} & \cdots & \frac{\partial^2 \, p(E|M,\Theta)}{\partial \, \theta\{\frac{M}{m_N}\}^2} \end{bmatrix}$$

The expressions for the partial derivatives with respect to $p(E|M, \Theta)$ are obtained by differentiating Eqn (15.1). For compactness of notation, we introduce S, n and θ as vectors.

$$S = [S(E|m_1), S(E|m_2), \ldots, S(E|m_n)]$$

$$n = [n(m_1), n(m_2), \ldots, n(m_n)]$$

$$\theta = \left[\theta\{\tfrac{M}{m_1}\}, \theta\{\tfrac{M}{m_2}\}, \ldots, \theta\{\tfrac{M}{m_n}\} \right]$$

The partial differentials are then given as

$$\frac{\partial \, p(E|M, \Theta)}{\partial \, \theta_i} = \frac{n_i S_i}{\sum_k n_k \theta_k} - \frac{n_i \sum_k S_k n_k \theta_k}{\left(\sum_k n_k \theta_k \right)^2}$$

$$\frac{\partial^2 \, p(E|M, \Theta)}{\partial \, \theta_i \partial \, \theta_j} = -\frac{n_i S_j + n_j S_i}{\left(\sum_k n_k \theta_k \right)^2} + \frac{n_i n_j \sum_k S_k n_k \theta_k}{\left(\sum_k n_k \theta_k \right)^3}$$

Reference Point: the Informed Transformation

The second-order Taylor series approximation requires a reference point for Θ in order to obtain expressions for each likelihood. The informed transformation makes use of the most credible assignment of Θ as a reference point based on prior probabilities. As an example, let us consider a three-mode system with the following priors: $p(m_1) = 0.5$, $p(m_2) = 0.25$ and $p(m_3) = 0.25$. Now let us say that there is a body of evidence belonging to the ambiguous mode $\{m_1, m_2\}$. If, according to the prior probabilities, mode m_1 is twice as probable as mode m_2 then the most credible allocation is

$$\hat{\theta}\{\tfrac{m_1}{\{m_1,m_2\}}\} = p(m_1|\{m_1,m_2\}) = \frac{0.5}{0.5 + 0.25} = 2/3$$

$$\hat{\theta}\{\tfrac{m_2}{\{m_1,m_2\}}\} = p(m_2|\{m_1,m_2\}) = \frac{0.25}{0.5 + 0.25} = 1/3$$

where $\hat{\theta}$ indicates a most credible estimate. Similarly, consider body of evidence belonging to the ambiguous mode $\{m_2, m_3\}$ where the priors of m_2 and m_3 are equal. The most credible allocation in this case is

$$\hat{\theta}\{\tfrac{m_2}{\{m_2,m_3\}}\} = p(m_2|\{m_2,m_3\}) = \frac{0.25}{0.25 + 0.25} = 1/2$$

$$\hat{\theta}\{\tfrac{m_3}{\{m_2,m_3\}}\} = p(m_3|\{m_2,m_3\}) = \frac{0.25}{0.25 + 0.25} = 1/2$$

This technique to obtain credible allocation is called the *informed transformation* as it is based on information from prior probabilities. In general, the informed transformation $\hat{\theta}\{\tfrac{M}{m}\}$ is given as

$$\hat{\theta}\{\tfrac{M}{m_k}\} = \frac{p(M)}{\sum\limits_{M \subset m_k} p(M)} \tag{15.4}$$

This transformation also yields the informed likelihood $\hat{p}(E|M)$ by substituting $\hat{\Theta}$ in for Θ in the likelihood expression in Eqn (15.1)

$$\hat{p}(E|M) = p(E|M, \hat{\Theta})$$

15.2.3   Second-order Bayesian Combination

Combination Method

Two types of combination need to be considered: the combination of multiple likelihoods and combination with a prior. In both cases, two second-order approximations are multiplied together and all terms that are third order or higher are discarded. The difference lies in the normalization. For combining likelihoods, normalization is not needed, but when combining with a prior using Bayes' rule, normalization must be performed.

The goal of combining likelihoods is to express the joint probability of two independent pieces of evidence.

$$p(E_1, E_2|M, \Theta) = p(E_1|M, \Theta)p(E_2|M, \Theta)$$

$$p(E_1|M, \Theta) \approx \hat{p}(E_1|M) + \boldsymbol{J}_1\Delta\Theta + \frac{1}{2}\Delta\Theta^T\boldsymbol{H}_1\Delta\Theta$$

$$p(E_2|M, \Theta) \approx \hat{p}(E_2|M) + \boldsymbol{J}_2\Delta\Theta + \frac{1}{2}\Delta\Theta^T\boldsymbol{H}_2\Delta\Theta$$

When ignoring terms of third order and higher, the joint probability must also be expressed as second order

$$p(E_1, E_2|M, \Theta) \approx \hat{p}(E_1, E_2|M) + \boldsymbol{J}_{12}\Delta\Theta + \frac{1}{2}\Delta\Theta^T \boldsymbol{H}_{12}\Delta\Theta$$

By multiplying the second-order approximations of $p(E_1|M, \Theta)$ and $p(E_2|M, \Theta)$ and then discarding terms that are third order and higher, one obtains a rule for updating key terms

$$\hat{p}(E_1, E_2|M) = \hat{p}(E_1|M)\hat{p}(E_2|M)$$

$$\boldsymbol{J}_{12} = \hat{p}(E_1|M)\boldsymbol{J}_2 + \hat{p}(E_2|M)\boldsymbol{J}_1$$

$$\boldsymbol{H}_{12} = \hat{p}(E_1|M)\boldsymbol{H}_2 + \hat{p}(E_2|M)\boldsymbol{H}_1 + \boldsymbol{J}_2^T \boldsymbol{J}_1 + \boldsymbol{J}_1^T \boldsymbol{J}_2$$

After the independent likelihoods are combined, they can be combined with a prior distribution $p(M|\Theta)$ in a similar manner in order to obtain the posterior estimate $p(M|E, \Theta)$. However, for combination with a prior, the normalization constant K needs to be accounted for. First, we define the prior

$$p(M|\Theta) = \hat{p}(M) + \boldsymbol{J}_P\Delta\Theta + \frac{1}{2}\Delta\Theta^T \boldsymbol{H}_P\Delta\Theta$$

In the static case, there is often no ambiguity in the prior, and in such a case $\boldsymbol{J}_P$ and $\boldsymbol{H}_P$ are zero matrices. When combining the likelihood with the prior, normalization is required

$$K = \sum_k \hat{p}(m_k)\hat{p}(E_1, E_2|m_k)$$

The second-order terms for the posterior probability are similar to the likelihood, except now the terms are normalized by the inverse of K

$$\hat{p}(M|E_1, E_2) = \frac{1}{K}\hat{p}(m_k)\hat{p}(E_1, E_2|m_k)$$

$$\boldsymbol{J}_F = \frac{1}{K}[\hat{p}(M)\boldsymbol{J}_{12} + \hat{p}(E_1, E_2|M)\boldsymbol{J}_P]$$

$$\boldsymbol{H}_F = \frac{1}{K}[\hat{p}(M)\boldsymbol{H}_{12} + \hat{p}(E_1, E_2|M)\boldsymbol{H}_P + \boldsymbol{J}_{12}^T \boldsymbol{J}_P + \boldsymbol{J}_P^T \boldsymbol{J}_{12}]$$

so that the second-order expression for the posterior probability is

$$p(M|E_1, E_2, \Theta) = \hat{p}(M|E_1, E_2) + \boldsymbol{J}_F\Delta\Theta + \frac{1}{2}\Delta\Theta^T \boldsymbol{H}_F\Delta\Theta$$

Diagnosis Methods: the Reference Point Method

The diagnosis is made by determining which posterior probability has the highest value. A simple posterior probability estimate is the second-order reference point, $\hat{p}(M|E_1, E_2)$, or the informed transformation. This value is already given in the second-order probability calculations and requires no information from $\boldsymbol{J}$ or $\boldsymbol{H}$. In fact, if the objective is to make a diagnosis based on the simple point estimate, $\boldsymbol{J}$ or $\boldsymbol{H}$ does not need to be calculated at all. One can simply use the Bayesian method based on the reference likelihoods $\hat{p}(E_1, E_2|M)$.

Diagnosis Methods: the Expected Value Method

The more complex and rigorous estimate is the expected value $E[p(M|E)]$, which assumes a distribution over the θ parameters and calculates the expected value of the posterior. The posterior expected value is given as

$$E[p(M|E)] = C + J^* E[\Theta] + \frac{1}{2}[E[\Theta]^T H_{OD} E[\Theta] + E[\Theta^2]^T H_D] \tag{15.5}$$

where H_{OD} is the Hessian H with the diagonal elements being set to zero, and H_D is a column vector of the diagonal elements of H. The parameters C and J^* are calculated as

$$C = \hat{p}(M|E) - J\hat{\Theta} + \frac{1}{2}\hat{\Theta}^T H \hat{\Theta}$$

$$J^* = (J - \hat{\Theta}^T H)$$

where $\hat{\Theta}$ is the same value as that given in Eqn (15.4). The terms $E[\Theta]$ and $E[\Theta^2]$ are column vectors with each element pertaining to $E[\theta\{\frac{M}{m_k}\}]$ and $E[\theta^2\{\frac{M}{m_k}\}]$. These expectations are calculated assuming that θ follows a Dirichlet distribution; for each term the expectation results are given as follows:

$$E[\theta\{\tfrac{M}{m_k}\}] = \frac{\alpha(\frac{M}{m_k})}{\sum\limits_{m_i \subset m_k} \alpha(\frac{m_i}{m_k})} \tag{15.6}$$

$$E[\theta^2\{\tfrac{M}{m_k}\}] = \frac{\alpha(\frac{M}{m_k}) + \alpha^2(\frac{M}{m_k})}{\left[\sum\limits_{m_i \subset m_k} \alpha(\frac{m_i}{m_k})\right] + \left[\sum\limits_{m_i \subset m_k} \alpha(\frac{m_i}{m_k})\right]^2} \tag{15.7}$$

where $\alpha(\frac{M}{m_k})$ represents the prior frequency of mode M happening given the ambiguous mode m_k. This value $\alpha(\frac{M}{m_k})$ can be calculated as

$$\alpha(\tfrac{M}{m_k}) = p(M|m_k)n(m_k) = \hat{\theta}\{\tfrac{M}{m_k}\}n(m_k)$$

Diagnosing by means of the expected value is advantageous when $n(m_k)$ is small, in other words where there is historical data with a large number of ambiguous modes and few data points. Small values of $n(m_k)$ lead to large variances of the Θ parameters. If there is a large number of data points for an ambiguous mode, the variances of Θ will be much smaller and the expected values will be nearly identical to the much simpler point estimates obtained from the informed transformation.

15.2.4   Optional Step: Separating Monitors into Independent Groups

Motivation

If the evidence E is multivariate with a large number of components, the likelihood function $p(E|M)$ is high-dimensional and can be difficult to estimate, so it is often desirable to break down E into independent groups. For example, if E has ten elements, and each element can take on two values, then the distribution would have ten dimensions with $2^{10} = 1024$ different possible values, resulting in 1042 discrete bins for this distribution. However, if elements in E can be divided into two independent groups (E_1 and E_2) with five elements each, we would end up estimating two distributions having five dimensions and $2^5 = 32$ bins each: a significant simplification over a 1024-bin distribution. The joint probability can be calculated through simple multiplication

$$p(E_1, E_2|M) = p(E_1|M)p(E_2|M) \qquad E_1|m \perp E_2|m$$

Note that evidence groups can change with each mode, so they only have to be conditionally independent given the mode. This results in the following Bayesian combination rule

$$p(M|E_1, E_2) = \frac{p(E_1, E_2|M)p(M)}{\sum_k p(E_1, E_2|m_k)p(m_k)}$$
$$= \frac{p(E_1|M)p(E_2|M)p(M)}{\sum_k p(E_1, E_2|m_k)p(m_k)}$$

where $p(E_1, E_2|m_k)$ can be broken down differently depending on m_k. Breaking down likelihoods into components gives added utility to the second-order combination rule; when ambiguous modes are present, the second-order rule is needed to calculate the joint probabilities and their respective ranges.

Dependency Metrics

In order to break down the likelihood into separate components, a dependency metric is required. A popular metric is the *mutual information criterion* (MIC).

$$I(X_1; X_2) = \int_{\mathcal{X}_1} \int_{\mathcal{X}_2} p(x_1, x_2) \log \left(\frac{p(x_1, x_2)}{p(x_1)p(x_2)} \right) dx_1 \, dx_2 \qquad (15.8)$$

where $\mathcal{X}_1$ and $\mathcal{X}_2$ represent the domain of the PDF $p(x_1, x_2)$. In this chapter, evidence components x_1 and x_2 are discrete, thus the MIC is calculated using summation instead of integration. The MIC yields a value of zero if X_1 and X_2 are independent, and a positive value if they are dependent (and ideally a value of infinity if they are perfectly dependent).

15.2.5 Grouping Methodology

The MIC values should first be arranged in a matrix, in a similar manner to covariance. In order to avoid redundancy, we only consider the lower off-diagonal portion of this matrix

$$MIC(\boldsymbol{X}) = \begin{bmatrix} 0 & 0 & 0 & \cdots & 0 \\ MIC(X_2, X_1) & 0 & 0 & \cdots & 0 \\ MIC(X_3, X_1) & MIC(X_3, X_2) & 0 & \cdots & 0 \\ \vdots & \vdots & \vdots & \ddots & \ddots \\ MIC(X_p, X_1) & MIC(X_p, X_2) & MIC(X_p, X_3) & \cdots & 0 \end{bmatrix}$$

By assuming independence between groups, elements of the MIC matrix will be assumed to be zero. In order to minimize the loss of information, elements of the MIC are sorted from largest to smallest. The grouping algorithm starts with the largest MIC value, which corresponds to X_a and X_b. The first group to be formed is thus $\{X_a, X_b\}$. The algorithm then proceeds to the next MIC value. For any newly drawn MIC value for $\{X_a, X_b\}$ one is faced with one of four different scenarios:

1. **Scenario:** Variables $\{X_a, X_b\}$ are not contained in any previous groups.
   Solution: Start a new group which contains $\{X_a, X_b\}$.
2. **Scenario:** X_a is contained in one of the previous groups but X_b is not.
   Solution: Add variable X_b to the group that contains $\{X_a\}$.
3. **Scenario:** X_a and X_b are contained in two different existing groups.
   Solution: Consider merging the two existing groups together as long as the merged group is not too large.
4. **Scenario:** X_a and X_b are both contained in the same existing group.
   Solution: Do nothing, as the $MIC(X_a, X_b)$ is already taken into account.

This is performed until either all MIC values are accounted for or the MIC values become small, for example $MIC < 0.01$.

15.3   Illustrative Example of Proposed Methodology

15.3.1   Introduction

We now go over a simple example of how to implement the proposed algorithm. Consider a control loop as shown in Figure 15.1. Let us assume that the sensor may be subject to bias, and that the valve may be subject to stiction. Consider two pieces of evidence in the form of monitors: E_1 is a bias monitor with outputs 'bias' and 'no bias', while E_2 is a stiction monitor with outputs 'stiction' and 'no stiction'. Positive results are given as (1) while negative results are given as (0).

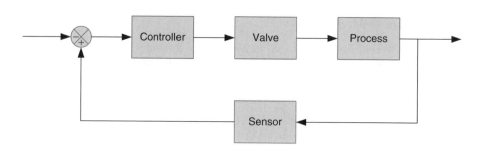

Figure 15.1   Typical control loop

15.3.2 Offline Step 1: Historical Data Collection

The first step is to go through the historical data and note the instances where each of the four possible modes occurs:

1. m_1 $[0,0]$ where bias and stiction do not occur
2. m_2 $[0,1]$ where bias does not occur but stiction does
3. m_3 $[1,0]$ where bias occurs but stiction does not
4. m_4 $[1,1]$ where both bias and stiction occur.

Data should be collected according to each of these modes. In certain instances, one or more of these ambiguous modes may also occur:

1. $\{m_1, m_3\}$ $[\times, 0]$ where bias is undetermined and stiction does not occur
2. $\{m_2, m_4\}$ $[\times, 1]$ where bias is undetermined and stiction occurs
3. $\{m_1, m_2\}$ $[0, \times]$ where bias does not occur and stiction is undetermined
4. $\{m_3, m_4\}$ $[1, \times]$ where bias occurs and stiction is undetermined
5. $\{m_1, m_2, m_3, m_4\}$ $[\times, \times]$ where both bias and stiction are undetermined.

Data should be collected for any of the ambiguous modes that appear in the data.

15.3.3 Offline Step 2: Mutual Information Criterion (Optional)

This step is optional and, considering the data is only two-dimensional, the benefit of assuming independence is insignificant in this case. However, for the sake of demonstration, the MIC matrix can be calculated as follows

$$MIC = \begin{bmatrix} 0 & 0 \\ MIC(1,2) & 0 \end{bmatrix}$$

where $MIC(1,2)$ is calculated as

$$MIC(1,2) = \sum_{E_1} \sum_{E_2} p(E_1, E_2 | m) \log \left(\frac{p(E_1, E_2 | M)}{p(E_1 | M) p(E_2 | M)} \right)$$

The MIC matrix must be calculated for every unambiguous mode. For each mode, the two pieces of evidence are either considered independent or dependent. For example, let us say that under Mode (1), the probability is distributed as given in Table 15.1. The MIC for

Table 15.1 Probability of evidence given Mode (1)

	$E = [0,0]$	$E = [0,1]$	$E = [1,0]$	$E = [1,1]$
$n(E|m_1)$	70	10	15	5
$p(E|m_1)$	0.7	0.1	0.15	0.05

Mode (1) is given as

$$MIC = p([0,0]|m_1)\log\left(\frac{p([0,0]|m_1)}{p([0,\times]|m_1)p([\times,0]|m_1)}\right)$$

$$+ p([0,1]|m_1)\log\left(\frac{p([0,1]|m_1)}{p([0,\times]|m_1)p([\times,1]|m_1)}\right)$$

$$+ p([1,0]|m_1)\log\left(\frac{p([1,0]|m_1)}{p([1,\times]|m_1)p([\times,0]|m_1)}\right)$$

$$+ p([1,1]|m_1)\log\left(\frac{p([1,1]|m_1)}{p([1,\times]|m_1)p([\times,1]|m_1)}\right)$$

$$= 0.7\log\left(\frac{0.7}{(0.8)(0.85)}\right) + 0.1\log\left(\frac{0.1}{(0.8)(0.15)}\right)$$

$$+ 0.15\log\left(\frac{0.15}{(0.2)(0.85)}\right) + 0.05\log\left(\frac{0.05}{(0.2)(0.15)}\right)$$

$$= 0.0088259$$

Because this number (0.0088259) is so small, the two monitors E_1 and E_2 can be considered independent under this mode.

15.3.4 Offline Step 3: Calculate Reference Values

The reference values $\hat{\Theta}$ can be calculated offline after the data has been collected; in particular, we work with the subsets $\hat{\theta}\{m_i\}$ of $\hat{\Theta}$. The term $\hat{\theta}\{\frac{m_i}{m_i \subset \bullet}\}$ is defined as a vector containing all modes that can support m_i, for example:

$$\hat{\theta}\{\tfrac{m_1}{m_1 \subset \bullet}\} = \left[\hat{\theta}\{\tfrac{m_1}{m_1}\}, \hat{\theta}\{\tfrac{m_1}{\{m_1,m_2\}}\}, \hat{\theta}\{\tfrac{m_1}{\{m_1,m_3\}}\}, \hat{\theta}\{\tfrac{m_1}{\{m_1,m_2,m_3,m_4\}}\}\right]^T$$

$$\hat{\theta}\{\tfrac{m_2}{m_2 \subset \bullet}\} = \left[\hat{\theta}\{\tfrac{m_2}{m_2}\}, \hat{\theta}\{\tfrac{m_2}{\{m_1,m_2\}}\}, \hat{\theta}\{\tfrac{m_2}{\{m_2,m_4\}}\}, \hat{\theta}\{\tfrac{m_2}{\{m_1,m_2,m_3,m_4\}}\}\right]^T$$

Note that $\hat{\theta}\{\frac{m_1}{m_1}\} = 1$ and $\hat{\theta}\{\frac{m_2}{m_2}\} = 1$ by definition. Each element of $\hat{\theta}\{\frac{m_i}{m_i \subset \bullet}\}$ is calculated as

$$\hat{\theta}\{\tfrac{m_i}{m_k}\} = \frac{n(m_i)}{\sum_{m_j \subseteq m_k} n(m_j)} m_i \subseteq m_k$$

For example, let us say that the prior probabilities of the four modes are as given in Table 15.2.

Table 15.2 Prior probabilities

	m_1	m_2	m_3	m_4
$p(E|m_1)$	0.6	0.15	0.15	0.1

Then parameters $\hat{\theta}\{\frac{m_i}{m_k}\}$ can be calculated, for example, as

$$\hat{\theta}\{\frac{m_1}{\{m_1,m_2\}}\} = \frac{p(m_1)}{p(m_1) + p(m_2)}$$

$$= \frac{0.6}{0.6 + 0.15} = 0.8$$

$$\hat{\theta}\{\frac{m_2}{\{m_1,m_2,m_3,m_4\}}\} = \frac{p(m_2)}{p(m_1) + p(m_2) + p(m_3) + p(m_4)}$$

$$= \frac{0.15}{0.6 + 0.15 + 0.15 + 0.1} = 0.15$$

However, if the term in the numerator is not contained in the denominator, the value of $\hat{\theta}\{\frac{m_i}{m_k}\}$ is zero. For example,

$$\hat{\theta}\{\frac{m_1}{\{m_2,m_4\}}\} = \frac{0}{p(m_2) + p(m_4)} = 0$$

This is because given the ambiguous mode $\{m_2, m_4\}$ the probability of m_1 is zero.

15.3.5   Online Step 1: Calculate Support

When a new piece of evidence $[E_1, E_2]$ becomes available, we calculate the support according to

$$S(E_1, E_2|\boldsymbol{m}_k) = \frac{n(E_1, E_2|\boldsymbol{m}_k)}{n(\boldsymbol{m}_k)}$$

For example, using the values in Table 15.1, if $E = [0,0]$ then $S(E_1, E_2|\boldsymbol{m}_1)$ would be calculated as

$$S([0,0]|\boldsymbol{m}_1) = \frac{70}{70 + 10 + 15 + 5} = 0.7$$

If E_1 and E_2 are considered independent given m_k, the support is calculated separately

$$S(E_1|\boldsymbol{m}_k) = \frac{n(E_1|\boldsymbol{m}_k)}{n(\boldsymbol{m}_k)} \qquad S(E_2|\boldsymbol{m}_k) = \frac{n(E_2|\boldsymbol{m}_k)}{n(\boldsymbol{m}_k)}$$

Using the same example, the probability given the independence assumption would be

$$S([0, \times]|\boldsymbol{m}_k) = \frac{70 + 10}{70 + 10 + 15 + 5} = 0.8$$

$$S([\times, 0]|\boldsymbol{m}_k) = \frac{70 + 15}{70 + 10 + 15 + 5} = 0.85$$

As a comparison with the first result, the joint probability given the independence assumption would be

$$S([0,0]|\boldsymbol{m}_1) = S([0, \times]|\boldsymbol{m}_k)S([\times, 0]|\boldsymbol{m}_k) = 0.8 * 0.85 = 0.68$$

which is fairly close to our original result, $S([0,0]|\boldsymbol{m}_1) = 0.7$, thus validating the independence assumption.

15.3.6   Online Step 2: Calculate Second-order Terms

The second-order terms are calculated as a linearization around $\hat{\Theta}$. The required terms are

1. $\hat{p}(E|M) = p(E|M, \hat{\Theta})$, required for probability boundaries and all diagnosis methods
2. $\left.\frac{\partial\, p(E|M,\Theta)}{\partial\, \theta_i}\right|_{\hat{\Theta}}$, required for probability boundaries and expected value diagnosis method
3. $\left.\frac{\partial^2\, p(E|M,\Theta)}{\partial\, \theta_i\, \partial\, \theta_j}\right|_{\hat{\Theta}}$, required for probability boundaries and expected value diagnosis method.

These terms can be calculated as follows:

$$p(E|M, \hat{\Theta}) = \frac{\sum_k S_k n_k \hat{\theta}_k}{\sum_k n_k \hat{\theta}_k}$$

$$\left.\frac{\partial\, p(E|M,\Theta)}{\partial\, \theta_i}\right|_{\hat{\Theta}} = \frac{n_i S_i}{\sum_k n_k \hat{\theta}_k} - \frac{n_i \sum_k S_k n_k \hat{\theta}_k}{\left(\sum_k n_k \hat{\theta}_k\right)^2}$$

$$\left.\frac{\partial^2\, p(E|M,\Theta)}{\partial\, \theta_i\, \partial\, \theta_j}\right|_{\hat{\Theta}} = -\frac{n_i S_j + n_j S_i}{\left(\sum_k n_k \hat{\theta}_k\right)^2} + \frac{n_i n_j \sum_k S_k n_k \hat{\theta}_k \{m\}}{\left(\sum_k n_k \hat{\theta}_k\right)^3}$$

where S and n are horizontal vectors containing the support and frequency of the modes that can support mode M, both ambiguous and unambiguous. In order to illustrate this, let us consider a more complete version of Table 15.1 shown below in Tables 15.3 and 15.4. Each element in S and n pertain to an element in $\hat{\theta}\{m_i\}$. For example, consider Mode (1); the resulting vectors S and n are obtained as

$$S = [S(E|m_1), S(E|\{m_1, m_2\}), S(E|\{m_1, m_3\}), S(E|\{m_1, m_2, m_3, m_4\})]$$
$$= [0.7, 0.5, 0.4, 0.28]$$
$$n = [n(m_1), n\{m_1, m_2\}, n\{m_1, m_3\}, n\{m_1, m_2, m_3, m_4\}]$$
$$= [100, 50, 50, 25]$$

Table 15.3   Frequency of modes containing m_1

	$E = [0, 0]$	$E = [0, 1]$	$E = [1, 0]$	$E = [1, 1]$
$n(E|m_1)$	70	10	15	5
$n(E|m_2)$	14	59	6	21
$n(E|m_3)$	13	7	58	22
$n(E|m_4)$	12	8	23	57
$n(E|m_1, m_2)$	25	15	7	3
$n(E|m_1, m_3)$	20	20	6	4
$n(E|m_2, m_4)$	19	17	7	7
$n(E|m_3, m_4)$	19	8	7	16
$n(E|m_1, m_2, m_3, m_4)$	7	6	6	6

Table 15.4 Support of modes containing m_1

	$E = [0,0]$	$E = [0,1]$	$E = [1,0]$	$E = [1,1]$
$S(E|m_1)$	0.7	0.1	0.15	0.05
$S(E|m_2)$	0.14	0.59	0.6	0.21
$S(E|m_3)$	0.13	0.07	0.58	0.22
$S(E|m_4)$	0.12	0.08	0.23	0.57
$S(E|m_1, m_2)$	0.5	0.3	0.14	0.06
$S(E|m_1, m_3)$	0.4	0.4	0.12	0.08
$n(E|m_2, m_4)$	0.38	0.34	0.14	0.14
$n(E|m_3, m_4)$	0.38	0.16	0.14	0.32
$S(E|m_1, m_2, m_3, m_4)$	0.28	0.24	0.24	0.24

The function value, as well as the first and second derivative expressions are evaluated at $\hat{\Theta}$. For m_1, the parameter vector $\hat{\theta}$ is given as:

$$\hat{\theta} = \left[1, \hat{\theta}\{\tfrac{m_1}{m_1,m_2}\}, \hat{\theta}\{\tfrac{m_1}{m_1,m_3}\}, \hat{\theta}\{\tfrac{m_1}{m_1,m_2,m_3,m_4}\}\right]$$

$$= [1, 0.6/0.75, 0.6/0.75, 0.6/1.0] = [1, 0.8, 0.8, 0.6]$$

Note that the terms J, H only pertain to the variable terms in Θ, thus, in the case of m_1, $\theta\{\tfrac{m_1}{m_1}\} = 1$ is not included because it is a constant.

$$\theta = \left[\theta\{\tfrac{m_1}{\{m_1,m_2\}}\}, \theta\{\tfrac{m_1}{\{m_1,m_3\}}\}, \theta\{\tfrac{m_1}{\{m_1,m_2,m_3,m_4\}}\}\right]$$

By taking derivatives according to the variable terms in Θ, the Jacobian and Hessian can be obtained along with the reference value. Note that the derivative can be taken for other terms that are not variable, but their corresponding elements in J and H will be zero. For the purposes of notation compactness, we will only be taking the derivative with respect to θ, the variable elements in Θ for the mode in question. For example, when m_1 is selected, J and H take the following form:

$$\hat{p}(E|m_1) = p(E|m_1, \hat{\Theta})$$

$$J_i^{m_1} = \left[\frac{\partial\, p(E|m_1,\Theta)}{\partial\, \theta_1} \quad \frac{\partial\, p(E|m_1,\Theta)}{\partial\, \theta_2} \quad \frac{\partial\, p(E|m_1,\Theta)}{\partial\, \theta_3}\right]_{\hat{\theta}}$$

$$H_i^{m_1} = \begin{bmatrix} \dfrac{\partial^2\, p(E|m_1,\Theta)}{\partial\, \theta_1^2} & \dfrac{\partial^2\, p(E|m_1,\Theta)}{\partial\, \theta_1 \partial\, \theta_2} & \dfrac{\partial^2\, p(E|m_1,\Theta)}{\partial\, \theta_1 \partial\, \theta_3} \\[2ex] \dfrac{\partial^2\, p(E|m_1,\Theta)}{\partial\, \theta_2 \partial\, \theta_1} & \dfrac{\partial^2\, p(E|m_1,\Theta)}{\partial\, \theta_2^2} & \dfrac{\partial^2\, p(E|m_1,\Theta)}{\partial\, \theta_2 \partial\, \theta_3} \\[2ex] \dfrac{\partial^2\, p(E|m_1,\Theta)}{\partial\, \theta_3 \partial\, \theta_1} & \dfrac{\partial^2\, p(E|m_1,\Theta)}{\partial\, \theta_3 \partial\, \theta_2} & \dfrac{\partial^2\, p(E|m_1,\Theta)}{\partial\, \theta_3^2} \end{bmatrix}_{\hat{\theta}}$$

where the superscript m_1 pertains to Mode (1). Using data from Tables 15.3 and 15.4, when $E = [0,0]$ is observed, the reference likelihood is:

$$p(E|m_1, \hat{\Theta}) = \frac{\sum_k S_k n_k \hat{\theta}_k}{\sum_k n_k \hat{\theta}_k}$$

$$= \frac{0.7 \times 100 \times 1 + 0.5 \times 50 \times 0.8 + 0.4 \times 50 \times 0.8 + 0.28 \times 25 \times 0.6}{100 \times 1 + 50 \times 0.8 + 50 \times 0.8 + 25 \times 0.6}$$

$$= \frac{70 + 20 + 16 + 4.2}{100 + 40 + 40 + 15} = 110.2/195 = 0.565$$

In a similar manner, the Jacobian terms are calculated as

$$\frac{\partial^2 p(E|m_1, \Theta)}{\partial \theta_i^2} = \frac{n_i S_i}{\sum_k n_k \hat{\theta}_k} - \frac{n_i \sum_k S_k n_k \hat{\theta}_k}{\left(\sum_k n_k \hat{\theta}_k\right)^2}$$

$$= \frac{n_i S_i}{195} - \frac{n_i \, 110.2}{195^2} = \frac{n_i (195 \, S_i - 110.2)}{195^2}$$

Thus the Jacobian can be expressed as

$$\boldsymbol{J}^{m_1} = \left[\frac{50(195(0.5) - 110.2)}{195^2} \quad \frac{50(195(0.4) - 110.2)}{195^2} \quad \frac{25(195(0.28) - 110.2)}{195^2} \right]$$

Likewise, the Hessian terms can be expressed as

$$\frac{\partial^2 p(E|m_1, \Theta)}{\partial \theta_i \, \partial \theta_j}\bigg|_{\hat{\theta}} = -\frac{n_i S_j + n_j S_i}{\left(\sum_k n_k \hat{\theta}_k\right)^2} + \frac{n_i n_j \sum_k S_k n_k \hat{\theta}_k \{m\}}{\left(\sum_k n_k \hat{\theta}_k\right)^3}$$

$$= -\frac{n_i S_j + n_j S_i}{195^2} + \frac{n_i n_j 110.2}{195^3}$$

$$= \frac{n_i n_j \left(\frac{110.2}{195} - \frac{S_j}{n_j} - \frac{S_i}{n_i}\right)}{195^2}$$

so that the Hessian takes the following form

$$\boldsymbol{H}^{m_1} = \begin{bmatrix} \frac{50 \times 50 \left(\frac{110.2}{195} - \frac{0.5}{50} - \frac{0.5}{50}\right)}{195^2} & \frac{50 \times 50 \left(\frac{110.2}{195} - \frac{0.4}{50} - \frac{0.5}{50}\right)}{195^2} & \frac{50 \times 25 \left(\frac{110.2}{195} - \frac{0.28}{25} - \frac{0.5}{50}\right)}{195^2} \\ \frac{50 \times 50 \left(\frac{110.2}{195} - \frac{0.5}{50} - \frac{0.4}{50}\right)}{195^2} & \frac{50 \times 50 \left(\frac{110.2}{195} - \frac{0.4}{50} - \frac{0.4}{50}\right)}{195^2} & \frac{50 \times 25 \left(\frac{110.2}{195} - \frac{0.28}{25} - \frac{0.4}{50}\right)}{195^2} \\ \frac{25 \times 50 \left(\frac{110.2}{195} - \frac{0.5}{50} - \frac{0.28}{25}\right)}{195^2} & \frac{25 \times 50 \left(\frac{110.2}{195} - \frac{0.4}{50} - \frac{0.28}{25}\right)}{195^2} & \frac{25 \times 25 \left(\frac{110.2}{195} - \frac{0.28}{25} - \frac{0.28}{25}\right)}{195^2} \end{bmatrix}$$

Note that these Jacobian and Hessian terms do not assume independence, and are calculated from joint probabilities of E_1, E_2; if independence is assumed, Jacobian and Hessian matrices have to be calculated for each independent piece of evidence.

15.3.7   Online Step 3: Perform Combinations

For some modes, E_1 and E_2 are considered independent; if this is the case for mode M, second-order terms exist for each piece of evidence. These terms must be combined before performing combination with the prior probabilities.

$$\hat{p}(E_1, E_2|M) = \hat{p}(E_1|M)\,\hat{p}(E_2|M)$$
$$\boldsymbol{J}_L = \hat{p}(E_1|M)\boldsymbol{J}_2 + \hat{p}(E_2|M)\boldsymbol{J}_1$$
$$\boldsymbol{H}_L = \hat{p}(E_1|M)\boldsymbol{H}_2 + \hat{p}(E_2|M)\boldsymbol{H}_1 + \boldsymbol{J}_2^T\boldsymbol{J}_1 + \boldsymbol{J}_1^T\boldsymbol{J}_2$$

where $\boldsymbol{J}_L$ and $\boldsymbol{H}_L$ are the Jacobian and Hessian of the overall likelihood. If E_1 and E_2 are considered dependent, then $\boldsymbol{J}_L$ and $\boldsymbol{H}_L$ are obtained directly from the previous step; no combination is required. These likelihood terms are then combined with the prior terms $p(M)$, $\boldsymbol{J}_P$ and $\boldsymbol{H}_P$, in order to obtain the posterior terms $p(M|F_1, F_2)$, $\boldsymbol{J}_F$ and $\boldsymbol{H}_F$

$$K = \sum_k \hat{p}(m_k)\hat{p}(E_1, E_2|m_k)$$

$$\hat{p}(M|E_1, E_2) = \frac{1}{K}\hat{p}(M)\hat{p}(E_1, E_2|M)$$

$$\boldsymbol{J}_F = \frac{1}{K}[\hat{p}(M)\boldsymbol{J}_L + \hat{p}(E_1, E_2|M)\boldsymbol{J}_P]$$

$$\boldsymbol{H}_F = \frac{1}{K}[\hat{p}(M)\boldsymbol{H}_L + \hat{p}(E_1, E_2|M)\boldsymbol{H}_P + \boldsymbol{J}_L^T\boldsymbol{J}_P + \boldsymbol{J}_P^T\boldsymbol{J}_L]$$

Keep in mind that priors focus on the frequency of unambiguous modes, thus $\boldsymbol{J}_F$ and $\boldsymbol{H}_F$ tend to be zero matrices.

In the case of our example, $\hat{p}(M)$ is given in Table 15.2; the likelihood reference $\hat{p}(E_1, E_2|m_1)$ has already been given as 0.565. By calculating the reference likelihoods for other modes, we get the likelihood vector $\hat{p}(E_1, E_2|M) = [0.565, 0.198, 0.184, 0.168]$. From this, the normalization constant is

$$K = \sum_k \hat{p}(m_1)\hat{p}(E_1, E_2|m_1)$$
$$= 0.565 \times 0.6 + 0.198 \times 0.15 + 0.184 \times 0.15 + 0.168 \times 0.1 = 0.413$$

Keeping in mind previous Jacobian and Hessian terms $\boldsymbol{J}^{m_1}, \boldsymbol{H}^{m_1}$, and that Jacobian and Hessian terms associated with prior probabilities are zero matrices, the posterior probability along with the Jacobian and Hessian matrices are obtained as

$$\hat{p}(m_1|E_1, E_2) = \frac{1}{0.413}(0.6 \times 0.565) = 0.821$$

$$\boldsymbol{J}_F^{m_1} = \frac{1}{0.413}(0.6\boldsymbol{J}^{m_1} + 0.565[\boldsymbol{0}]) = 1.45\,\boldsymbol{J}^{m_1}$$
$$= \begin{bmatrix} -0.0228 & -0.0579 & -0.0500 \end{bmatrix}$$

$$\boldsymbol{H}_F^{m_1} = \frac{(0.6\boldsymbol{H}^{m_1} + 0.565[\boldsymbol{0}] + (\boldsymbol{J}^{m_1})^T\boldsymbol{0} + \boldsymbol{0}^T\boldsymbol{J}^{m_1})}{0.413} = 1.45\,\boldsymbol{H}^{m_1}$$
$$= \begin{bmatrix} 0.0490 & 0.0492 & 0.0244 \\ 0.0492 & 0.0493 & 0.0245 \\ 0.0244 & 0.0245 & 0.0122 \end{bmatrix}$$

15.3.8   Online Step 4: Make a Diagnosis

Diagnosis using the Point Estimate $\hat{p}(M|E_1, E_2)$

A diagnosis can be made using the reference posterior probabilities $\hat{p}(M|E_1, E_2)$. This is the simplest method available and does not require the calculation of J_F and H_F. The diagnosis is made by selecting the mode with largest corresponding value of $\hat{p}(M|E_1, E_2)$. In our example, the set of posterior probabilities is found to be

$$\hat{p}(M|E_1, E_2) = \frac{1}{0.413} \begin{bmatrix} 0.565 \times 0.6 \\ 0.198 \times 0.15 \\ 0.184 \times 0.15 \\ 0.168 \times 0.1 \end{bmatrix} = \begin{bmatrix} 0.8207 \\ 0.0719 \\ 0.0668 \\ 0.0406 \end{bmatrix}$$

From this example, one would diagnose Mode (1).

Diagnosis using the Expected Value of $p(M|E_1, E_2)$

Instead of assuming a point value for $p(M|E_1, E_2)$ using $\hat{\Theta}$, we could assume a distribution over the possible values of Θ in order to calculate the expected value of $p(M|E_1, E_2)$. Consider Mode (1) with its corresponding values J_F and H_F along with the corresponding parameter vector θ from Θ, which can be varied

$$\theta = \left[\theta\{\tfrac{m_1}{\{m_1, m_2\}}\}, \theta\{\tfrac{m_1}{\{m_1, m_3\}}\}, \theta\{\tfrac{m_1}{\{m_1, m_2, m_3, m_4\}}\} \right]^T$$

$$= [\theta_1, \theta_2, \theta_3]^T$$

Note that because $\theta\{\tfrac{m_1}{m_1}\} = 1$ is not variable but constant, it is now excluded from θ. The vector of expected values and expected squared values are given as

$$E[\theta] = [E(\theta_1), E(\theta_2), E(\theta_3)]^T$$

$$E[\theta^2] = [E(\theta_1^2), E(\theta_2^2), E(\theta_3^2)]^T$$

These expected values are calculated using Eqn (15.6) and (15.7)

$$E[\theta\{\tfrac{m_1}{m_k}\}] = \frac{\alpha(\tfrac{m_1}{m_k})}{\sum\limits_{m_i \subset m_k} \alpha(\tfrac{m_1}{m_k})}$$

$$E[\theta^2\{\tfrac{m_1}{m_k}\}] = \frac{\alpha(\tfrac{m_1}{m_k}) + \alpha^2(\tfrac{m_1}{m_k})}{\left[\sum\limits_{m_i \subset m_k} \alpha(\tfrac{m_1}{m_k})\right] + \left[\sum\limits_{m_i \subset m_k} \alpha(\tfrac{m_1}{m_k})\right]^2}$$

where $\alpha(\tfrac{m_i}{m_k})$ is calculated using

$$\alpha(\tfrac{m_i}{m_k}) = \hat{\theta}\{\tfrac{m_i}{m_k}\} n(m_k)$$

As an example, let us consider the prior probabilities in Table 15.2 and the mode frequencies in Table 15.3. The resulting α parameters would be

$$\alpha(\tfrac{m_1}{m_1,m_2}) = \hat{\theta}\{\tfrac{m_1}{m_1,m_2}\}n(m_1,m_2)$$

$$= 0.8 \times 50 = 40$$

The value for α conveys a degree of certainty on $\hat{\theta}\{\tfrac{m_1}{m_1,m_2}\}$, where $\alpha \to \infty$ indicates complete certainty. If one wishes to reduce certainty on α it is possible to scale the values for α by a value less than 1; scaling by zero conveys complete uncertainty. Using these values for α, the expectation $E(\theta_1)$ is calculated as

$$E(\boldsymbol{\theta}_1) = E[\theta\{\tfrac{m_1}{m_1,m_2}\}] = \frac{\alpha(\tfrac{m_1}{m_1,m_2})}{\alpha(\tfrac{m_1}{m_1,m_2}) + \alpha(\tfrac{m_2}{m_1,m_2})}$$

$$= \frac{40}{40+10} = 0.8$$

$$E(\boldsymbol{\theta}_1^2) = E[\theta^2\{\tfrac{m_1}{m_1,m_2}\}] = \frac{\alpha(\tfrac{m_1}{m_1,m_2}) + \alpha^2(\tfrac{m_1}{m_1,m_2})}{[\alpha(\tfrac{m_1}{m_1,m_2}) + \alpha(\tfrac{m_1}{m_1,m_2})] + [\alpha(\tfrac{m_1}{m_1,m_2}) + \alpha(\tfrac{m_1}{m_1,m_2})]^2}$$

$$= \frac{40 + 40^2}{50 + 50^2} = 0.643$$

When this technique is applied to m_1 using the other ambiguous modes $\{m_1, m_3\}$, $\{m_1, m_2, m_3, m_4\}$ the elements of $E(\boldsymbol{\theta})$ and $E(\boldsymbol{\theta}^2)$ are given as follows:

$$E(\boldsymbol{\theta}) = [0.8, 0.8, 0.6]^T$$

$$E(\boldsymbol{\theta}^2) = [0.643, 0.643, 0.369]^T$$

The next step is to obtain all the terms in the expression for the second-order expectation

$$E[p(M|E)] = C + \boldsymbol{J}^* E[\boldsymbol{\theta}] + \frac{1}{2}[E[\boldsymbol{\theta}]^T \boldsymbol{H}_{OD} E[\boldsymbol{\theta}] + E[\boldsymbol{\theta}^2]^T \boldsymbol{H}_D] \tag{15.9}$$

The first terms C and $\boldsymbol{J}^*$ are obtained using the reference parameter values $\hat{\boldsymbol{\theta}}$. Note that because the nonvariable terms in $\hat{\boldsymbol{\theta}}$ have zero Jacobian and Hessian elements, they are omitted.

$$C = \hat{p}(M|E) - \boldsymbol{J}\hat{\theta} + \frac{1}{2}\hat{\theta}^T \boldsymbol{H}\hat{\theta}$$

$$= 0.821 - \begin{bmatrix} -0.0228 \\ -0.0579 \\ -0.0500 \end{bmatrix}^T \begin{bmatrix} 0.8 \\ 0.8 \\ 0.6 \end{bmatrix}$$

$$+ \frac{1}{2}\begin{bmatrix} 0.8 \\ 0.8 \\ 0.6 \end{bmatrix}^T \begin{bmatrix} 0.0490 & 0.0492 & 0.0244 \\ 0.0492 & 0.0493 & 0.0245 \\ 0.0244 & 0.0245 & 0.0122 \end{bmatrix} \begin{bmatrix} 0.8 \\ 0.8 \\ 0.6 \end{bmatrix}$$

$$= 0.725$$

$$J^* = J - \hat{\theta}^T H$$

$$= \begin{bmatrix} -0.0228 \\ -0.0579 \\ -0.0500 \end{bmatrix}^T - \begin{bmatrix} 0.8 \\ 0.8 \\ 0.6 \end{bmatrix}^T \begin{bmatrix} 0.0490 & 0.0492 & 0.0244 \\ 0.0492 & 0.0493 & 0.0245 \\ 0.0244 & 0.0245 & 0.0122 \end{bmatrix}$$

$$= \begin{bmatrix} -0.0910 & -0.1263 & -0.0840 \end{bmatrix}$$

Next, H_{OD} is obtained by subtracting the main diagonal of H from H. Meanwhile, H_D is a vector of the main diagonal of H that was removed from H_{OD}. For our example, H_{OD} and H_D are given as

$$H^1_{OD} = \begin{bmatrix} 0 & 0.0492 & 0.0244 \\ 0.0492 & 0 & 0.0245 \\ 0.0244 & 0.0245 & 0 \end{bmatrix}$$

$$H^1_D = \begin{bmatrix} 0.0490 & 0.0493 & 0.0122 \end{bmatrix}$$

With all of the expectation of vectors and matrix terms defined, the expected value $E[p(M|E)]$ can be calculated according to Eqn 15.5.

$$E[p(M|E)] = C + J^* E[\theta] + \frac{1}{2}[E[\theta]^T H_{OD} E[\theta] + E[\theta^2]^T H_D]$$

$$= 0.821$$

which is almost identical to the posterior probability of Mode (1) obtained from the point estimate.

Diagnosis using the Probability Ranges of $p(M|E_1, E_2)$

While the simple method and the expectation method yield point estimates of the probability, the second-order approximation can also be used to obtain probability ranges. The probability is a function of Θ given by

$$\Delta\Theta = \Theta - \hat{\Theta}$$

$$p(M|E, \Theta) = \hat{p}(M|E) + J^m_F \Delta\Theta + \frac{1}{2}\Delta\Theta^T H^m_F \Delta\Theta$$

The belief or lower-bound probability is obtained by minimizing $p(M|E, \Theta)$ subject to the constraint $0 \leq \Theta \leq 1$. The plausibility or upper-bound probability is obtained by maximizing $p(M|E, \Theta)$ subject to the same constraints. Because this is a constrained quadratic expression, the maximization and minimization problems can be solved using standard quadratic programming techniques.

While the actual diagnosis can be obtained using either the point estimate or expected value of $p(M|E_1, E_2)$, the belief and plausibility can give additional information about the uncertainty of the diagnosis. Consider the hypothetical result in Figure 15.2, where point estimate is given by a dotted line and the uncertainty regions are given in a lighter shade of gray. In this

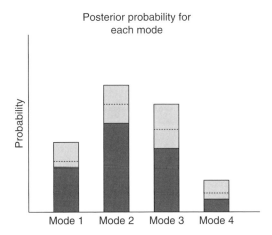

Figure 15.2   An illustration of diagnosis results with uncertainty region

example, the uncertainty region for Mode (2) and Mode (3) overlap. Thus while the system is most likely operating at Mode (2), it would be unwise to rule out Mode (3), as under certain circumstances, Mode (3) could be more probable than Mode (2).

15.4   Simulated Case

The proposed second-order method was tested on the simulated Tennessee Eastman problem. In this simulation, data was masked as ambiguous based on its resemblance toward other modes. In the masking process, distributions were estimated for each mode and a likelihood ratio threshold was set, so that if another mode was likely enough, the data point was classified as ambiguous. For example, consider the data from Mode (1). If there is a data point d_i, consisting of evidence, where the likelihood ratio R between mode k and Mode 1 was large enough

$$R = \frac{p(d_i|m_k)}{p(d_i|m_1)} > \text{Threshold}$$

then the mode m_k was added, rendering the mode associated with d_i ambiguous. By manipulating the threshold, certain amounts of data can be classified as ambiguous. For this simulation case, data was abundant and the monitors showed fairly clear results. In order to demonstrate the effectiveness of the second-order method, 50% of data was deliberately removed from the history, and noise was added.

First, we took a look at the average probability boundaries given for the system. Here, second-order posterior terms were calculated and boundaries were determined via quadratic programming. The first set of figures (see Figure 15.3) had an ambiguity threshold set so that 30% of the data had modes with ambiguity. In the second set of figures (see Figure 15.4) each had 70% of the data belonging to ambiguous modes. From these figures, one can see that Modes 2 and 4 were the easiest to diagnose, with the true probability being notably higher, and with small probability boundaries. Modes 3 and 5 were the most difficult, with much larger

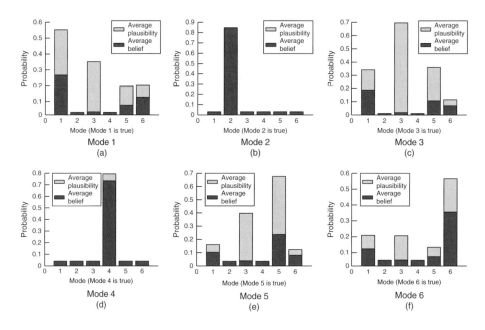

Figure 15.3   Probability bounds at 30% ambiguity

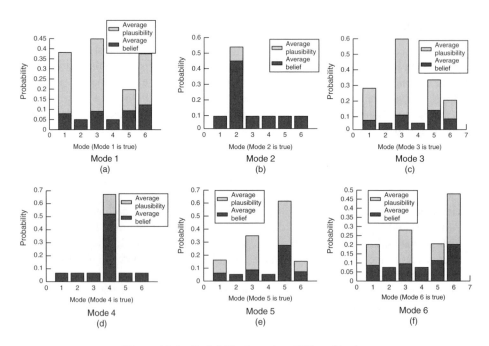

Figure 15.4   Probability bounds at 70% ambiguity

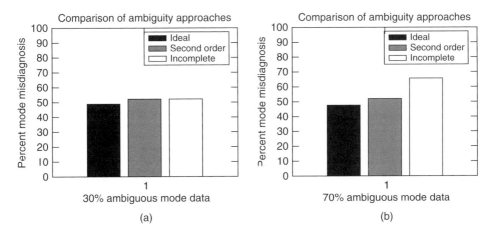

Figure 15.5   Tennessee Eastman problem mode-diagnosis error

probability boundaries and a tendency toward mutual confusion. When the amount of ambiguous mode data was increased from 30% to 70%, probability ranges grew, primarily from Modes 1 and 6. However, there was a slight increase in probability boundaries from Modes 2 and 4. Probability boundaries for Modes 3 and 5 were already quite large, with 30% ambiguous mode data; increasing the amount of ambiguous mode data from 30% to 70% had little effect.

After evaluating the probability boundaries, we proceed to evaluating overall diagnosis performance. For the sake of simplicity, diagnosis was based on the point-estimate method, from the informed transformation $\hat{p}(M|E) = p(M|E, \hat{\Theta})$. In Figure 15.5, mode diagnosis performance was assessed. The performance metric used is the percentage of misdiagnosed modes. Three different methods were compared:

1. **Ideal case:** This refers to the case where ambiguity is not present. The original Bayesian method was performed on data before masking techniques were applied.
2. **Second-order method:** This refers to the case where the second-order Bayesian method is performed on data containing ambiguous modes.
3. **Incomplete Bayesian method:** This refers to the case where ambiguous modes is present in the data, but they were ignored so that the original Bayesian method could be performed on the remaining dataset.

The second-order method was compared to the *incomplete Bayesian method*, where masked data was simply discarded, and the *ideal case*, where masked data was known. In Figure 15.5(a), the three methods performed very similarly, so ignoring the 30% of the data that was ambiguous did not significantly affect the result. However, when the amount of ambiguous data increased to 70%, the incomplete Bayesian method ignores a significant amount of data. Thus when there was more ambiguous data, the performance improvement brought on by the second-order method was significant.

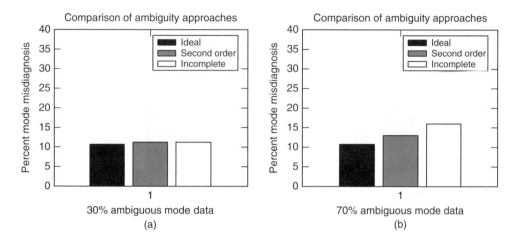

Figure 15.6 Tennessee Eastman component-diagnosis error

In addition to diagnosing modes, the state of each individual component was diagnosed. For this system, the component states were:

1. A/C feed ratio step change (stream 4)
2. B composition step change (stream 4)
3. C header pressure loss
4. A, B, C feed composition step change (stream 4)
5. D feed temperature (stream 2)
6. Reactor cooling-water inlet temperature
7. Sticky reactor cooling-water valve

In many cases, if a mode is incorrectly diagnosed, the correct mode is similar, usually differing by one or two components. Consequently, component diagnosis tends to exhibit better performance, with fewer false results. Unsurprisingly, when diagnosing components, the second-order method showed a performance improvement that was similar to the improvement shown when diagnosing modes. The results are shown in Figure 15.6.

15.5 Bench-scale Case

This method was also applied to the bench-scale hybrid tank system, in which tank leaks and sensor bias are to be detected. Again, data was masked based on its proximity to other modes and diagnosis was based on the point estimate given by the informed transformation $\hat{p}(M|E) = p(M|E, \hat{\Theta})$.

Mode-diagnosis results are available in Figure 15.7. When diagnosing modes, performance seemed to be relatively unchanged, even when 70% of the data was missing; this robustness toward missing data was again likely due to the abundance of data. In contrast, Figure 15.8 shows results when diagnosing components, where it can be seen that the second-order

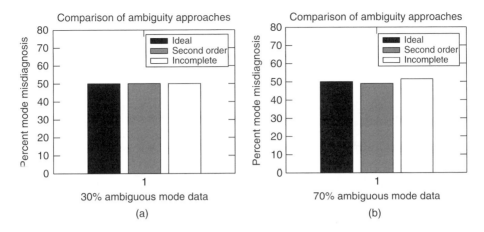

Figure 15.7 Hybrid tank system mode-diagnosis error

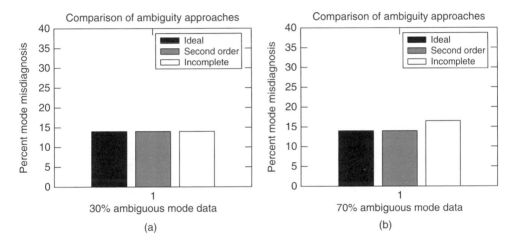

Figure 15.8 Hybrid tank system component-diagnosis error

approach shows a slightly more significant improvement. While the frequency of diagnosing the correct mode may not have significantly improved when the second order method was applied, it appears that the incorrectly diagnosed modes bear more resemblance to the true mode.

15.6   Industrial-scale Case

Finally, the method was tested on industrial data, where the on and off conditions were to be detected for each subsystem. Data was masked by taking into account proximity with other modes, and again, diagnosis results were based on the point estimate of the posterior probability $\hat{p}(M|E) = p(M|E, \hat{\Theta})$.

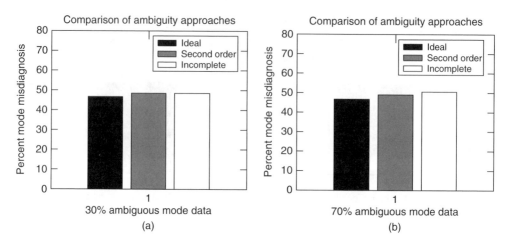

Figure 15.9   Industrial system mode-diagnosis error

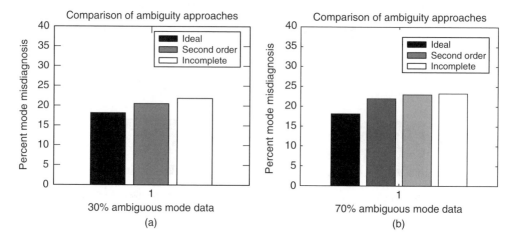

Figure 15.10   Industrial system component-diagnosis error

Mode-diagnosis results are shown in Figure 15.9. In a similar manner to the experimental system, diagnosis results are fairly uniform for both 30% ambiguous mode data and 70% ambiguous mode data. When diagnosing components, however, the improvements for the second-order method appear to be more pronounced (Figure 15.10). Thus while a similar number of modes may still be incorrectly diagnosed, it is evident that the incorrect modes tend to resemble the correct mode more closely.

15.7   Notes and References

The original likelihood problem expressed in Eqn (15.1) was described in detail in Gonzalez and Huang (2013b), which was the first paper in a two-part series. The problem of combining multiple evidence sources was elaborated in the second part of Gonzalez and Huang (2013a), which describes the second-order combination method in detail. Readers are referred to these two papers for more details.

References

Gonzalez R and Huang B 2013a Control loop diagnosis from historical data containing ambiguous operating modes: Part 2. information synthesis based on proportional parameterization. *Journal of Process Control* **23**(4), 1441–1454.

Gonzalez R and Huang B 2013b Control loop diagnosis with ambiguous historical operating modes: Part 1. a proportional parametrization approach. *Journal of Process Control* **23**(4), 585–597.

16

Accounting for Ambiguous Modes in Historical Data: A Dempster–Shafer Approach

16.1 Introduction

Dempster–Shafer theory has long been considered a more general alternative to Bayesian combination since it directly accounts for ambiguity and has had a wide variety of adaptations in the literature. However, Dempster–Shafer theory does not lend itself to making inference from history-based likelihoods containing ambiguous modes. Because of this, it needs to be generalized in order to better fit this task.

As in Bayesian combination, the goal is to combine information from independent sources of evidence, including prior probabilities. The application of generalized Dempster–Shafer theory takes the same form as the second-order method discussed in Chapter 15. This chapter gives details on how to:

1. set up a method for calculating likelihoods given θ parameters $p(E|M,\Theta))$
2. calculate the generalized basic belief assignment (GBBA) for both priors and likelihoods
3. combine the GBBAs to obtain a final diagnosis result
4. group monitors into independent groups (optional).

16.2 Algorithm

16.2.1 Parametrized Likelihoods

Likelihoods are parametrized in the same way as in Chapter 15

$$p(E|M,\Theta) = \frac{\sum_{\boldsymbol{m}_k \geq m} \theta\{\frac{m}{\boldsymbol{m}_k}\} S(E|\boldsymbol{m}_k) n(\boldsymbol{m}_k)}{\sum_{\boldsymbol{m}_k \geq m} \theta\{\frac{m}{\boldsymbol{m}_k}\} n(\boldsymbol{m}_k)} \tag{16.1}$$

Process Control System Fault Diagnosis: A Bayesian Approach, First Edition. Ruben Gonzalez, Fei Qi and Biao Huang.
© 2016 John Wiley & Sons, Ltd. Published 2016 by John Wiley & Sons, Ltd.

where $n(\boldsymbol{m}_k)$ is the number of times $\boldsymbol{m}_k$ appears in the history, and $S(E|\boldsymbol{m}_k)$ is the support function for evidence E given $\boldsymbol{m}_k$, which for discrete data is calculated as

$$S(E|\boldsymbol{m}_k) = \frac{n(E|\boldsymbol{m}_k)}{n(\boldsymbol{m}_k)}$$

In this chapter, Eqn (16.1) is used as a basis for calculating the GBBAs.

16.2.2   Basic Belief Assignments

Traditional Basic Belief Assignments

The basic belief assignment (BBA) is a function with respect to the mode and is denoted $S(\boldsymbol{m}_k)$; it is similar to probability except that it can yield support to ambiguous modes as well.[1] The probability $p(M|E)$ is calculated using $S(\boldsymbol{m}|E)$ according to

$$p(M|E,\Theta) = \sum_{\boldsymbol{m}_k \supseteq \boldsymbol{m}} \theta\{\tfrac{\boldsymbol{m}}{\boldsymbol{m}_k}\} S(\boldsymbol{m}_k|E) \qquad (16.2)$$

where the support function (or BBA) $S(\boldsymbol{m}|E)$ is given as

$$S(\boldsymbol{m}|E) = \frac{n(\boldsymbol{m}, E)}{n(E)} \qquad (16.3)$$

Here, $n(\boldsymbol{m}, E)$ is the number of times the mode $\boldsymbol{m}$ and evidence E jointly occur, while $n(E)$ is the total number of times evidence E occurs. In addition to probability, Dempster–Shafer theory also concerns itself with belief (the lower-bound probability) and plausibility (the upper-bound probability). Because $p(M|E,\Theta)$ is linear with respect to θ and the coefficients $S(\boldsymbol{m})$ on θ are positive, $p(M|E,\Theta)$ is maximized by setting θ to 1 whenever possible and is minimized by setting θ to 0 whenever possible.

$$Bel(M|E,\Theta) = \min_{\Theta}[p(M|E,\Theta)] = \sum_{\boldsymbol{m}_k \subseteq \boldsymbol{m}} S(\boldsymbol{m}_k|E)$$

$$Pl(M|E,\Theta) = \max_{\Theta}[p(M|E,\Theta)] = \sum_{\boldsymbol{m}_k \supseteq \boldsymbol{m}} S(\boldsymbol{m}_k|E)$$

Unfortunately, the form of Eqn (16.2) is too restrictive to adequately represent the base problem in Eqn (16.1). The two most detrimental restrictions of Eqn (16.2) lie in the terms $S(\boldsymbol{m})$, which function as coefficients on Θ. The first restriction is that terms in $S(\boldsymbol{m})$ are positive, while the second restriction is that terms must be identical for all unambiguous modes contained in $\boldsymbol{m}$. The reasons for relaxing these restrictions have been discussed in Chapter 7, but they mainly boil down to the fact that Eqn (16.1) is nonlinear with respect to Θ. The task at hand is to obtain a generalized expression of the BBA that better represents the problem presented in Eqn (16.1). This can be done by allowing the BBA to have negative support, and to have different amounts of support for different modes in $\boldsymbol{m}$.

---

[1] The boldface $\boldsymbol{m}$ is used to represent potential ambiguity.

Generalized basic belief assignments

The GBBA, denoted G, is a matrix that can approximate the expression of $p(E|M, \Theta)$ in Eqn (16.2)

$$p(E|M, \Theta) = G[:, m]^T \Theta[:, m] \qquad (16.4)$$

where Θ is the matrix form of Θ and $\Theta[:, m]$ is the column of Θ that pertains to the specific mode m. Likewise, $G[:, m]^T$ is the column of G pertaining to m. The expression in Eqn (16.4) is a first-order approximation of Eqn (16.1).

As an example for the structure of G, let us consider a three-mode system m_1, m_2, m_3 with ambiguous modes $\{m_1, m_2\}$, $\{m_2, m_3\}$, $\{m_1, m_3\}$. The structure of G and Θ are given as

$$
G =
\begin{array}{c|ccc}
 & m_1 & m_2 & m_3 \\
\hline
m_1 & G\{\frac{m_1}{m_1}\} & 0 & 0 \\
m_2 & 0 & G\{\frac{m_2}{m_2}\} & 0 \\
m_3 & 0 & 0 & G\{\frac{m_3}{m_3}\} \\
\{m_1, m_2\} & G\{\frac{m_1}{m_1, m_2}\} & G\{\frac{m_2}{m_1, m_2}\} & 0 \\
\{m_1, m_3\} & G\{\frac{m_1}{m_1, m_3}\} & 0 & G\{\frac{m_3}{m_1, m_3}\} \\
\{m_2, m_3\} & 0 & G\{\frac{m_2}{m_2, m_3}\} & G\{\frac{m_3}{m_2, m_3}\}
\end{array}
$$

$$
\Theta =
\begin{array}{c|ccc}
 & m_1 & m_2 & m_3 \\
\hline
m_1 & 1 & 0 & 0 \\
m_2 & 0 & 1 & 0 \\
m_3 & 0 & 0 & 1 \\
\{m_1, m_2\} & \theta\{\frac{m_1}{m_1, m_2}\} & \theta\{\frac{m_2}{m_1, m_2}\} & 0 \\
\{m_1, m_3\} & \theta\{\frac{m_1}{m_1, m_3}\} & 0 & \theta\{\frac{m_3}{m_1, m_3}\} \\
\{m_2, m_3\} & 0 & \theta\{\frac{m_2}{m_2, m_3}\} & \theta\{\frac{m_3}{m_2, m_3}\}
\end{array}
$$

Any value of Θ that is not set to 0 or 1 is considered to be unknown or *flexible*. The approximation for $p(E|M, \Theta)$ is then given as

$$p(E|M_1, \Theta) = G[:, m_1]^T \Theta[:, m_1]$$
$$= G\{\tfrac{m_1}{m_1}\}\theta\{\tfrac{m_1}{m_1}\} + G\{\tfrac{m_1}{m_1, m_2}\}\theta\{\tfrac{m_1}{m_1, m_2}\} + G\{\tfrac{m_1}{m_1, m_3}\}G\{\tfrac{m_1}{m_1, m_3}\}$$

The individual terms of G can be calculated according to the following heuristic

$$
G[k, i] =
\begin{cases}
0 & m_i \cap m_k = \emptyset \\
\tilde{p}(E|m_i) & m_i = m_k \\
\frac{\partial p(E|m_i)}{\partial \theta[k, i]} & m_i \subset m_k
\end{cases}
\qquad (16.5)
$$

where

$$\tilde{p}(E|M) = p(E|M, \hat{\Theta}) - \sum_{\boldsymbol{m}_k \supset M} \hat{\theta}\{\tfrac{M}{\boldsymbol{m}_k}\} \frac{\partial \, p(E|M, \boldsymbol{\Theta})}{\partial \, \theta\{\tfrac{M}{\boldsymbol{m}_k}\}}\bigg|_{\hat{\Theta}}$$

$$\frac{\partial p(E|M)}{\partial \boldsymbol{\Theta}[k, i]} = \frac{\partial \, p(E|M, \boldsymbol{\Theta})}{\partial \, \theta\{\tfrac{M}{\boldsymbol{m}_k}\}}\bigg|_{\hat{\Theta}}$$

Here, $\hat{\Theta}$ is the reference value of Θ which, for the generalized Dempster–Shafer method, is set to the *inclusive value* of Θ.

$$\hat{\Theta} = \Theta^*$$

where

- Θ^* is the *inclusive value*, which is defined by setting all flexible values to 1
- Θ_* is the *exclusive value*, which is defined by setting all flexible values to 0.

When G is defined in this manner, Eqn (16.4) is a first-order Taylor series approximation of $p(E|m_1, \Theta)$.

Belief and plausibility can also be calculated from Eqn (16.4) by minimizing and maximizing $p(E|m_1, \Theta)$.

$$Bel(E|M) = \min_{\Theta} \boldsymbol{G}[:, m]^T \boldsymbol{\Theta}[:, m]$$

$$Pl(E|M) = \max_{\Theta} \boldsymbol{G}[:, m]^T \boldsymbol{\Theta}[:, m]$$

where Θ is the set of variable elements in $\Theta[:, m]$; the elements not automatically set to 1 or 0 by logic. Again, the values in Θ are constrained to be between 0 and 1. However, one should note that the elements in G are no longer positive. Consequently, when calculating belief and plausibility, the minimum is no longer obtained by setting all flexible values in Θ to zero; likewise, the maximum is no longer maximized by setting all flexible values in Θ to one. Instead, the belief and plausibility should be obtained using linear programming methods.

16.2.3   The Generalized Dempster's Rule of Combination

In the same way as Bayesian methods, the generalized Dempster's rule of combination can be used to combine information from multiple pieces of evidence, including prior probabilities. Originally, Dempster's rule of combination is given for BBAs in the following form:

$$S(\boldsymbol{m}_k) = \frac{1}{1 - K} \sum_{\boldsymbol{m}_k = \boldsymbol{m}_i \cap \boldsymbol{m}_j \neq \emptyset} S(\boldsymbol{m}_i) S(\boldsymbol{m}_j) \qquad (16.6)$$

$$1 - K = \sum_{\boldsymbol{m}_k = \boldsymbol{m}_i \cap \boldsymbol{m}_j \neq \emptyset} S(\boldsymbol{m}_i) S(\boldsymbol{m}_j) \qquad (16.7)$$

where K is a normalization constant that ensures that the terms $S(\boldsymbol{m}_k)$ sum to unity. In a similar manner, the generalized Dempster's rule of combination is applied to the rows of $\boldsymbol{G}$ that pertain to $\boldsymbol{m}_k$, or equivalently, $G[\boldsymbol{m}_k, :]$

$$G_{12}[\boldsymbol{m}_k, :] = \frac{1}{1-K} \sum_{\boldsymbol{m}_k = \boldsymbol{m}_i \cap \boldsymbol{m}_j \neq \emptyset} G_1[\boldsymbol{m}_i, :] \circ G_2[\boldsymbol{m}_j, :] \tag{16.8}$$

$$1 - K = \sum_{\boldsymbol{m}_k = \boldsymbol{m}_i \cap \boldsymbol{m}_j \neq \emptyset} \mathrm{mean}_{x \neq 0}(G[\boldsymbol{m}_i, :] \circ G[\boldsymbol{m}_j, :]) \tag{16.9}$$

where $X \circ Y$ denotes the Hadamard (or element-wise) product between X and Y, while $\mathrm{mean}_{x \neq 0}(X)$ is the mean of the nonzero values of X.

Short-cut Combination Rule with Bayesian GBBAs

When the generalized Dempster's rule is applied and at least one of the two GBBAs is Bayesian, having no support to ambiguous modes, the resulting GBBA will also be Bayesian. In such a case, it is easier to apply a short-cut rule that will yield the exact answer with much less computational burden. Here we can consider the case where $\boldsymbol{G}_1$ is a Bayesian prior, expressed as $P_1(M)$, and where $\boldsymbol{G}_2$ is an arbitrary GBBA. The resulting GBBA ($\boldsymbol{G}_{12}$) can be expressed as

$$\boldsymbol{G}_{12}(m_i, m_i) = \frac{1}{1-K} P_1(m_i) In_2(m_i) \tag{16.10}$$

$$\boldsymbol{G}_{12}(m_i, m_{j \neq i}) = 0$$

$$1 - K = \sum_m P_1(M) In_2(M) \tag{16.11}$$

where $In_2(m_i)$ is the inclusive probability of m_i expressed as

$$In_2(m_i) = \sum_{\boldsymbol{m}_k \supseteq \boldsymbol{m}_i} \boldsymbol{G}_2[\boldsymbol{m}_k, m_i] = \boldsymbol{G}_2[:, m_i]^T \boldsymbol{\Theta}^*[:, m_i]$$

The end result is $\boldsymbol{G}_{12}(m_i, m_i)$, being a diagonal matrix with the main diagonals representing Bayesian posterior probabilities. The short-cut combination rule is particularly useful in a dynamic setting, where successive combinations will yield Bayesian posteriors pretty quickly; it is better to simply use a Bayesian prior in the first step and use the short-cut rule for every successive combination.

16.3   Example of Proposed Methodology

16.3.1   Introduction

To demonstrate how to use the generalized Dempster's rule, we consider in Figure 16.1 the same control-loop example, and the same modes and evidence, as in Chapter 15.

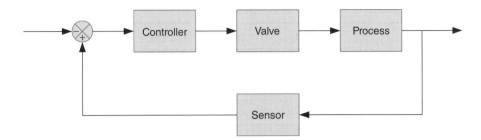

Figure 16.1   Typical control loop

16.3.2   Offline Step 1: Historical Data Collection

Again, the first step is to go through the historical data and note the instances where each of the four possible modes occurs:

1. m_1 $[0, 0]$ where bias and stiction do not occur
2. m_2 $[0, 1]$ where bias does not occur but stiction does
3. m_3 $[1, 0]$ where bias occurs but stiction does not
4. m_4 $[1, 1]$ where both bias and stiction occur.

Data should be collected according to each mode. In certain instances, one of these ambiguous modes may also occur:

1. $\{m_1, m_2\}$ $[\times, 0]$ where bias does not occur and stiction is undetermined
2. $\{m_1, m_3\}$ $[\times, 1]$ where bias is undetermined and stiction does not occur
3. $\{m_2, m_4\}$ $[0, \times]$ where bias is undetermined and stiction occurs
4. $\{m_3, m_4\}$ $[1, \times]$ where bias occurs and stiction is undetermined
5. $\{m_1, m_2, m_3, m_4\}$ $[\times, \times]$ where both bias and stiction are undetermined.

Such ambiguous cases should be classified under one of these ambiguous modes.

16.3.3   Offline Step 2: Mutual Information Criterion (Optional)

As in Chapter 15, this step is optional. One can reduce the dimensionality of the problem by assuming independence where the MIC yields a result close to zero, such as less than 0.01.

$$MIC = \begin{bmatrix} 0 & 0 \\ MIC(1, 2) & 0 \end{bmatrix}$$

where, for example, $MIC(1, 2)$ is calculated as

$$MIC(1, 2) = \sum_{E_1} \sum_{E_2} p(E_1, E_2 | m) \log \left(\frac{p(E_1, E_2 | M)}{p(E_1 | M) p(E_2 | M)} \right)$$

As in the second-order Bayesian method, the MIC matrix must be calculated for every unambiguous mode. For each mode, the two pieces of evidence are either considered independent or dependent. For example, let us say that under Mode (1), the probability is distributed as given in Table 16.1:

Table 16.1 Probability of evidence given Mode (1)

	$E = [0,0]$	$E = [0,1]$	$E = [1,0]$	$E = [1,1]$
$n(E\|m_1)$	70	10	15	5
$p(E\|m_1)$	0.7	0.1	0.15	0.05

The MIC for Mode (1) is given as

$$MIC = p([0,0]|m_1) \log \left(\frac{p([0,0]|m_1)}{p([0,\times]|m_1)p([\times,0]|m_1)} \right)$$

$$+ p([0,1]|m_1) \log \left(\frac{p([0,1]|m_1)}{p([0,\times]|m_1)p([\times,1]|m_1)} \right)$$

$$+ p([1,0]|m_1) \log \left(\frac{p([1,0]|m_1)}{p([1,\times]|m_1)p([\times,0]|m_1)} \right)$$

$$+ p([1,1]|m_1) \log \left(\frac{p([1,1]|m_1)}{p([1,\times]|m_1)p([\times,1]|m_1)} \right)$$

$$= 0.7 \log \left(\frac{0.7}{(0.8)(0.85)} \right) + 0.1 \log \left(\frac{0.1}{(0.8)(0.15)} \right)$$

$$+ 0.15 \log \left(\frac{0.15}{(0.2)(0.85)} \right) + 0.05 \log \left(\frac{0.05}{(0.2)(0.15)} \right)$$

$$= 0.0088259$$

Because this number (0.0088259) is so small, the two monitors E_1 and E_2 can be considered independent under this mode.

16.3.4 Offline Step 3: Calculate Reference Value

GBBAs use inclusive Θ values (Θ^*) as a reference for Θ in the same manner as the second-order Bayesian method uses the informed values ($\hat{\Theta}$). The boldface notation indicates that Θ^* takes the form of a matrix; in this illustration, Θ^* takes the following form:

$$\Theta^* = \begin{bmatrix}
1 & 0 & 0 & 0 \\
0 & 1 & 0 & 0 \\
0 & 0 & 1 & 0 \\
0 & 0 & 0 & 1 \\
\theta^*\{\frac{m_1}{m_1,m_2}\} & \theta^*\{\frac{m_2}{m_1,m_2}\} & 0 & 0 \\
\theta^*\{\frac{m_1}{m_1,m_3}\} & 0 & \theta^*\{\frac{m_3}{m_1,m_3}\} & 0 \\
0 & \theta^*\{\frac{m_2}{m_2,m_4}\} & 0 & \theta^*\{\frac{m_4}{m_2,m_4}\} \\
0 & 0 & \theta^*\{\frac{m_3}{m_3,m_4}\} & \theta^*\{\frac{m_4}{m_3,m_4}\} \\
\theta^*\{\frac{m_4}{m_1,\dots,m_4}\} & \theta^*\{\frac{m_4}{m_1,\dots,m_4}\} & \theta^*\{\frac{m_4}{m_1,\dots,m_4}\} & \theta^*\{\frac{m_4}{m_1,\dots,m_4}\}
\end{bmatrix}$$

where each element of Θ^* is calculated as

$$\Theta^*[i,j] = 1 \qquad\qquad \boldsymbol{m}_i \cap \boldsymbol{m}_k \neq \emptyset$$
$$\Theta^*[i,j] = 0 \qquad\qquad \boldsymbol{m}_i \cap \boldsymbol{m}_k = \emptyset$$

resulting in

$$\Theta^* = \begin{bmatrix}
1 & 0 & 0 & 0 \\
0 & 1 & 0 & 0 \\
0 & 0 & 1 & 0 \\
0 & 0 & 0 & 1 \\
1 & 1 & 0 & 0 \\
1 & 0 & 1 & 0 \\
0 & 1 & 0 & 1 \\
0 & 0 & 1 & 1 \\
1 & 1 & 1 & 1
\end{bmatrix}$$

16.3.5   Online Step 1: Calculate Support

For this application, we assume that the priors are unambiguous, so that the short-cut method can be applied. When a new piece of evidence $[E_1, E_2]$ becomes available, we calculate the support according to

$$S(E_1, E_2 | \boldsymbol{m}_k) = \frac{n(E_1, E_2 | \boldsymbol{m}_k)}{n(\boldsymbol{m}_k)}$$

If E_1 and E_2 are considered independent given m_k, the support can be calculated separately:

$$S(E_1 | \boldsymbol{m}_k) = \frac{n(E_1 | \boldsymbol{m}_k)}{n(\boldsymbol{m}_k)} \qquad S(E_2 | \boldsymbol{m}_k) = \frac{n(E_2 | \boldsymbol{m}_k)}{n(\boldsymbol{m}_k)}$$

The support can be arranged in matrix form as follows:

$$\boldsymbol{S} = \begin{bmatrix}
S(E|m_1) & 0 & 0 & 0 \\
0 & S(E|m_2) & 0 & 0 \\
0 & 0 & S(E|m_3) & 0 \\
0 & 0 & 0 & S(E|m_4) \\
S(E|m_1, m_2) & S(E|m_1, m_2) & 0 & 0 \\
S(E|m_1, m_3) & 0 & S(E|m_1, m_3) & 0 \\
0 & S(E|m_2, m_4) & 0 & S(E|m_2, m_4) \\
0 & 0 & S(E|m_3, m_4) & S(E|m_3, m_4) \\
S(E|m_1, \ldots, m_4) & S(E|m_1, \ldots, m_4) & S(E|m_1, \ldots, m_4) & S(E|m_1, \ldots, m_4)
\end{bmatrix}$$

where each column corresponds to an unambiguous mode, and each row corresponds to an ambiguous mode. Note that for some modes (hence columns), evidence independence is assumed, and therefore some columns have multiple entries to correspond to multiple pieces of evidence.

For example, consider the data collected in Table 16.2 with the support calculated in Table 16.3.

Table 16.2 Frequency of modes containing m_1

	$E = [0,0]$	$E = [0,1]$	$E = [1,0]$	$E = [1,1]$
$n(E\|m_1)$	70	10	15	5
$n(E\|m_2)$	14	59	6	21
$n(E\|m_3)$	13	7	58	22
$n(E\|m_4)$	12	8	23	57
$n(E\|m_1, m_2)$	25	15	7	3
$n(E\|m_1, m_3)$	20	20	6	4
$n(E\|m_2, m_4)$	19	17	7	7
$n(E\|m_3, m_4)$	19	8	7	16
$n(E\|m_1, m_2, m_3, m_4)$	7	6	6	6

Table 16.3 Support of modes containing m_1

	$E = [0,0]$	$E = [0,1]$	$E = [1,0]$	$E = [1,1]$
$S(E\|m_1)$	0.7	0.1	0.15	0.05
$S(E\|m_2)$	0.14	0.59	0.6	0.21
$S(E\|m_3)$	0.13	0.07	0.58	0.22
$S(E\|m_4)$	0.12	0.08	0.23	0.57
$S(E\|m_1, m_2)$	0.5	0.3	0.14	0.06
$S(E\|m_1, m_3)$	0.4	0.4	0.12	0.08
$n(E\|m_2, m_4)$	0.38	0.34	0.14	0.14
$n(E\|m_3, m_4)$	0.38	0.16	0.14	0.32
$S(E\|m_1, m_2, m_3, m_4)$	0.28	0.24	0.24	0.24

If $E = [0,1]$ is observed, the support matrix S takes on the following form:

$$
S = \begin{bmatrix}
0.1 & 0 & 0 & 0 \\
0 & 0.59 & 0 & 0 \\
0 & 0 & 0.07 & 0 \\
0 & 0 & 0 & 0.08 \\
0.3 & 0.3 & 0 & 0 \\
0.4 & 0 & 0.4 & 0 \\
0 & 0.34 & 0 & 0.34 \\
0 & 0 & 0.16 & 0.16 \\
0.24 & 0.24 & 0.24 & 0.24
\end{bmatrix}
$$

16.3.6   Online Step 2: Calculate the GBBA

The GBBA is calculated by taking the derivative of Eqn (16.2), which in vector form is written as

$$
p(E|m_i, \Theta) = \frac{[S[:, m_i] \circ n]^T \Theta[:, m_i]}{n^T \Theta[:, m_i]}
$$

where n is the vertical vector of mode frequencies:

$$n = [n(m_1), n(m_2), \ldots, n(m_1, m_2, m_3, m_4)]^T$$

In the case of Table 16.2, n is given as

$$n = [100, 100, 100, 100, 50, 50, 50, 50, 25]^T$$

In MATLAB, given the variables S (in the form of S), Theta (in the form of Θ) and n (in the form of n), probabilities can be calculated using the following code:

```
1   n_m = length(S(1,:));
2   P = zeros(n_m,1);
3   for m = 1:n_m
4       Num = (S(:,m).*n)'*Theta(:,m);
5       Den = n'*Theta(:,m);
6           P(m) = Num/Den;
7   end
```

In a similar manner, we can calculate the derivatives as

$$\left. \frac{\partial\, p(E|m_i, \Theta)}{\partial\Theta[k, m_i]} \right|_{\hat\Theta} = \frac{n[k]S[k, m_i]}{n^T\Theta^*[:, m_i]} - \frac{n[k](n \circ S[:, m_i])\Theta^*[:, m_i]}{(n^T\Theta^*[:, m])^2}$$

which, in MATLAB, can be obtained using

```
1   n = sum(N,2); %Find number of samples for each mode
2   [n_M,n_m] = size(S);
3   amb = (n_m+1):n_M; % indices of ambiguous modes
4   for m = 1:n_m
5       Num = (S(:,m).*n)'*Theta(:,m); %Likelihood numerator
6       Den = n'*Theta(:,m); %Likelihood denominator
7       P(m,1) = Num/Den; %Inclusive Likelihood
8
9       for k = amb % Only consider ambiguous modes for dP
10          if ModeStruct(k,m) == 1
11              %Abridged code which used Num and Den
12              dP(k-n_m,m) = n(k)*S(k,m)/Den - n(k)*Num/(Den^2);
13          end
14      end
15  end
```

After calculating the derivatives, the GBBA will need reference probabilities. The reference probability is calculated as

$$\tilde p(E|m_i, \Theta^*) = p(E|m_i, \Theta^*) - \sum_{m_k \supset m_i} \left. \frac{\partial\, p(E|m_i, \Theta)}{\partial\Theta[k, m_i]} \right|_{\hat\Theta} \Theta^*[k, m_i]$$

As an example, for Mode (1), $\tilde{p}(E|m_1, \Theta^*)$ is given as

$$\tilde{p}(E|m_1, \Theta^*) = p(E|m_1, \Theta^*) - \left.\frac{\partial p(E|m_1, \Theta)}{\partial \Theta[5,1]}\right|_{\hat{\Theta}} \Theta^*[5,1]$$

$$- \left.\frac{\partial p(E|m_1, \Theta)}{\partial \Theta[6,1]}\right|_{\hat{\Theta}} \Theta^*[6,1] - \left.\frac{\partial p(E|m_1, \Theta)}{\partial \Theta[9,1]}\right|_{\hat{\Theta}} \Theta^*[9,1]$$

where Θ^* is the matrix given in Offline Step 3. In MATLAB, the reference probability Pr can be obtained using the following code

```
1  for m = 1:n_m
2      Pr(m,1) = P(m,1) - dP(:,m)'*Theta(amb,m);
3  end
```

Using these inputs, the resulting GBBA is given as

$$G = \begin{bmatrix} \tilde{p}(E|m_1, \Theta^*) & 0 & 0 & 0 \\ 0 & \tilde{p}(E|m_2, \Theta^*) & 0 & 0 \\ 0 & 0 & \tilde{p}(E|m_3, \Theta^*) & 0 \\ 0 & 0 & 0 & \tilde{p}(E|m_4, \Theta^*) \\ \left.\frac{\partial p(E|m_1,\Theta)}{\partial \Theta[5,1]}\right|_{\hat{\Theta}} & \left.\frac{\partial p(E|m_2,\Theta)}{\partial \Theta[5,2]}\right|_{\hat{\Theta}} & 0 & 0 \\ \left.\frac{\partial p(E|m_1,\Theta)}{\partial \Theta[6,1]}\right|_{\hat{\Theta}} & 0 & \left.\frac{\partial p(E|m_3,\Theta)}{\partial \Theta[6,3]}\right|_{\hat{\Theta}} & 0 \\ 0 & \left.\frac{\partial p(E|m_2,\Theta)}{\partial \Theta[7,2]}\right|_{\hat{\Theta}} & 0 & \left.\frac{\partial p(E|m_4\Theta)}{\partial \Theta[7,4]}\right|_{\hat{\Theta}} \\ 0 & 0 & \left.\frac{\partial p(E|m_3,\Theta)}{\partial \Theta[8,3]}\right|_{\hat{\Theta}} & \left.\frac{\partial p(E|m_4,\Theta)}{\partial \Theta[8,4]}\right|_{\hat{\Theta}} \\ \left.\frac{\partial p(E|m_1,\Theta)}{\partial \Theta[9,1]}\right|_{\hat{\Theta}} & \left.\frac{\partial p(E|m_2,\Theta)}{\partial \Theta[9,2]}\right|_{\hat{\Theta}} & \left.\frac{\partial p(E|m_3,\Theta)}{\partial \Theta[9,3]}\right|_{\hat{\Theta}} & \left.\frac{\partial p(E|m_4,\Theta)}{\partial \Theta[9,4]}\right|_{\hat{\Theta}} \end{bmatrix}$$

In MATLAB, obtaining the GBBA is a simple matrix concatenation

```
1      GBBA = [diag(Pr);dP]
```

In our example, the GBBA is found to be

$$G = \begin{bmatrix} 0.16 & 0 & 0 & 0 \\ 0 & 0.57 & 0 & 0 \\ 0 & 0 & 0.08 & 0 \\ 0 & 0 & 0 & 0.10 \\ 0.02 & -0.07 & 0 & 0 \\ 0.05 & 0 & 0.10 & 0 \\ 0 & -0.06 & 0 & 0.07 \\ 0 & 0 & 0.02 & 0.01 \\ 0.00 & -0.05 & 0.02 & 0.02 \end{bmatrix}$$

16.3.7   Online Step 3: Combine BBAs and Diagnose

In this example, the prior probability is unambiguous. Thus the short-cut method can be used for combination. The first step to the short-cut method is to calculate the inclusive probability $P^*(E|M, \Theta)$

$$p(E|M, \Theta^*) = G[:, M]^T \Theta^*[:, M]$$

where Θ^* is the inclusive probability, and where all nonzero values are set to 1

$$\Theta^* = \begin{bmatrix} 1 & 0 & 0 & 0 \\ 0 & 1 & 0 & 0 \\ 0 & 0 & 1 & 0 \\ 0 & 0 & 0 & 1 \\ 1 & 1 & 0 & 0 \\ 1 & 0 & 1 & 0 \\ 0 & 1 & 0 & 1 \\ 0 & 0 & 1 & 1 \\ 1 & 1 & 1 & 1 \end{bmatrix}$$

The posterior probability can be obtained using the following short-cut expression

$$p(M|E) = \frac{1}{K} p(E|M, \Theta^*) p(M) = \frac{1}{K} In(E|M) p(M)$$

$$K = \sum_k p(E|m_k, \Theta^*) p(m_k)$$

After the BBAs are combined, the posterior probability $p(M|E)$ has a single value for each mode (it is not variable with respect to Θ), thus one simply diagnoses the mode based on the highest posterior.

16.4   Simulated Case

The proposed second-order method was tested on the simulated Tennessee Eastman problem. In the same manner as the second-order Bayesian method, data was masked as ambiguous based on its resemblance toward other modes. If 30% of the data is classified as ambiguous, then for every mode, it classifies the 70% of the most likely data for this mode as unambiguous, but each remaining data point will have at least one additional mode associated with it, based on its proximity toward other modes.

Results for the generalized Dempster–Shafer method are compared to:

- the ideal scenario, in which no modes are ambiguous
- the incomplete Bayesian method, which ignores ambiguous mode data
- the second-order method.

The results in Figure 16.2 compare diagnosis results based on modes, while those in Figure 16.3 compare diagnosis results based on the individual components that make up the modes.

The results for all four scenarios were nearly identical when 30% of the data belonged to ambiguous modes, but more significant improvements were exhibited when 70% of the data

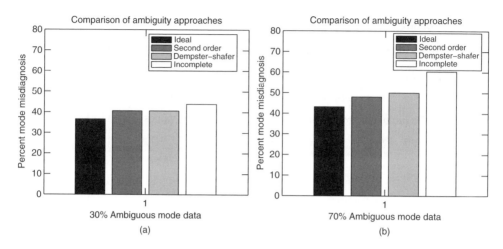

Figure 16.2   Tennessee Eastman problem mode-diagnosis error

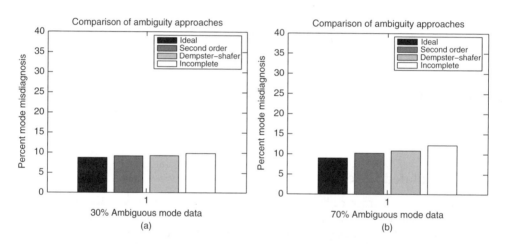

Figure 16.3   Tennessee Eastman problem component-diagnosis error

came from ambiguous modes. These improvements showed that data was beginning to become more scarce, as performance for the incomplete Bayesian method, in which ambiguous mode data is simply ignored, was beginning to decline. The results of the second-order method and the generalized Dempster–Shafer method were very similar.

16.5   Bench-scale Case

The generalized Dempster–Shafer method was also applied to the bench-scale hybrid tank system. Diagnosis based on modes is shown in Figure 16.4 while diagnosis based on component is

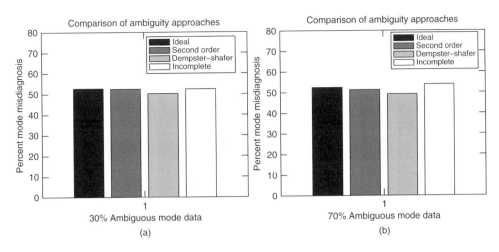

Figure 16.4   Hybrid tank system mode-diagnosis error

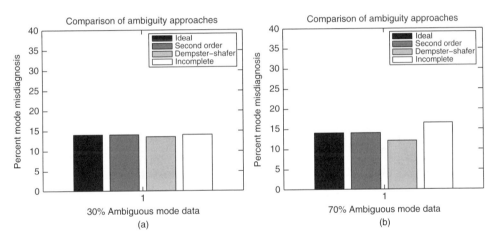

Figure 16.5   Hybrid tank system component-diagnosis error

shown in Figure 16.5. Performance in this case showed that the generalized Dempster–Shafer method represented a slight improvement over the second-order method.

The main difference between the second-order method and the generalized Dempster–Shafer method is that the second-order method distributes ambiguous mode data over the specific modes in accordance with prior probabilities. However, the generalized Dempster–Shafer method includes all possible ambiguous mode data in each specific mode. If the actual ambiguous mode distribution is significantly different from what is expected from prior probabilities, the generalized Dempster–Shafer method will likely yield better results than the second-order method.

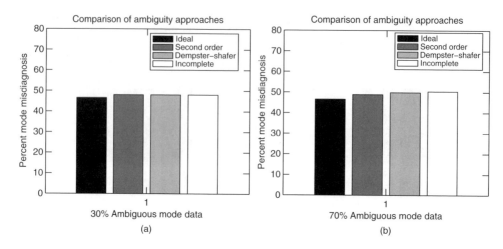

Figure 16.6   Industrial system mode-diagnosis error

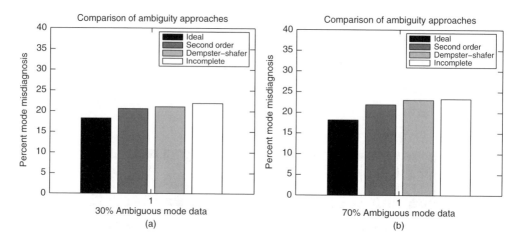

Figure 16.7   Industrial system component-diagnosis error

16.6   Industrial System

Finally, the generalized Dempster–Shafer method was applied to the industrial system along-side the second-order method. In contrast to the experimental system, the second-order method had a slightly better performance than the generalized Dempster–Shafer method when applied to the industrial system. In this case, the second-order method yields better results most likely because the real proportions Θ are similar to the estimated values Θ^* given by the prior probabilities. Results for mode diagnosis are shown in Figure 16.6, while results for component diagnosis are shown in Figure 16.7.

Due to the abundance of data, excluding ambiguous mode data did not have a strong effect when 30% of the data belonged to ambiguous modes, but differences became more pronounced when 70% of the data came from ambiguous modes. There was also a more pronounced change with respect to diagnosing components in contrast to diagnosing modes.

16.7   Notes and References

The bulk of this material can be accessed from proceedings of the 16th International Conference on Information Fusion (Gonzalez and Huang 2013b). For detailed reasons as to why traditional Dempster–Shafer theory docs not suit this problem, readers are referred to Gonzalez and Huang (2013a). The fundamentals of Dempster–Shafer theory can be found in Shafer (1976), while Cobb and Shenoy (2006) gives the rationale behind the traditional plausibility short-cut method, a precursor to the inclusive probability short-cut method described in this chapter.

References

Cobb B and Shenoy P 2006 On the plausibility transformation method for translating belief function models to probability models. *International Journal of Approximate Reasoning* **41**(3), 314–330.

Gonzalez R and Huang B 2013a Control loop diagnosis with ambiguous historical operating modes: Part 1. a proportional parametrization approach. *Journal of Process Control* **23**(4), 585–597.

Gonzalez R and Huang B 2013b Data-driven diagnosis with ambiguous hypotheses in historical data: A generalized Dempster–Shafer approach. *Information Fusion (FUSION), 16th International Conference on*, pp. 2139–2144.

Shafer G 1976 *A Mathematical Theory of Evidence*. Princeton University Press.

17

Making use of Continuous Evidence through Kernel Density Estimation

17.1 Introduction

In previous chapters of Part II, we have considered evidence that takes on a discrete form, such as alarms, which were obtained by discretizing a continuous performance metric. By classifying a continuous performance index into low-resolution bins, information is lost. However, using methods that can deal with continuous data preserves this information and, as a result, continuous-data methods can significantly improve performance.

Kernel density estimation (sometimes called the Parzen–Rosenblatt window method) is a popular method for estimating probability density functions from data. It is a nonparametric method that places a 'kernel function', centred around each data point, so that adding the kernel functions results in a smoothed probability density estimate. Kernel density estimation is nonparametric and thus does not assume a predefined shape for the distribution, allowing the distribution estimate to naturally follow the shape of the data distribution, regardless of the shape it takes.

While kernel density estimation does not contain shape parameters, it does contain a crucial smoothing parameter called the bandwidth. A variety of techniques exist for estimating bandwidths, and this chapter will discuss two of the most popular approaches. In addition to the required techniques for kernel densities, there are complementary techniques that can help increase the accuracy of a kernel density estimate. This chapter focuses on how to perform the following techniques:

- kernel density estimation
- bandwidth selection
- dimension reduction
- handling missing data.

Process Control System Fault Diagnosis: A Bayesian Approach, First Edition. Ruben Gonzalez, Fei Qi and Biao Huang.
© 2016 John Wiley & Sons, Ltd. Published 2016 by John Wiley & Sons, Ltd.

17.2   Algorithm

17.2.1   Kernel Density Estimation

The goal of kernel density estimation is to estimate a density function $f(x)$ using kernels $K(x)$, which are centred around each data point in the historical dataset D. The kernel $K(x)$ can be any function that integrates to unity; it is preferable for $K(x)$ to be a density function. From a multivariate dataset D with n entries, a kernel density estimate from the kernel $K(x)$ is obtained using the following sum:

$$f(x) \approx \frac{1}{n} \sum_{i=1}^{n} \frac{1}{|H|^{1/2}} K_H(H^{1/2}(x - D_i)) \tag{17.1}$$

Due to its many desirable mathematical properties, a popular choice of kernel density estimate is the multivariate Gaussian kernel

$$K_H(z) = \frac{1}{\sqrt{(2\pi)^d}} \exp\left(z^T z\right)$$

where d is the dimensionality of the data. Using this kernel results in the following kernel density estimate

$$f(x) \approx \frac{1}{n} \sum_{i=1}^{n} \frac{1}{\sqrt{(2\pi)^d |H|}} \exp\left([x - D_i]^T H^{-1}[x - D_i]\right) \tag{17.2}$$

17.2.2   Bandwidth Selection

The smoothness of a kernel density estimate is affected by the parameter H in the same manner that histograms are affected by bin width. When the bin width increases, the histogram becomes smoother, but as the bin size decreases, the histogram takes a rougher shape. Similarly, shrinking the bandwidth H will result in a jagged kernel density estimate, while increasing H will result in a smoother one. Many different methods for selecting bandwidths exist, and bandwidth selection is still a fairly active area of research (Duong and Hazelton 2003, 2005). However, a small selection of the more popular methods are presented in this section.

Optimal bandwidth for normal distributions

Selecting a bandwidth to estimate normal distributions is a mature subject within the literature, and the result is well established. Based on the AMISE criterion (discussed in Section 8.3.3), the optimal bandwidth for estimating a normal distribution with normal kernels is given as

$$H_N = \left(\frac{4}{n(d+2)}\right)^{\frac{2}{d+4}} \Sigma \tag{17.3}$$

where Σ is the covariance of the normal distribution, d is the dimension of the data and n is the number of sampled data points. In practice, the sample covariance S can be used in place of Σ. While this bandwidth is optimal for normal distributions, for many other distributions this bandwidth can be larger than optimal, resulting in an over-smoothed distribution, especially if the distribution has more than one mode. Engineering judgement can be exercised by inspecting the data to see if the shape deviates significantly from normal.

In higher dimensions, the direction of the data can change in different locations. Thus it is often safer to use the main diagonal of the covariance estimate S instead of the full matrix, as S will stretch the covariance matrix in a certain direction. Using the main diagonal will eliminate directional preference.

17.2.3   Adaptive Bandwidths

One problem often encountered in kernel density estimation is that when a single bandwidth is used, peaks tend to be over-smoothed, while tails tend to be under-smoothed. Adaptive bandwidth estimation is a common solution for this problem. In the adaptive bandwidth estimation problem, a pilot density function is estimated in order to give a rough probability for all the data points. These probabilities are used to scale the bandwidth. Intuitively, larger probabilities result in narrower bandwidths and smaller probabilities result in wider ones.

In the first step, the pilot density is estimated using the optimal normal bandwidth H_N

$$\hat{f}_H^p(x) = \frac{1}{n} \sum_{i=1}^{n} \frac{1}{\sqrt{(2\pi)^d |H_N|}} \exp\left([x - D_i]^T H_N^{-1} [x - D_i]\right)$$

Then, we calculate a geometric mean probability g to be used as a standardization constant.

$$\log(g) = \frac{1}{n} \sum_i \log[\hat{f}_H^p(D[i])] \tag{17.4}$$

Next, we obtain a bandwidth matrix scalar λ_i for each data point:

$$\lambda_i = \left(\frac{\hat{f}_H^p(D[i])}{g}\right)^\alpha$$

where the parameter α is a user-defined parameter. For practical purposes, it is most often set to $\alpha = 0.5$ for moderate sample sizes, say 100 sample points in the univariate case, but should be reduced for larger sample sizes. The parameter λ_i is used to scale the bandwidth as follows:

$$H_i = \lambda_i^{-2/d} H_N \tag{17.5}$$

A possible adaptation of this step is to use a local covariance estimate S_i to account for the local direction of the data. In this adaptation, we obtain a sample covariance estimate S_i using the data closest to $D[i]$ – for example the closest 25% of the data, or using a weighting function for the covariance estimator – and scale it according to the following expression:

$$H_i = \lambda_i^{-2/d} \frac{|H_N|^{1/d}}{|S_i|^{1/d}} S_i \tag{17.6}$$

Finally, the kernel density estimate is given as

$$\hat{f}(x) = \frac{1}{n} \sum_{i=1}^{n} \frac{1}{\sqrt{(2\pi)^d |H_i|}} \exp \left([x - D_i]^T H_i^{-1} [x - D_i] \right) \tag{17.7}$$

17.2.4 Optional Step: Dimension Reduction by Multiplying Independent Likelihoods

Kernel density methods, like discrete methods, tend to suffer from performance degradation when dimensionality is increased. For the discrete method, estimation difficulty increases exponentially with respect to the number of monitoring inputs. For example, if 2 monitors were used, each having 3 different states, there would be $2^3 = 9$ bins required for estimation. Now if 15 monitors were used, the number of bins for estimation would balloon to over 14 million. Kernel density methods are more efficient at approximating densities, and thus difficulty may not grow as fast as discrete/histogram methods, but dimensionality can still be a problem.

If monitors are highly correlated, data tends to exhibit lower-dimensionality behaviour, but independent monitors are more problematic. However, independent monitors have a convenient solution. If certain monitors, or groups of monitors, can be considered independent, $E_1 \perp E_2$, then their higher-dimensional joint likelihood $p(E_1, E_2|M)$ can be calculated by multiplying the lower-dimensional individual likelihoods $p(E_1|M)$, $p(E_2|M)$ together.

$$p(E_1, E_2|M) = p(E_1|M)p(E_2|M)$$

Assuming independence and verifying these assumptions through the MIC was discussed in Chapter 15. The procedure largely remains the same in this chapter, except that now probability functions $p(\pi_1), p(\pi_2), p(\pi_1, \pi_2)$ can be expressed as kernel density estimates

$$I(x_1; x_2) = \int_{x_1} \int_{x_2} p(x_1, x_2) \log \left(\frac{p(x_1, x_2)}{p(x_1)p(x_2)} \right) dx_1 \ dx_2 \tag{17.8}$$

which can be integrated numerically. When the MIC is applied to kernel density estimates, values less than 0.05 are considered negligible, while values greater than 0.2 are considered significant.

17.2.5 Optional Step: Creating Independence via Independent Component Analysis

Independent component analysis (ICA) assumes that data was generated by a linear combination of independent variables; it is similar to principal component analysis (PCA) except that the latent variable t is not restricted to be Gaussian. Thus, ICA can be used effectively on data that is not multivariate normal. Nevertheless, there still is a restriction, in that observations y are assumed to be linear combinations of latent variables t

$$y - \mu = At$$

The goal of ICA is to obtain the transformation matrix A or, more importantly, its inverse $W = A^{-1}$. While many algorithms for ICA exist – some are even based on the MIC – the fixed-point algorithm has been shown to exhibit both accuracy and speed. Software for this algorithm has been previously developed under the MATLAB and Octave platforms and is available at the following website: http://research.ics.aalto.fi/ica/fastica/.

ICA has been shown to be effective when observation inputs consist of raw or relatively unprocessed data. However, if the data consists of monitors that were tuned to be sensitive to underlying problem sources, ICA can reduce this sensitivity and possibly result in poorer diagnostic performance.

17.2.6   Optional Step: Replacing Missing Values

When values are missing for a kernel density estimation, one cannot use the marginalization methods that are used for discrete monitors. However, being able to assume independence among smaller groups of observations makes missing values less of an issue. For example, consider a system with nine observations $y_1, y_2, \ldots, y_9$. If it were possible to break them down into three independent groups $[y_1, y_2, y_3] \perp [y_4, y_5, y_6] \perp [y_7, y_8, y_9]$, then if y_7 is missing, we would only have to discard y_8 and y_9. If independence was not assumed, then observations $y_1, \ldots, y_6$ would also have to be discarded along with y_8 and y_9.

If discarding missing data significantly reduces data reliability, then missing data entries have to be estimated. This can be done by applying the following steps:

1. Estimate the kernel density function using only complete data entries.
2. Estimate the expected values of the incomplete data entries.
3. Using the estimated data entries, estimate the kernel density function which includes the new data points.
4. Repeat the log likelihood maximization based on the new kernel density function.
5. Repeat steps (3) and (4) until the likelihood converges.

This approach resembles the EM algorithm, except that the missing values must be explicitly calculated instead of its statistics. Such an approach is required because the kernel density function is nonparametric.

After obtaining the kernel density estimate, one can use *kernel density regression* to replace the missing values. There are two main techniques for kernel density regression: the *zeroth-order (Nadaraya–Watson) method* and the *first-order method*.

1. **Zeroth-order (Nadaraya–Watson) method:** For a point with known elements x and missing elements y, we look at the historical record of data having both known X and Y values. The value of y is calculated by a weighted average of the historical Y values, the weighting based on the proximity of their X components to x.

$$E(y) = \frac{\sum_{i=1}^{n} K\left(H_x^{-1/2}[X_i - x]\right) Y_i}{\sum_{i=1}^{n} K\left(H_x^{-1/2}[X_i - x]\right)} \tag{17.9}$$

2. **First-order method:** The zeroth-order method tends to have a flat bias of the function, particularly around peaks in y and around the edges of the data. In order to correct this bias, the first-order method is proposed. Instead of taking a weighted average of Y, weighted linear regression on Y is performed. Given the point x, y, with known elements x and unknown elements y, we first obtain the regressor variable Z as

$$Z_i = \begin{bmatrix} 1 \\ X_i - x \end{bmatrix}$$

The weighted linear regression is then performed as

$$\begin{bmatrix} E(y) \\ \hat{\beta}(x) \end{bmatrix} = \left[\sum_{i=1}^{n} K(H_x^{-1/2}[X_i - x]) Z_i Z_i^T \right]^{-1}$$

$$\times \left[\sum_{i=1}^{n} K(H_x^{-1/2}[X_i - x]) Z_i Y_i \right] \qquad (17.10)$$

The result of the regression is a vector, with $E(y)$ as the first element (corresponding to $Z_i = 1$), and with the remaining elements $\hat{\beta}(x)$ being ignored.

After finding $E(y)$ for all missing data points, the completed data points can be added to the data. For a second iteration, because estimates are obtained for missing values, one can use the kernel density estimate from the completed data history to estimate $E(y)$ again. The steps of updating the kernel density estimate and filling in missing values can be repeated until the likelihood converges.

17.3   Example of Proposed Methodology

Again, we use the control-loop example presented in Chapter 15, with the same modes and evidence considered. This is shown in Figure 17.1. The methodology of kernel density estimation is data-intensive and cannot be easily summarized; thus the example will be illustrated through MATLAB/Octave code.

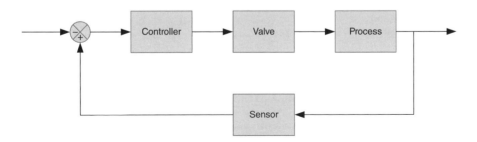

Figure 17.1   Typical control loop

Kernel density estimation functions

The code below is an optimized MATLAB/Octave implementation of kernel density estimation, which works in arbitrary dimensions. The main feature is that Cholesky decomposition is used for the bandwidth so that it only needs to be performed once offline.

```
1   function DensityInfo = fKernelEstimateNorm(X)
2   [n,d] = size(X);
3   %A diagonal covariance matrix tends yield more stable results
4   S = diag(var(X));
5   %Optimal bandwidth for Gaussian estimation
6   H = (4/(n*(d+2)))^(2/(d+4))*S;
7
8   %Inversion from cholesky decomposition (for speed and forced symmetry)
9   HRi = eye(d)/chol(H);
10
11  %we use the log of normalization constant for the kernel density to avoid NaNs
12  BWMd = det(H);
13  k = -0.5*log( (2*pi)^d*BWMd ) - log(n);
14
15  DensityInfo.bwm = H;          %bandwidth matrix
16  DensityInfo.bwmZt = HRi';     %transformation to Z
17  DensityInfo.logk = k;     %log of KDE normalizing constant
18  DensityInfo.data = X;         %Data
```

The `fKernelEstimateNorm` command is used to set up the bandwidth parameters in a way to allow efficient kernel density estimation from the `fKernelDensity`, given as

```
1   function fx = fKernelDensity(x,DensityInfo)
2   %x is multivariate random variable realization (row vector)
3   %multiple rows in x will result in multiple probabilities
4
5   Data = DensityInfo.data;                %KDE data
6   k(1,:) = DensityInfo.logk;     %Log of KDE normalization
7   Zt = DensityInfo.bwmZt;         %Transformation to Mahalanobis distance
8
9
10  %This function estimates multiple probabilities from multiple rows in x
11  %Data has nD rows, each are horizonal entries of x
12  [nX,p] = size(x);
13  [nD,¬] = size(Data);
14
15  %Calculate kernel density (can take both static and variable kernels)
16  fx = zeros(nX,1);
17  for n = 1:nX
18      Xd = ( Data - ones(nD,1)*x(n,:) )'; %Raw distances
19      DM = sum( (Zt*Xd).^2,1);     %Mahalanobis distances
20      Pe = k - 0.5*DM;     %Exponent of probability
21      fx(n,1) = sum(exp(Pe));     %Probability density
22  end
```

where '¬' is the 'not' operator, which is written as '~' in MATLAB. This function treats x as a set of row vectors, so that multiple rows will yield multiple likelihoods. Note that the log-normalization constant is added in the exponent; this is less likely to result in numerical errors if k and DM are large. For the sake of speed, it is key to ensure that the exponent $\exp$ (Pe) is evaluated for the entire vector Pe, as this is much faster than evaluating elements of Pe one at a time due to the parallel computation methods invoked by MATLAB and Octave for vectors.

17.3.1 Offline Step 1: Historical Data Collection

As in previous cases, the first step is to go through the historical data and note the instances where each of the four possible modes occurs:

1. m_1 $[0,0]$ where bias and stiction do not occur
2. m_2 $[0,1]$ where bias does not occur but stiction does
3. m_3 $[1,0]$ where bias occurs but stiction does not
4. m_4 $[1,1]$ where both bias and stiction occur.

In this chapter, we assume that no ambiguous modes exist, thus Bayesian methods are used. Nevertheless, because the only aspect that changes is how the likelihood function $p(E|M)$ is calculated (as kernel density estimates are used) kernel density estimates can be combined with the second-order Bayesian method or the Dempster–Shafer method in order to handle ambiguity.

If we consider the MATLAB cell variable Data, which contains data for each mode, we can set up the kernel density estimates with the following code:

```
1        for m = 1:length(Data)
2                KDE(m) = fKernelEstimateNorm(Data{m})
3        end
```

Offline Step 2: Replacing Missing Values (Optional)

If elements of E are missing, they cannot be used for learning unless the missing values are replaced. The data is separated into two sections $\mathcal{D}_m = [\mathcal{D}_c, \mathcal{D}_{ic}]$. In MATLAB, it is assumed that missing elements are represented by the value NaN; for example

$$\mathcal{D}_m = \begin{bmatrix} 4.5 & 2.8 & 0.3 & 1.2 \\ 2.7 & \text{NaN} & 1.0 & 0.9 \\ 5.6 & 3.0 & 0.6 & 0.11 \\ 4.5 & 2.7 & 0.4 & 1.3 \\ 3.6 & 2.8 & 0.5 & \text{NaN} \end{bmatrix}$$

Each row represents a historical piece of evidence. From this example, rows 2 and 5 would be moved to the set D_{ic}, while rows 1, 3 and 4 would be moved to the set D_c. This sorting can easily be done in MATLAB; consider the dataset for mode M denoted by Data{m} that has some NaN entries in various rows. Sorting is done using the following code:

```
1   vNaN = sum(Data{m},2); %rows with NaN will sum to NaN
2   indIC = find(isnan(vNaN)); %identify rows with NaN
3   indC = find(¬isnan(vNaN)); %identify rows without NaN
4
5   %Place incomplete and complete data into appropriate data sets
6   Dc = Data{m}(indC,:);
7   Dic = Data{m}(indIC,:);
8   DcF = Dc;
9   DicF = Dic;
```

Note that DcF represents the data with missing sets filled in, but on the first iteration the missing data are *not* filled in. For each incomplete data entry, the expected value of the missing entries can be obtained using kernel density regression.

```
1   for i = 1:size(Dic,1)
2           dic = Dic(i,:);
3           Xind = find(¬isnan(dic));
4           Yind = find(isnan(dic));
5           KDE = fKernelEstimateNorm(DcF(:,Xind));
6
7           dic(i,Yind) = fKernelRegFirst(dic(Xind),DcF(:,Yind),KDE)
8           DicF(i,:) = dic;
9   end
```

The function fKernelRegFirst returns the expected value of the missing elements Yind using the historical values of the missing component DcF(:,Yind) and the kernel density estimate KDE, which is obtained using historical values of the available component DcF(:,Xind). While the basics of the function fKernelRegFirst have already been described, a number of safeguards and efficiency-increasing steps are included, making the code quite lengthy. Because of this, details of fKernelRegFirst are set out in Appendix 17.A. Once estimation has been performed over all elements in Dic, one can add the completed elements to the data.

```
1   DcF = [Dc,DicF]
```

Then, one can estimate the missing elements Dic again with the new completed Dc dataset.

17.3.2 Offline Step 3: Mutual Information Criterion (Optional)

As in Chapters 15 and 16, this step is optional. The kernel density estimate is used to obtain the probability terms $p(E_1|M)p(E_2|M)$ and $p(E_1, E_2|M)$; if MIC(1,2) is greater than some threshold (0.01), it would be better to assume that E_1 and E_2 were dependent.

$$MIC = \begin{bmatrix} 0 & 0 \\ MIC(1,2) & 0 \end{bmatrix}$$

where $MIC(1,2)$ is calculated as

$$MIC(1,2) = \int_{E_1} \int_{E_2} p(E_1, E_2|m) \log\left(\frac{p(E_1, E_2|M)}{p(E_1|M)p(E_2|M)}\right) dE_1 \ dE_2$$

In MATLAB, the MIC can be calculated using numerical integration. Consider the example above, where D is the historical record of the bivariate random variable E. In MATLAB, the mutual information term can be expressed as:

```
1   function MI = fMI(X,Y,KDE,KDEx,KDEy)
2       %X and Y are usually scalars, but quad2d prefers the ability to have matrix ...
            inputs/outputs
3       [ni,nj] = size(X);
4       MI = zeros(ni,nj);
5       for i = 1:ni
6           for j = 1:nj
7               p12 = fKernelDensity([X(i,j),Y(i,j)],KDE);
8               p1 = fKernelDensity(X(i,j),KDEx);
9               p2 = fKernelDensity(Y(i,j),KDEy);
10              MI(i,j) = p12*log( p12/(p1*p2) );
11          end
12      end
13  end
```

Note that if there are regions where p12, p1, p2 are very small (and approach zero), there is a risk that MI will be undefined. It is useful to note that, due to the logarithmic term, if p12 approaches zero, then MI also approaches zero. Thus, for reliability, statements made in the code should be inserted to reflect this fact.

The MIC is obtained by integrating over the aforementioned mutual information term:

```
1   function MIC = fMIC(D)
2       x = D(:,1);
3       y = D(:,2);
4
5       %Kernel density for univariate and bivariate data sets
6       [KDE]  =  fKernelEstimateNorm(D);
7       [KDEx] =  fKernelEstimateNorm(x);
8       [KDEy] =  fKernelEstimateNorm(y);
9
10      %Set integration boundaries
11      minx = min(x) - 0.1*std(x);
12      maxx = max(x) + 0.1*std(x);
13      miny = min(y) - 0.1*std(y);
14      maxy = max(y) + 0.1*std(y);
15
16      %hMI is a function to be integrated
17      hMI = @(X,Y) fMI(X,Y,KDE,KDEx,KDEy);
18
19      %Integrate to obtain MIC
20      MIC = quad2d(hMI,minx,maxx,miny,maxy);
21
22  end
```

The MIC can be arranged in a lower triangular matrix for grouping:

```
1    MICmatrix = zeros(length{Data{m});
2    ne = length(Data{m}(1,:)); %number of evidence sources
3    for j = 1:ne
4        for i = (j+1):ne
5            MICmatrix(i,j) = fMIC(Data{m}(:,i,j))
6        end
7    end
```

When grouping, one prioritizes the elements of the MIC matrix that have large values. If the value in the MIC matrix is less than 0.05, the corresponding two pieces of evidence (by row and column) can be considered independent.

17.3.3   Offline Step 4: Independent Component Analysis (Optional)

If process data were used directly, applying ICA to transform the data may be advantageous. The fixed-point algorithm (Fast ICA: Hyvärinen *et al.* 2005) can be applied in order to solve for A in the expression

$$y = At + \epsilon$$

so that the new monitor inputs π^* are calculated as

$$D_m^* = A^{-1}D_m = WD_m$$

The transformed data D_m^* can be used as input instead of the original data D_m.

The Fast ICA package is used to produce the information necessary for ICA: the mean and the transformation matrix W.

```
1    Transform.mean = mean(data);
2    [¬,¬,W] = fastica(data','stabilization','on');
3    Transform.W = W;
```

The argument $'stabilization', 'on'$ helps the algorithm achieve more reliable results.

17.3.4   Offline Step 5: Obtain Bandwidths

There are two main options for bandwidth matrices; the optimal bandwidth for Gaussian distributions and the adaptive optimal bandwidth. The first option is applied to all data points, and for each mode the bandwidth matrix is given a value of

$$H_m = \left(\frac{4}{n(d+2)} \right)^{\frac{2}{d+4}} \Sigma_m \tag{17.11}$$

so that the kernel density estimate is given as

$$\hat{f}^p_{H_m}(x) = \frac{1}{n} \sum_{i=1}^{n} \frac{1}{\sqrt{(2\pi)^d |H_m|}} \exp\left([x - D_i]^T H_m^{-1}[x - D_i] \right)$$

where n is the number of data points in $\mathcal{D}_m$, D_i is the ith data point in $\mathcal{D}_m$, d is the dimension of the data in $\mathcal{D}_m$ and Σ_m is the covariance matrix of $\mathcal{D}_m$. The equivalent bandwidth selection step in MATLAB has already been given as:

```
1        for m = 1:length(Data)
2               KDE(m)  = fKernelEstimateNorm(Data{m});
3        end
```

so that likelihood probabilities given x are written as

```
1    for m = 1:length(DATA)
2      L(m)  = fKernelDensity(x,KDE(m));
3    end
```

As an alternative to the optimal normal bandwidth, one can use the adaptive bandwidths H_m^k, which are different for every kth element.

$$\log(g) = \frac{1}{n} \sum_i \log[\hat{f}^p_H(D_i)]$$

$$\lambda_i = \frac{\hat{f}^p_H(D_i)}{g}$$

$$H_m[i] = \lambda_i^{-1} H_m$$

so that the kernel density estimate is given as

$$\hat{f}^p_{H_m}(x) = \frac{1}{n} \sum_{i=1}^{n} \frac{1}{\sqrt{(2\pi)^d |H_m[i]|}} \exp\left([x - D_i]^T H_m[i]^{-1}[x - D_i] \right)$$

When implementing in MATLAB, new commands have to be created:

- fKernelEstimateNorm
- fKernelDensityAdaptive

The adaptive bandwidth selector is given as:

```
1    function [DensityInfo]  = fKernelEstimateAdaptive(X)
2
3    %Generate initial bandwidth matrix using optimal normal method
4    [n,p]  = size(X);
```

```
5   KDE = fKernelEstimateNorm(X);
6   H0 = KDE.bwm;
7
8   %obtain scaling parameters Lam
9   fx = fKernelDensity(X,KDE);
10  g = exp( mean( log(fx)) );
11  Lam = fx/g;
12
13  %Scale bandwidth for individual points
14  for i = 1:n
15      BWMi = H0/Lam(i);
16      BWM(:,:,i) = BWMi;
17      Zt(:,:,i)  = chol(BWMi,'lower')\eye(p);
18      dBWM(i,:)  = det(BWMi);
19  end
20  %log of normalizing constant
21  logk = -0.5*log((2*pi)^p*dBWM) - log(n);
22
23  %Save results as a structure
24  DensityInfo.bwm = BWM;
25  DensityInfo.bwmZt = Zt;
26  DensityInfo.logk = logk;
27  DensityInfo.data = X;
28
29  end
```

while the adaptive kernel density estimator is given as:

```
1   function fx = fKernelDensityAdaptive(x,DensityInfo)
2   %Unload key parameters
3   Data       = DensityInfo.data;
4   Zt         = DensityInfo.bwmZt;
5   k(1,:)     = DensityInfo.logk;
6
7   %Data has nD rows, each are horizonal entries of x
8   [nX,p] = size(x);
9   [nD,¬] = size(Data);
10
11  %Calculate kernel density (can take both static and variable kernels)
12  fx = zeros(nX,1);
13  kz(1,1,:) = k;
14  for n = 1:nX
15      %find distances (distances are nD vertical vectors)
16      Xd = (Data - ones(nD,1)*x(n,:))';
17      %Multiply each transformation by the appropriate distance
18      for zi = 1:nD
19              Z(:,:,zi) = Zt(:,:,zi)*Xd(:,zi);
20      end
21      Pe(:,1) = kz - 0.5*sum(Z.^2,1); %exponent of probability
22      fx(n,1) = sum(exp(Pe));
23  end
```

If likelihoods contain multiple independent groups, one has to construct the groups first. For example, let us say that under Mode 1, the independent groups are $[1, 3, 4]$ and $[2, 5]$ but for Mode 2, all evidence is dependent.

```
1  groups1{1} = [1,3,4];
2  groups1{2} = [2,5];
3  Mgroups{1} = groups1;
4
5  groups2{1} = [1,2,3,4,5];
6  Mgroups{2} = groups2;
```

The optimal normal bandwidths are then obtained as:

```
1  cKDE = cell(length(Mgroups),1);
2
3  for m = 1:length(Mgroups)
4      groups = Mgroups{m};
5      for c = 1:length(groups)
6          %use only data from relevent groups for each KDE
7          cKDE{m}(c) = fKernelEstimateNorm( Data{m}(:,groups{c}) );
8      end
9  end
```

Note that one could also use adaptive bandwidths here.

17.3.5   Online Step 1: Calculate Likelihood of New Data

When this method is performed online, new evidence from monitors E is used as input to calculate the probability. When kernel density estimation is applied

$$p(E|M) = \frac{1}{n} \sum_{i=1}^{n} \frac{1}{\sqrt{(2\pi)^d |H_m|}} \exp\left([e - D_i]]^T H_m^{-1}[e - D_i]\right)$$

where $H_m[i]$ is used instead of H_m if the adaptive kernel density estimation method is used. Again, probabilities can be calculated using the fKernelDensity command:

```
1  for m = 1:length(KDE)
2      L(m) = fKernelDensity(x,KDE(m));
3  end
```

Note that if the evidence is separated into independent components, there is a likelihood $p(E_c|M)$ for every component c. These component likelihoods have to be multiplied together in order to obtain the net likelihood.

```
1   for m = 1:length(Data)
2       groups = Mgroups{m};
3       L(m) = 1;
4           for c = 1:length(groups)
5               L(m) = L(m)* fKernelDensity( x(groups{c}),KDE{m}(c) );
6           end
7       end
8   end
```

17.3.6   Online Step 2: Calculate Posterior Probability

The posterior probability is obtained by combining the likelihood with a prior according to Bayes' rule

$$p(M|E) = \frac{p(E|M)p(M)}{\sum_k p(E|M_k)p(m_k)}$$

Note that if there is more than one component,

$$p(E|M) = p(E_1|M) \times p(E_2|M) \times \ldots \times p(E_{n_c}|M)$$

In MATLAB, it is convenient to set the prior $p(m) = $ P as a vertical vector along with the likelihood $p(E|m) = $ L. In this way, the vector of posteriors can be calculated as:

```
1          P = (L.*P)/(L'*P);
```

17.3.7   Online Step 3: Make a Diagnosis

The diagnosis is made by selecting the mode with the highest posterior probability.

17.4   Simulated Case

The kernel density estimation method is used on the Tennessee Eastman problem, a popular benchmark simulation system. As in previous applications, each fault (or problem source) is simulated, with some of the data being used for learning, while other parts of the data are used for validation. The first set of results (in Figure 17.2) compare the diagnosis results from the discrete method against the optimal normal and adaptive kernel density estimation methods. From this figure, kernel density estimation performs significantly better than the discrete methods both for diagnosing modes and problem sources. However, the performance of adaptive and optimal Gaussian kernel density estimation methods are similar.

In addition, we consider how different grouping methods affect diagnosis rates. Four different grouping approaches are considered.

1. The *lumped* approach does not assume independence, and considers all evidence as a single multivariate variable.
2. The *independent* approach is the complete opposite; it assumes that all pieces of evidence are independent, so that likelihoods are calculated for each piece of evidence and the joint likelihood is combined by multiplying results together.

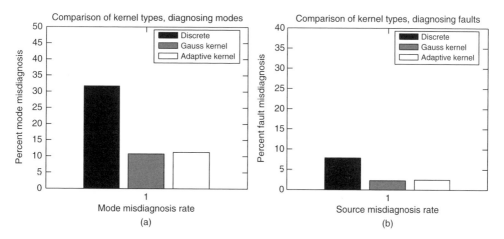

Figure 17.2   Tennessee Eastman problem: discrete vs. kernel density estimation

3. The *grouped* approach is a compromise between the lumped and independent approaches. It uses the MIC as an indicator to determine which variables are dependent and which are not. The grouping algorithms discussed in this chapter (as well as Chapters 15 and 16) are used to separate evidence into independent groups. Likelihoods are calculated for each group and are multiplied together in order to obtain a joint likelihood.
4. Finally, the ICA *transformed* method is used to transform the evidence into independent components, so that likelihoods are obtained for each component; the joint likelihood is again obtained through multiplying individual likelihoods together.

Chapters 15 and 16 have not compared grouping approaches for the discrete method, thus we compare grouping approaches for the discrete method first. The grouping comparison for the discrete method is shown in Figure 17.3. The grouping approaches are also compared for

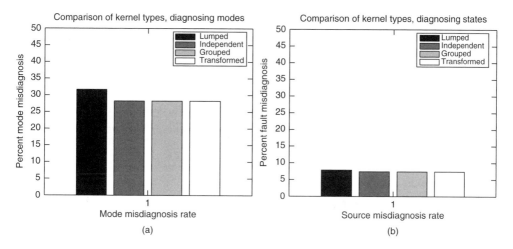

Figure 17.3   Grouping approaches for discrete method

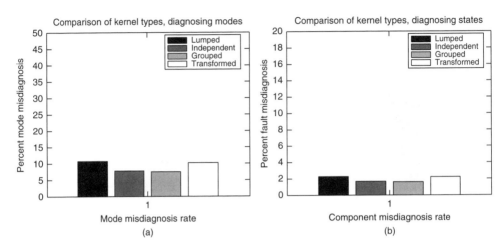

Figure 17.4   Grouping approaches for kernel density method

the kernel density estimation method; the comparison for kernel density estimation is shown in Figure 17.4. Because the performance of optimal Gaussian and adaptive kernel density estimation is similar, the optimal Gaussian method is used here as it is the simpler of the two.

Figures 17.3 and 17.4 display the mode and component misdiagnosis rates. Diagnosis can be made for each of the components as well as the total mode, and the percentage of incorrect diagnosis rates are displayed in the appropriate figures. Note that modes are harder to diagnose than components (as they contain more information). Thus mode misdiagnosis rates tend to be higher. From Figures 17.3 and 17.4 one can see that different grouping techniques have a negligible effect on diagnosis performance, regardless of whether discrete or kernel density estimation is used.

17.5   Bench-scale Case

The kernel density estimation method was also used on the hybrid tank system and compared with discrete results. Comparison between discrete and kernel density estimation methods is shown in Figure 17.5. Again, for both mode and component diagnosis, kernel density methods exhibit a very significant improvement over the discrete method. In addition, the adaptive and optimal Gaussian kernel density estimation methods perform similarly yet again.

Grouping methods for the lab-scale system are also compared. As seen in Figure 17.6, for the discrete method, the independent, grouped and transformed approaches all exhibit a significant improvement over the lumped method. However, for the kernel density estimation method, all grouping approaches perform similarly (as seen in Figure 17.7).

17.6   Industrial-scale Case

Finally, the kernel density methods were compared to the discrete methods for the industrial system. In Figure 17.8, yet again we see a significant improvement when kernel density estimation methods are used instead of discrete methods.

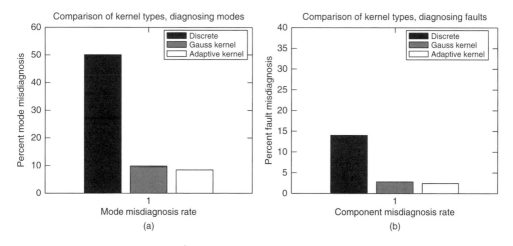

Figure 17.5   Hybrid tank problem: discrete vs. kernel density estimation

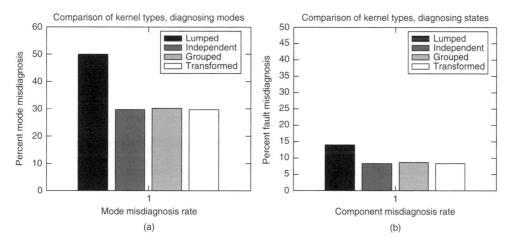

Figure 17.6   Grouping approaches for discrete method

   Different grouping approaches were tried on the discrete method when applied to the industrial system and the results are shown in Figures 17.9 and 17.10. Here, the independent, grouped and transformed approaches outperformed the lumped approach; because process measurements were directly used and a large number of instruments were applied, the transformed approach performed slightly better than the other methods. However, it should be noted that the transformed method is application-specific and can result in worse performance for applications to which it is not well-suited. When the different grouping methods were applied to the optimal Gaussian kernel density estimation method, the other approaches still exhibit superior performance over the lumped method.

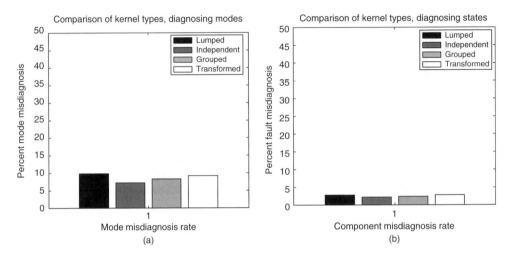

Figure 17.7   Grouping approaches for kernel density method

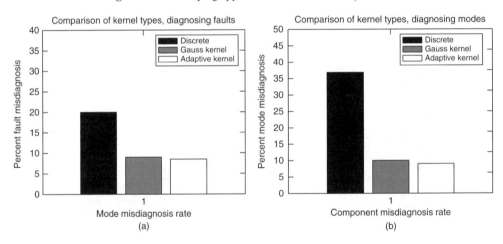

Figure 17.8   Solids-handling problem: discrete vs. kernel density estimation

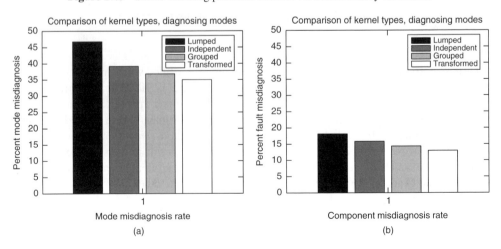

Figure 17.9   Grouping approaches for discrete method

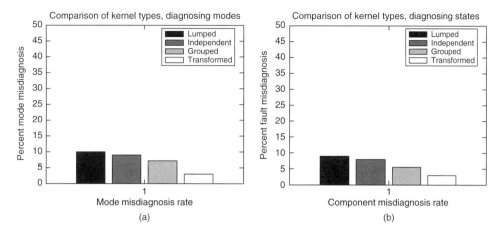

Figure 17.10   Grouping approaches for kernel density method

17.7   Notes and References

Contents in this chapter are mostly based on the previous work in Gonzalez *et al.* (2014); however, more detailed information about kernel density estimation can be found in Wand and Jones (1995). In addition, details on the independent component analysis technique used in this chapter can be found in Hyvarinen and Oja (1997), while information and application of the mutual information criterion can be found in Tourassi *et al.* (2001).

References

Duong T and Hazelton ML 2003 Plug-in bandwidth matrices for bivariate kernel density estimation. *Journal of Non-parametric Statistics* **15**, 17–30.

Duong T and Hazelton ML 2005 Cross-validation bandwidth matrices for multivariate kernel density estimation. *Scandinavian Journal of Statistics* **32**, 485–506.

Hyvärinen A, Karhunen J and Oja E 2005 *The FastICA package for MATLAB*. Helsinki Univ. of Technology, Espoo, Finland.

Gonzalez R and Huang B 2014 Control-loop diagnosis using continuous evidence through kernel density estimation. *Journal of Process Control* **24**(5), 640–651.

Hyvarinen A and Oja E 1997 A fast fixed-point algorithm for independent component analysis. *Neural Computation* **9**, 1483–1492.

Tourassi GD., Frenderick ED, Markey MK and Floyd CE 2001 Application of the mutual information criterion for feature selection in computer-aided diagnosis. *Medical Physics* **28**(12), 2394–2402.

Wand MP and Jones MC 1995 *Kernel Smoothing*. Chapman & Hall/CRC.

Appendix

17.A Code for Kernel Density Regression

While kernel density regression has been presented in Chapter 17 in a relatively simple manner, there are a number of safeguards and efficiency schemes that need to be put in place. As such, the code is too complex to be considered in standard chapter material and is instead, presented in this appendix.

17.A.1 Kernel Density Regression

This section contains the overall code for kernel density regression – it works for both standard and adaptive kernels – with both:

- zeroth order (fKernelRegFirst(x,Y,KDEx))
- first order (fKernelRegZeroth(x,Y,KDEx)).

being considered. In order to increase speed, a three-dimensional matrix toolbox was constructed and various functions from that toolbox were used. The functions are explained in Section 17.A.2. Such functions are relatively easy to identify as they start with the character z.

Zeroth-order Kernel Density Regression

```
1   function y = fKernelRegZeroth(x,Y,KDEx)
2
3   X  = KDEx.data;
4   Zt = KDEx.bwmZt;
5
6   %Data has nD rows, each are horizonal entries of x
7   [nX,p] = size(x);
8   [nD,¬] = size(X);
9   [¬,¬,nz] = size(Zt);
10  y = zeros(nX,length(Y(1,:)));
11
12  %Calculate kernel density (can take both static and variable kernels)
13  if nz == 1 %The usual case, for non-adaptive kernels
14      clear('Weight');
15      for n = 1:nX
16          Xd = (X - ones(nD,1)*x(n,:))';
17          Z = Zt*Xd;
18          Weight(1,:) = exp( - 0.5*(sum(Z.^2,1)) );
19          y(n,:) = (Weight*Y)/sum(Weight);
20      end
21  else %If the kernel is adaptive, we have to multiply by different matrices
22      clear('Weight')
23      for n = 1:nX
24          %entries are column vectors strung out depth-wise
25          Xd = zcols((X - ones(nD,1)*x(n,:))');
```

```
26          Z = z_matmultiply(Zt,Xd);
27          Weight(1,:) = exp(- 0.5*sum(Z.^2,1));
28          y(n,:) = (Weight*Y)/sum(Weight);
29      end
30  end
```

First-order Kernel Density Regression

```
1   function y = fKernelRegFirst(x,Y,KDEx)
2
3   X  = KDEx.data;
4   Zt = KDEx.bwmZt;
5
6   %Data has nD rows, each are horizonal entries of x
7   [nX,p]  = size(x);
8   [nD,py] = size(Y);
9   [¬,¬,nz] = size(Zt);
10  y = zeros(nX,py);
11
12  %Calculate kernel density (can take both static and variable kernels)
13  if nz == 1 %The usual case, for non-adaptive kernels
14      clear('Weight');
15      for n = 1:nX
16          %Obtain Weights
17          Xd = (X - ones(nD,1)*x(n,:))';
18          Z = Zt*Xd;
19          Weight(1,1,:) = exp( - 0.5*(sum(Z.^2,1)) );
20
21          %Obtain regression denominator
22          Z = [ones(1,1,nD);zcols(Xd)];
23          oZ = z_matmultiply(Z,z_transpose(Z));
24          Den = sum( z_matmultiply(Weight,oZ), 3);
25
26          %Obtain regression numerator
27          oY = z_matmultiply(Z,zrows(Y));
28          Num = sum( z_matmultiply(Weight,oY), 3);
29
30          B = Den\Num;
31
32          y(n,:) = B(1,:);
33      end
34  else %If the kernel is adaptive, we have to multiply by different matrices
35      clear('Weight')
36      for n = 1:nX
37          %entries are column vectors strung out depth-wise
38          Xd = zcols((X - ones(nD,1)*x(n,:))');
39          Z = z_matmultiply(Zt,Xd);
40          Weight(1,1,:) = exp(- 0.5*sum(Z.^2,1));
41
42          %Obtain regression denominator
43          Z = [ones(1,1,nD);Xd];
44          oZ = z_matmultiply(Z,z_transpose(Z));
45          Den = sum( z_matmultiply(Weight,oZ), 3);
```

```
46
47          %Obtain regression numerator
48          oY = z_matmultiply(Z,zrows(Y));
49          Num = sum( z_matmultiply(Weight,oY), 3);
50
51          B = Den\Num;
52
53          y(n,:) = B(1,:);
54      end
55  end
```

17.A.2   Three-dimensional Matrix Toolbox

MATLAB matrix operations only support two-dimensional matrices, but higher-dimensional equivalents can be implemented through element-wise multiplication and summation.

Multiplying Matrices

In this work, matrix manipulation must be done repeatedly for each element of data used in the kernel density estimate. Consider Figure 17.11. On one side there is a typical two-dimensional matrix multiplication, but on the other side, there are two three-dimensional matrices, where two-dimensional matrix multiplication is to be repeated over the z-axis.

The simple way to do this is to use a for loop

```
1   [nA1,¬,nA3] = size(A);
2   [¬,nB2,¬] = size(B);
3   C = zeros(nA1,nB2,nA3);
4   for z = 1:length(A(1,1,:))
5           C(:,:,z) = A(:,:,z)*B(:,:,z);
6   end
```

However, this method can speed up by using parallel for loops parfor if the toolbox is available:

```
1   C = z_matmultiply(A,B)
```

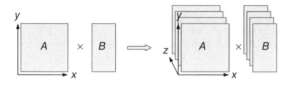

Figure 17.11   Function z_matmultiply

which is executed as:

```
1   function C = z_matmultiply(A,B)
2   [¬,¬,oc]  = size(A);
3   [mc,nc]  = size(A(:,:,1)*B(:,:,1));
4
5   C = zeros([mc,nc,oc]);
6   for z = 1:oc
7       C(:,:,z)  = A(:,:,z)*B(:,:,z);
8   end
9
10  end
```

or more efficiently as,

```
1   function C = z_matmultiply(A,B)
2   [¬,¬,oc]  = size(A);
3   [mc,nc]  = size(A(:,:,1)*B(:,:,1));
4
5   C = zeros([mc,nc,oc]);
6   parfor z = 1:oc
7       C(:,:,z)  = A(:,:,z)*B(:,:,z);
8   end
9
10  end
```

Rearranging Matrices

Some code has also been used for rearranging three-dimensional matrices. For example, the transposition of all depth-wise matrices (as seen in Figure 17.12) is obtained using the function z_transpose

```
1   function Xt = z_transpose(X)
2   Xt = permute(X,[2,1,3]);
3   end
```

In addition, matrices can be converted into n depth-wise columns or n depth-wise rows using zcols and zrows as shown in Figure 17.13.

Figure 17.12   Function z_transpose

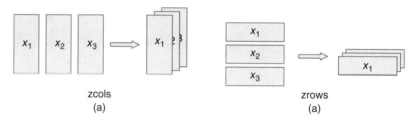

Figure 17.13   Converting matrices depth-wise

```
1   function Azr = zrows(A)
2   Azr = permute(A,[3,2,1]);
3   end
4
5   function Azc = zcols(A)
6   Azc = permute(A,[1,3,2]);
7   end
```

18

Dynamic Application of Continuous Evidence and Ambiguous Mode Solutions

18.1 Introduction

Chapter 13 as well as earlier work (Qi and Huang 2010, 2011) addressed the topic of taking into account mode and evidence dependency when implementing Bayesian diagnosis techniques online. However, the dynamic techniques that were previously introduced were specific to unambiguous modes and discrete evidence. This chapter aims to tie up previous loose ends so that the autodependent nature of modes and evidence can be taken into account.

When considering the material in this book, the existing autodependent mode procedure in Chapter 13 is only affected by the introduction of ambiguous modes; introduction of continuous evidence will not affect the autodependent mode procedure. Likewise, introducing ambiguous modes will not affect the autodependent evidence procedure. Thus this chapter proposes two solutions that can be applied independently:

- taking into account autodependent modes with ambiguous modes in history
- taking into account autodependent continuous evidence.

18.2 Algorithm for Autodependent Modes

In this section, we consider the probability of each unambiguous mode at time t, and its propagation over time, as illustrated in Figure 18.1. At time t, the probability of the mode set M is predicted from time $t - 1$. The prediction is made using a transformation by the probability transition matrix A. After prediction, it is updated by the evidence obtained at time t (denoted E^t). Updating through evidence is done using Bayes' rule. The predict–update procedure is applied recursively so that diagnosis is made by probabilities that are constantly being updated.

Process Control System Fault Diagnosis: A Bayesian Approach, First Edition. Ruben Gonzalez, Fei Qi and Biao Huang.
© 2016 John Wiley & Sons, Ltd. Published 2016 by John Wiley & Sons, Ltd.

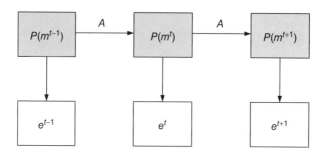

Figure 18.1   Mode autodependence

18.2.1   Transition Probability Matrix

The probability transition matrix A is used to predict the probabilities of each mode at the next
time step:

$$p(M^t) = Ap(M^{t-1})$$

where $p(M^t)$ and $p(M^{t-1})$ are column vectors of probability. The probability transition
matrix A can be constructed as follows:

$$A = \begin{bmatrix} p(m_1^t|m_1^{t-1}) & p(m_1^t|m_2^{t-1}) & \cdots & p(m_1^t|m_n^{t-1}) \\ p(m_2^t|m_1^{t-1}) & p(m_2^t|m_2^{t-1}) & \cdots & p(m_2^t|m_n^{t-1}) \\ \vdots & \vdots & \ddots & \vdots \\ p(m_n^t|m_1^{t-1}) & p(m_n^t|m_2^{t-1}) & \cdots & p(m_n^t|m_n^{t-1}) \end{bmatrix}$$

where $p(m_i^t|m_j^{t-1})$ is the probability of mode i occurring given mode j occurring at the pre-
vious time instant. The transition probability matrix is constructed in such a manner that each
column must sum to unity.

18.2.2   Review of Second-order Method

In Chapter 15, where the second-order Bayesian rule of combination was introduced, the likelihood was expressed in terms of the parameter set Θ.

$$p(E|M, \Theta) = \frac{S(E|M)n(M) + \sum\limits_{m_k \supset M} \theta\{\frac{M}{m_k}\}S(E|m_k)n(m_k)}{n(M) + \sum\limits_{m_k \supset M} \theta\{\frac{M}{m_k}\}n(m_k)}$$

The second-order Bayesian method re-expressed the likelihood in terms of a second-order
approximation

$$\Delta\Theta = \Theta - \hat{\Theta}$$

$$p(E|M, \Theta) = \hat{p}(E|M) + J_L^m \Delta\Theta + \frac{1}{2}\Delta\Theta^T H_L^m \Delta\Theta$$

where $\boldsymbol{J}_L^m$ is the likelihood Jacobian for mode M and $\boldsymbol{H}_L^m$ is the likelihood Hessian for mode M. In a similar manner, the prior probability was expressed as

$$p(M|\Theta) = \hat{p}(M) + \boldsymbol{J}_R^m \Delta\Theta + \frac{1}{2}\Delta\Theta^T \boldsymbol{H}_R^m \Delta\Theta$$

Here, if using prior probabilities from experts, it is common that priors do not have any ambiguity; in such cases, $\boldsymbol{J}_R^m$ and $\boldsymbol{H}_R^m$ are set to be zero matrices. By using the second-order rule of combination (introduced in Chapter 15), new Jacobians and Hessians can be calculated:

$$p(M|E,\Theta) = \hat{p}(M|E) + \boldsymbol{J}_P^m \Delta\Theta + \frac{1}{2}\Delta\Theta^T \boldsymbol{H}_P^m \Delta\Theta$$

where the second-order terms are calculated as

$$\hat{p}(M|E) = \frac{1}{K}\hat{p}(E|M)\hat{p}(M) \tag{18.1}$$

$$\boldsymbol{J}_P^m = \frac{1}{K}[\hat{p}(M)\boldsymbol{J}_L^m + \hat{p}(E|M)\boldsymbol{J}_R^m]$$

$$\boldsymbol{H}_P^m = \frac{1}{K}[\hat{p}(M)\boldsymbol{H}_L^m + \hat{p}(E|M)\boldsymbol{H}_R^m + (\boldsymbol{J}_L^m)^T(\boldsymbol{J}_R^m) + (\boldsymbol{J}_R^m)^T(\boldsymbol{J}_L^m)]$$

$$K = \sum_{k=1}^{n_m} \hat{p}(E|m_k)\hat{p}(m_k)$$

18.2.3   Second-order Probability Transition Rule

By merit of the probability transition matrix, the probability of mode k can be expressed as

$$p(m_k^t|E^{t-1}) = \sum_{i=1}^{n} A(k,i)p(m_i^{t-1}|E^{t-1}) \tag{18.2}$$

where, by applying this transformation, the posterior $p(m_i^{t-1}|E)$ at time $t-1$ is transformed into the prior $p(m_k^t)$ for time t. In this chapter, we apply the rule in Eqn (18.1) to the second-order probability expressions developed in Chapter 15 to create a second-order probability transition rule. First, let us consider the parameterized probability

$$p(m_k^t|\Theta) = \hat{p}(M^k) + \boldsymbol{J}_R^{m_k} \Delta\Theta + \frac{1}{2}\Delta\Theta^T \boldsymbol{H}_R^{m_k} \Delta\Theta$$

By applying the probability transition rule in Eqn (18.2) to the parametrized probability above, one can collect terms and obtain a new probability transition rule set for the reference probability, the Jacobian and Hessian, respectively.

$$\hat{p}(m_k^t) = \sum_{i=1}^{n} A(k,i)\hat{p}(m_i^{t-1}|E) \tag{18.3}$$

$$\boldsymbol{J}_R^{m_k} = \sum_{i=1}^{n} A(k,i)\boldsymbol{J}_R^{m_i} \tag{18.4}$$

$$\boldsymbol{H}_R^{m_k} = \sum_{i=1}^{n} A(k,i)\boldsymbol{H}_R^{m_i} \tag{18.5}$$

Thus, when transitioning to the next time step, the probability transition rules in Eqns (18.3–18.5) are applied to each reference probability, Jacobian and Hessian, respectively, in order to obtain the prior for the next time step. Updating can occur in the usual manner – the second-order Bayesian combination rule – but keep in mind that Jacobians and Hessians for priors no longer tend to be zero matrices.

18.3 Algorithm for Dynamic Continuous Evidence and Autodependent Modes

18.3.1 Algorithm for Dynamic Continuous Evidence

The problem of autodependent discrete evidence has been dealt with previously by Qi and Huang (2011). The goal was to take into account evidence dependence as indicated in Figure 18.2. Here, the likelihood must contain previous evidence $p(E^t|E^{t-1}, M)$, which can be combined with prior probabilities $p(M)$ to obtain a posterior $p(M|E^t, E^{t-1})$

$$p(m_i^t|E^t, E^{t-1}) = \frac{p(E^t|E^{t-1}, m_i^t)p(m_i^t)}{\sum_k p(E^t|E^{t-1}, m_i^t)p(m_i^t)}$$

The key challenge is to estimate the likelihood $p(E^t|E^{t-1}, M)$.

Review of Discrete Evidence Solution

In Chapter 2 it was shown that the likelihood for autodependent evidence can be estimated as

$$p(E^t|E^{t-1}, m_k) = \frac{n(E^t, E^{t-1}, m_k)}{n(E^{t-1}, m_k)} \tag{18.6}$$

where $n(E^t, E^{t-1}, m_k)$ is the number of times E^t, E^{t-1} and m_k jointly occur in the history, while $n(E^{t-1}, m_k)$ is the number of times $n(E^{t-1})$ and m_k jointly occur in the history. This solution also includes the ability to use prior samples, and these prior samples can be simply combined with the historical data.

One of the challenges was that introducing autodependence increased the evidence space, as we had to condition on both E^t and E^{t-1}. Correlation ratio tests were proposed to determine if some dependencies could be neglected, thus narrowing the evidence space by independence assumptions, although such assumptions cannot be made if test results are not favourable.

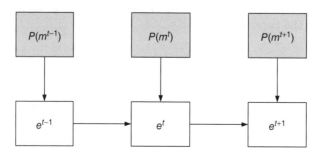

Figure 18.2   Evidence autodependence

For continuous data, the MIC is used as the independence-testing method instead of the correlation ratio test, as it functions for both discrete and continuous evidence.

Continuous Evidence Solution

One problem with the solution in Eqn (18.6) is that one cannot obtain $n(E^t \cap E^{t-1} \cap m_k)$ directly from continuous data. In Chapter 17, the kernel density estimate was used to approximate the likelihood

$$p(E|M) = \frac{n(E, M)}{n(M)}$$

$$p(E|M) \approx \frac{1}{N_K} \sum_{i=1}^{n} \frac{1}{|H|} K(H^{-1/2}E) = \hat{f}(E|M)$$

Thus the conditioning over continuous evidence $n(E, M)$ was smoothed over using a kernel density estimate. In order to condition on both E^t and E^{t-1} the rule of conditioning is applied:

$$p(Y|X) = \frac{p(X \cap Y)}{p(X)}$$

By applying the rule of conditioning to the kernel density estimation,

$$p(E^t|E^{t-1}, M^t) = \frac{p(E^t, E^{t-1}|M^t)}{p(E^{t-1}|M^t)} \tag{18.7}$$

From this result, one can see that two kernel density estimates are required:

1. The joint present and past evidence $p(E^t, E^{t-1}|M^t)$
2. The past evidence $p(E^{t-1}|M^t)$, which is also equal to $p(E^t|M^t)$

Thus for dynamic evidence one can estimate a kernel density function $p(E^t|M^t)$ as before, but an additional step is required: the kernel density estimate of the present and past observations $p(E^t, E^{t-1}|M^t)$ must also be estimated. To obtain the posterior, simply combine this new likelihood with a prior using Bayes' theorem

$$p(E^t|E^{t-1}, m_i^t) = \frac{p(E^t, E^{t-1}|m_i^t)}{p(E^{t-1}|m_i^t)}$$

$$p(m_i^t|E^t, E^{t-1}) = \frac{p(E^t|E^{t-1}, m_i^t)p(m_i^t)}{\sum_k p(E^t|E^{t-1}, m_k^t)p(m_k^t)}$$

Dimensionality Reduction

As was mentioned in Chapter 17, dimensionality can be an issue when dealing with a large number of instruments. However, dimensionality becomes an even more significant issue when taking dynamic evidence into account, as $p(E^t|E^{t-1}, m_i^t)$ has twice the dimensionality as $p(E|m_i^t)$. Thus dimensionality reduction techniques, such as grouping using the MIC, become an even more relevant practice.

18.3.2 Combining both Solutions

The second-order probability transition rules in Eqns (18.3–18.5) and the continuous autodependent solution in Eqn (18.7) are complementary solutions that can be independently applied. When applying both solutions, the corresponding Bayesian network diagram takes the form shown in Figure 18.3.

When applying both methods, one obtains the prior probability from the previous posterior according to the second-order transition rule:

$$\hat{p}(m_k^t) = \sum_{i=1}^{n} A(k, i)\hat{p}(m_i^{t-1}|E)$$

$$\boldsymbol{J}_R^{m_k} = \sum_{i=1}^{n} A(k, i)\boldsymbol{J}_R^{m_i}$$

$$\boldsymbol{H}_R^{m_k} = \sum_{i=1}^{n} A(k, i)\boldsymbol{H}_R^{m_i}$$

$$p(m_k^t|\Theta) = \hat{p}(m_k^t) + \boldsymbol{J}_R^{m_k}\Delta\Theta + \frac{1}{2}\Delta\Theta^T \boldsymbol{H}_R^{m_k}\Delta\Theta$$

Then, the second-order terms $\hat{p}(E^t|E^{t-1}, m_k^t)$, $\boldsymbol{J}_L^{m_k}$ and $\boldsymbol{H}_L^{m_k}$ are calculated for the likelihood expression

$$p(E^t|E^{t-1}, M^t, \Theta) = \frac{S(E|M^t)n(M^t) + \displaystyle\sum_{m_k \supset M^t} \theta\{\frac{M^t}{m_k}\}S(E^t|E^{t-1}, \boldsymbol{m}_k)n(\boldsymbol{m}_k)}{n(M^t) + \displaystyle\sum_{m_k \supset M^t} \theta\{\frac{M^t}{m_k}\}n(\boldsymbol{m}_k)}$$

where the support $S(E^t|E^{t-1}, \boldsymbol{m}_k)$ is calculated as

$$S(E^t|E^{t-1}, \boldsymbol{m}_k) = \frac{\hat{f}(E^t, E^{t-1}|\boldsymbol{m}_k)}{\hat{f}(E^t|\boldsymbol{m}_k)}$$

where $\boldsymbol{m}_k$ is a potentially ambiguous mode. The notation $\hat{f}(E^t, E^{t-1}|\boldsymbol{m}_k)$ indicates that the kernel density estimate uses evidence from times t and $t - 1$ collected under the mode

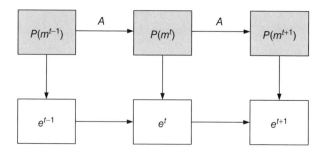

Figure 18.3   Evidence and mode autodependence

m_k, which can either be ambiguous or unambiguous. When values of $S(E^t|E^{t-1}, m_k)$ are obtained, the terms $\hat{p}(E^t|E^{t-1}, m_k^t)$, $\boldsymbol{J}_L^{m_k}$ and $\boldsymbol{H}_L^{m_k}$ can be obtained by taking derivatives of the resulting expression for $p(E^t|E^{t-1}, M^t, \Theta)$.

Once the second-order terms $\hat{p}(E^t|E^{t-1}, m_k^t)$, $\boldsymbol{J}_L^{m_k}$ and $\boldsymbol{H}_L^{m_k}$ have been obtained, they are combined with the previously obtained $\hat{p}(m_k^t|E^t, E^{t-1})$ $\boldsymbol{J}_R^{m_k}$ and $\boldsymbol{H}_R^{m_k}$ terms using the second-order Bayesian combination rule.

18.3.3 Comments on Usefulness

Mode Autodependence

The effectiveness of taking account of mode and evidence autodependence can vary quite significantly depending on the application. Taking autodependence of modes into account is particularly effective when modes change relatively slowly and if evidence is noisy. Noisy evidence can lead to relatively frequent false diagnosis results, but if modes change relatively slowly, taking mode autodependence into account creates a time-weighted average on the diagnosis results. However, mode autodependence will create a diagnosis tool that is sluggish to respond when mode changes, as a strong prior would have been created from the previous mode over time.

If taking mode autodependence into account results in a system that is too sluggish, one can replace the transition matrix A with an exponent to the power of n, which represents an acceleration factor.

$$A = A^n$$

For example, if $n = 2$, then it is assumed that the modes switch twice as frequently as previously thought. As n grows larger, the system responds more quickly to change. In fact, if $n \to \infty$, the resulting probability transition matrix will yield a flat prior for every time step which has the fastest possible reaction to a change in the process.

Evidence autodependence

Evidence autodependence tends to make the Bayesian diagnosis method slow to detect changes in the mode. In cases of strong evidence autodependence, the evidence trajectory follows a clear pattern in transition regions where values of evidence tend to slowly drift toward typical values for the new mode. This slow transition is often due to filtering or averaging done by the monitor. By documenting these transition regions, one now has a transition pattern that can be used to more quickly recognize changes in the operating mode; this is particularly beneficial if fast detection is desired. If one is more concerned about diagnosing slow-changing modes over longer periods, the benefits of accounting for evidence autodependence are much less significant.

In summary, taking account autodependence in the evidence is most effective for systems that have strongly autodependent (or slow-responding) evidence and frequently changing modes. Primarily, this is because of the time-sensitive nature of diagnosing systems with rapidly changing modes. Furthermore, taking autodependence into account is most manageable when the possible operating modes are few, as it requires data to be collected from transition regions between all modes.

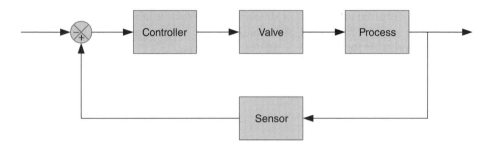

Figure 18.4   Typical control loop

18.4   Example of Proposed Methodology

18.4.1   Introduction

Again, for the tutorial, we consider the control-loop system shown in Figure 18.4.

18.4.2   Offline Step 1: Historical Data Collection

The first step is to go through the historical data and note the instances where each of the four possible modes occurs, and collect data belonging to the mode;

1. m_1 $[0, 0]$ where bias and stiction do not occur
2. m_2 $[0, 1]$ where bias does not occur but stiction does
3. m_3 $[1, 0]$ where bias occurs but stiction does not
4. m_4 $[1, 1]$ where both bias and stiction occur.

If ambiguous modes exist in the data, one must also collect data according to the ambiguous modes that appear in the history. For this system, the possible ambiguous modes are:

1. $\{m_1, m_3\}$ $[\times, 0]$ where bias is undetermined and stiction does not occur
2. $\{m_2, m_4\}$ $[\times, 1]$ where bias is undetermined and stiction occurs
3. $\{m_1, m_2\}$ $[0, \times]$ where bias does not occur and stiction is undetermined
4. $\{m_3, m_4\}$ $[1, \times]$ where bias occurs and stiction is undetermined
5. $\{m_1, m_2, m_3, m_4\}$ $[\times, \times]$ where both bias and stiction are undetermined.

Because the methods deal with kernel density estimation and other computationally intensive techniques, the application will be given in terms of MATLAB code.

18.4.3   Offline Step 2: Create Temporal Data

Let us consider the MATLAB cell variable `Data`, which has all data collected into modes, both unambiguous and ambiguous. We can construct a new dataset that includes the data from the previous time step (called the one-step temporal data). This is done by copying the data, deleting the first row of one set, and deleting the last row of the second set, then combining the two datasets.

```
1   for m = 1:length{Data}
2       e1 = Data{m};
3       e2 = e1;
4       e1(end,:) = []; %Delete last row of previous evidence
5       e2(1,:) = [];    %Delete first row of current evidence
6       DataT{m} = [e1,e2];
7   end
```

18.4.4   Offline Step 3: Mutual Information Criterion (Optional, but Recommended)

As in the kernel density estimation procedure, the MIC is used to group evidence into independent groups. However, grouping must now be done for two sets of data: the original data `Data` and the temporal data `DataT`:

```
1    MICmatrix = zeros(length{Data{m}});
2    ne = length(Data{m}(1,:)); %number of evidence sources
3    for j = 1:ne
4        for i = (j+1):ne
5            MICmatrix(i,j) = fMIC(Data{m}(:,i,j))
6        end
7    end
8
9    MICmatrixT = zeros(length{DataT{m}});
10   ne = length(DataT{m}(1,:)); %number of evidence sources
11   for j = 1:ne
12       for i = (j+1):ne
13           MICmatrixT(i,j) = fMIC(DataT{m}(:,i,j))
14       end
15   end
```

where `fMIC` is the function described in Chapter 17. Using the mutual information criterion matrices, the original data, and the one-step temporal data can be grouped into roughly independent groups `Groups` for original data and `GroupsT` for one-step temporal. The MATLAB cell variables `Groups` and `GroupsT` contain cell arrays `groups` for each mode. The cell array `groups` pertains to a specific mode and contains vectors that consist of grouped instruments; each vector represents a group.

    If one does not wish to break the evidence down into independent groups, it is still recommendable to evaluate the MIC in order to evaluate autodependence. For autodependence, the MIC is evaluated for two datasets: x_0, which is the original dataset, and x_1, which is the same as x_0 except that it is shifted by one time sample. When evaluating the MIC for kernel densities, if the MIC is less than 0.1, autodependence may be ignored.

Offline Step 4: Kernel Density Bandwidths for Original and One-step Temporal Data

Kernel density estimation is done in the same manner as in Chapter 17, where bandwidth and data are stored in KDE. However, now ambiguous modes also need to be estimated, and kernel density estimates now exist for both original and one-step temporal data.

```
1    for M = 1:length{Data}
2        %Kernel Density Estimate for original data
3        groups = Groups{M};
4        for g = 1:length{groups}
5            KDE{M}(g) = fKernelEstimateNorm(Data{M}(:,groups{g}));
6        end
7
8        %Kernel Density Estimate for one-step temporal data
9        groups = GroupsT{M};
10       for g = 1:length{groups}
11           KDE{M}(g) = fKernelEstimateNorm(DataT{M}(:,groups{g}));
12       end
13   end
```

18.4.5   Offline Step 5: Calculate Reference Values

Again, we make use of the parameter matrix $\hat{\Theta}$ in the same manner as the second-order Bayesian method.

$$
\hat{\Theta} =
\begin{bmatrix}
1 & 0 & 0 & 0 \\
0 & 1 & 0 & 0 \\
0 & 0 & 1 & 0 \\
0 & 0 & 0 & 1 \\
\hat{\theta}\{\frac{m_1}{m_1,m_2}\} & \hat{\theta}\{\frac{m_2}{m_1,m_2}\} & 0 & 0 \\
\hat{\theta}\{\frac{m_1}{m_1,m_3}\} & 0 & \hat{\theta}\{\frac{m_3}{m_1,m_3}\} & 0 \\
0 & \hat{\theta}\{\frac{m_2}{m_2,m_4}\} & 0 & \hat{\theta}\{\frac{m_4}{m_2,m_4}\} \\
0 & 0 & \hat{\theta}\{\frac{m_3}{m_3,m_4}\} & \hat{\theta}\{\frac{m_4}{m_3,m_4}\} \\
\hat{\theta}\{\frac{m_4}{m_1,m_2,m_3,m_4}\} & \hat{\theta}\{\frac{m_4}{m_1,m_2,m_3,m_4}\} & \hat{\theta}\{\frac{m_4}{m_1,m_2,m_3,m_4}\} & \hat{\theta}\{\frac{m_4}{m_1,m_2,m_3,m_4}\}
\end{bmatrix}
$$

18.4.6   Online Step 1: Obtain Prior Second-order Terms

One obtains prior second-order terms for this step from the posterior second-order terms in the previous step.

```
1    %number of historical modes and unambiguous modes
2    [nM,nm] = size(ThetaHat);
3
4    for m = 1:nm
5        Prior(m).probability = 0;
6        Prior(m).jacobian = zeros(1,(nA*nm));
```

```
7        Prior(m).hessian = zeros((nA*nm),(nA*nm));

8

9        for k = 1:nm
10           Prior(m).probability = Prior(m).probability + ...
                 Posterior(m).probability*A(m,k);
11           Prior(m).jacobian = Prior(m).jacobian + Posterior(k).jacobian*A(m,k);
12           Prior(m).hessian = Prior(m).hessian + Posterior(k).hessian*A(m,k);
13        end
14     end
```

18.4.7   Online Step 2: Calculate Support

When a new observation e has been made, one can calculate support for each mode. If multiple independent groups are present, the kernel density estimates can be multiplied together:

```
1   %we have observed current evidence e1
2   %we collected previous evidence e0
3   eT = [e0,e1];
4   S = ones(length(Data),1);
5   for M = 1:length(Data)
6       groups = Groups{M};
7       for g = 1:length(groups)
8           %Numerator and Denominator P(e1|e0,m)=P(e0,e1|m)/P(e0|m)
9           PN = fKernelDensity(eT(groups{g}),KDET{M}(g));
10          PD = fKernelDensity(e0(groups{g}),KDE{M}(g));

12          %If groups are independent, estimate support by multiplying
13          S(M) = S(M)*(PN/PD);
14      end
15  end
```

18.4.8   Online Step 3: Calculate Second-order Terms

Using the calculated support values (S), one can obtain the second-order terms as was previously done in Chapter 17.

```
1   function Lik = O2Terms(S,N,ThetaHat)
2       [nM,nm] = size(ThetaHat);
3       nA = nM - nm;

5       %Theta parameters for ambiguous modes
6       ThetaA = ThetaHat( (nm+1):end,:);

8       for m = 1:nm
9           %Initialize second-order terms
10          Lik(m).probability = 0;
11          Lik(m).jacobian = zeros(1,(nA*nm));
12          Lik(m).hessian = zeros((nA*nm),(nA*nm));
```

```
13
14          %Obtain numerators and denominators for the likelihood
15          SN = S.*N;   %Multiplication of S and N, useful for later
16          Num = SN'*ThetaHat(:,m);
17          Den = N'*ThetaHat(:,m);
18
19          Jac = zeros(1,nA);
20          Hes = zeros(nA,nA);
21
22          %Find indices where Theta values are not forced to be zero
23          Ind = find((ThetaA(:,m)≠0))';
24          for i = Ind
25              Jac(1,i) = SN(i)/Den - Num*N(i)/(Den^2);
26          end
27
28          for i = Ind
29              for j = Ind
30                  Hes(i,j) = -(N(i)*SN(j)+N(j)*SN(i))/(Den^2) + ...
                        2*Num*N(i)*N(j)/(Den^3);
31              end
32          end
33
34          %Find indices relevant to the current mode
35          ActiveInd = ((m-1)*nA+1):(m*nA);
36
37          Lik(m).probability = Num/Den;
38          Lik(m).jacobian(ActiveInd) = Jac;
39          Lik(m).hessian(ActiveInd,ActiveInd) = Hes;
40
41      end
```

18.4.9   Online Step 4: Combining Prior and Likelihood Terms

After obtaining the prior and likelihood probabilities, we can perform the second-order rule of combination in order to obtain the posterior second-order terms.

```
1   function Post = SecondOrderComb(Prior,Lik)
2   nm = length(Prior); %number of modes
3   %Normalization Constant
4   for m = 1:nm
5       K = K + Prior(m).probability * Lik(m).probability;
6   end
7
8   for m = 1:nm
9           PP = Prior(m).probability;
10          PL = Lik(m).probability;
11          JP = Prior(m).jacobian;
12          JL = Lik(m).jacobian;
13          HP = Prior(m).hessian;
14          HL = Lik(m).hessian;
15
```

```
16        Post(m).probability = 1/K*(PP*PL);
17        Post(m).jacobian = 1/K*(JP*PL + JL*PP);
18        Post(m).hessian = 1/K*(HP*PL + HL*PP + JL'*JP + JP'*JL);
19   end
```

The diagnosis can be finalised by selecting the posterior likelihood – `Post(m).probability` – that has the maximum value (the point estimate method) or one can also use the expected value method, which makes use of the second-order terms to calculate an expected value, as mentioned in Chapter 17.

18.5   Simulated Case

To create autodependent modes during the simulation, the mode was switched at random with a pre-specified switching probability. Autodependence in evidence, however, is a function of the monitors and cannot be easily created. Nevertheless, evidence autodependence can be easily reduced by sampling monitor data at a slower rate. Thus, the highest amounts of autodependence in the evidence are present when monitor data is sampled at its native frequency.

Four methods of interest were tested on the Tennessee Eastman simulation. The first method did not use any autodependent techniques, the second method only took into account autodependent modes, the third method only took into account autodependent evidence and the final method took into account autodependence in both modes and evidence. When the dynamic evidence was considered, the MIC was evaluated in order to determine if autodependence was strong enough to be considered.

Results for the four techniques can be seen in Figure 18.5. For the Tennessee Eastman simulation, it was found that autodependence in evidence was quite weak, as can be seen in the results. However, even before the results were obtained, it was noticed that the MIC values for autodependence averaged at around 0.05, which is a relatively weak value for autodependence;

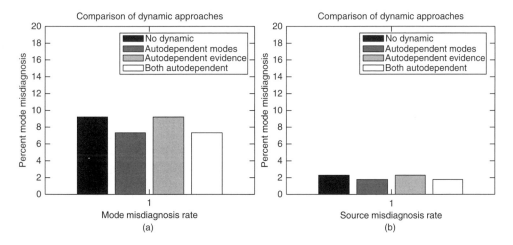

Figure 18.5   Comparison of dynamic methods

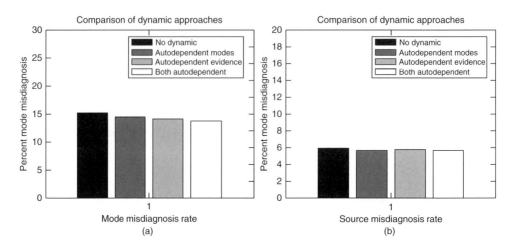

Figure 18.6   Comparison of dynamic methods

recall that if the MIC is less than 0.1, autodependence is ignored. The low autodependence in evidence was probably caused by a slow sampling rate.

In contrast to mode autodependence, however, modes tended to change relatively slowly, thus modes tended to have strong autodependence in the simulation. As seen in Figure 18.5, taking mode autodependence into account led to a significant performance improvement.

18.6   Bench-scale Case

The bench-scale system exhibited rather different behaviour from the simulated system. In the bench-scale system, the modes were set to switch relatively quickly and data was sampled at a slower rate. The MIC values for autodependence tended toward values of 0.08, which is closer to the threshold where autodependence is taken into account. From Figure 18.6, one can see that taking mode and evidence autodependence into account has a modest effect on the performance.

18.7   Industrial-scale Case

For the industrial system, data was sampled relatively quickly, but only small groups of data were selected, as modes changed extremely infrequently. In order to better assess conditions where modes were switching, validation data for each mode was divided into groups, and these groups were shuffled so that it would appear that modes switched more frequently. The results can be seen in Figure 18.7. In this case, one can see that accounting for autodependence in the evidence had a stronger effect than accounting for autodependence in the modes. The efficacy of accounting for evidence autodependence is mainly due to the fact that autodependence was relatively strong – MIC values averaged 0.2 – and that modes switched relatively quickly.

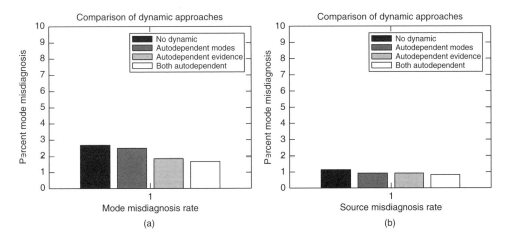

Figure 18.7   Comparison of dynamic methods

18.8   Notes and References

This chapter is a dynamic application of the material in Gonzalez and Huang (2013) and Gonzalez and Huang (2014). This dynamic application is a generalization of the material in Qi and Huang (2010).

References

Gonzalez R and Huang B 2013 Control loop diagnosis from historical data containing ambiguous operating modes: Part 2. information synthesis based on proportional parameterization. *Journal of Process Control* **23**(4), 1441–1454.

Gonzalez R and Huang B 2014 Control-loop diagnosis using continuous evidence through kernel density estimation. *Journal of Process Control* **24**(5), 640–651.

Qi F and Huang B 2010 Dynamic Bayesian approach for control loop diagnosis with underlying mode dependency. *Industrial and Engineering Chemistry Research* **49**, 8613–8623.

Qi F and Huang B 2011 Bayesian methods for control loop diagnosis in the presence of temporal dependent evidences. *Automatica* **47**, 1349–1356.

Index

Process Control System Fault Diagnosis: A Bayesian Approach, First Edition. Ruben Gonzalez, Fei Qi and Biao Huang.
© 2016 John Wiley & Sons, Ltd. Published 2016 by John Wiley & Sons, Ltd.